Plantation-Based Land Restoration

Vimal Chandra Pandey
Editor

Plantation-Based Land Restoration

Springer

Editor
Vimal Chandra Pandey
CSIR - National Botanical Research Institute
Lucknow, Uttar Pradesh, India

ISBN 978-981-95-2859-2 ISBN 978-981-95-2860-8 (eBook)
https://doi.org/10.1007/978-981-95-2860-8

This Springer imprint is published by the registered company Springer Nature Singapore Pte Ltd.
The registered company address is: 152 Beach Road, #21-01/04 Gateway East, Singapore 189721, Singapore

Preface

The upper earth's surface is a nonrenewable resource, and source of food for human life. Hence, land health is very important to promote sustainable livelihoods. Land degradation is one of the major concerns worldwide. There are some major reasons such as wind erosion, water erosion, waterlogging, salinization, lowering of the water table, soil fertility decline as well as soil pollution. Due to urbanization, industrialization, and population growth, land contamination is an ever-increasing issue across the nations. Land pollution is a type of land degradation that is the result of either anthropogenic activities or natural activities. The major pollutants of land contamination are heavy metals, metalloids, radioactive elements, pesticides, petroleum hydrocarbons, polyaromatic hydrocarbons, other organic pollutants, etc. In urban and peri-urban areas, irrigation of agricultural land with polluted water is also a reason for land pollution. Therefore, land security is an utmost important concern for the future sustainability of humans. Its restoration and management are important issues worldwide to protect land quality and functions. Land pollution means the degradation of the earth's surface. Contaminated land comes under the category of degraded/stressed land. Hence, the restoration of stressed land is essential for regaining biodiversity and ecosystem services and thereby attaining United Nations Sustainable Development Goals (UN SDGs). This fact showed the need to boost research toward land restoration. Plant-based land restoration has undergone a huge development.

Therefore, growing plantations on stressed land has a lot of potentials to secure the upper earth's surface through the land restoration programs targeted during the United Nations Decade of Ecosystem Restoration (2021–2030). This book will explore the growing plantations on stressed land (i.e., polluted land, degraded land, marginal land, nutrient-poor land, mine spoil, fly ash deposits, red mud dumpsites, neglected and underutilized land, etc.) which is very necessary to recover land and ecosystem services. Phytomanagement of stressed lands through economically valuable plants is in high demand to conserve nature and should be aligned with UN SDGs. Restoring stressed lands through growing plantations is one of the most popular holistic approaches due to its cost-effectiveness and sustainability. This book will cover a wide range of economically valuable and resilient plants for the restoration of stressed lands. The present book will also cover a wide range of areas including biomass production, land restoration, water security, and phytoproducts production from stressed land.

The core idea behind the compilation of this book is to draw chapters from eminent scientists across the globe, benefitted by their practical experience. This book will be a remarkable asset for researchers, students and scientists, policymakers, practitioners, stakeholders, and entrepreneurs to understand "Plantation-Based Land Restoration." The present book will offer how transforming stressed land into productive lands through promising and affordable practices of plantation for providing livelihood security to the people.

Lucknow, Uttar Pradesh, India Vimal Chandra Pandey

Acknowledgments

I sincerely pay gratitude to Sanchi Bhimrajka (Acquisition Editor) and *Aishwarya Thyagarajan* (Production Editor) from Springer Nature for their excellent support, guidance, and coordination during the production of this fascinating project. I thank all the contributors for their excellent chapter contributions. I also thank all the reviewers for their careful and insightful review of the book chapters. And last but not least, I must thank my family for endless support and encouragement.

Contents

Editor and Contributors

About the Editor

Vimal Chandra Pandey is an applied research scientist, internationally recognized for his research in the area of phytomanagement of polluted sites. His research focuses mainly on the remediation and management of polluted lands, using ecologically and socioeconomically valuable plants, to regain ecosystem services and support a bio-based economy as phytoproducts. His research interests also lie in fostering phytoremediation for utilizing polluted lands and thereby attaining the United Nations Sustainable Development Goals. His phytoremediation work has led to the extension of phytoremediation beyond its traditional application. He is now engaged in exploring commercial phytoremediation with the least risk, minimum input, and low maintenance. He worked at CSIR-National Botanical Research Institute, Babasaheb Bhimrao Ambedkar University, and the Council of Science and Technology, Uttar Pradesh, Lucknow, India. He has published over 121 scientific articles and book chapters in peer-reviewed journals and books.

Contributors

Neetu Arya Indian Institute of Forest Management, Bhopal, Madhya Pradesh, India
Forest Research Institute Deemed to be University, Dehradun, India

Arnab Banerjee Department of Environmental Science, Sant Gahira Guru Vishwavidyalaya, Ambikapur, Chhattisgarh, India

S. S. Bargali Department of Botany, D.S.B. Campus, Kumaun University, Nainital, Uttarakhand, India

Sanjib Kumar Behera ICAR-Indian Institute of Soil Science, Bhopal, Madhya Pradesh, India

V. Mani Chandana ICAR-Indian Agricultural Research Institute, New Delhi, India

ICAR-National Academy of Agricultural Research Management, Hyderabad, India

D. R. Chaudhary Central Salt and Marine Chemicals Research Institute, Bhavnagar, India

Gowardhan Kumar Chouhan Plant-Microbe Interaction Lab, Institute of Environment and Sustainable Development, Banaras Hindu University, Varanasi, Uttar Pradesh, India

Diganta B. Das Department of Chemical Engineering, Loughborough University, Loughborough, UK

C. K. Dotaniya Department of Soil Science & Agricultural Chemistry, SKRAU, Bikaner, India

M. L. Dotaniya ICAR-Directorate of Rapeseed-Mustard Research, Bharatpur, India

R. K. Doutaniya Department of Agronomy, SKN College of Agriculture, Jobner, Rajasthan, India

Joystu Dutta Department of Environmental Science, Sant Gahira Guru University Sarguja, Ambikapur, Chhattisgarh, India

IUCN Commission on Ecosystem Management (CEM), Gland, Switzerland

Sheikh Adil Edrisi Thapar School of Liberal Arts & Sciences, Thapar Institute of Engineering & Technology, Patiala, Punjab, India

Poornima Garg Department of Botany, Indira Gandhi National Tribal University, Amarkantak, Madhya Pradesh, India

Anand Kumar Gaurav Plant-Microbe Interaction Lab, Institute of Environment and Sustainable Development, Banaras Hindu University, Varanasi, Uttar Pradesh, India

Tushar Goyal Plant-Microbe Interaction Lab, Institute of Environment and Sustainable Development, Banaras Hindu University, Varanasi, Uttar Pradesh, India

Animekh Hazarika Department of Ecology and Environmental Science, Assam University, Silchar, Assam, India

M. Jagadesh ICAR-National Academy of Agricultural Research Management, Hyderabad, India

Manoj Kumar Jhariya Department of Farm Forestry, Sant Gahira Guru Vishwavidyalaya, Ambikapur, Chhattisgarh, India

Chandra Prakash Kala Faculty of Technical Forestry, Indian Institute of Forest Management, Bhopal, MP, India

Rayapati Karthik ICAR-National Academy of Agricultural Research Management, Hyderabad, India

Professor Jayashankar Telangana Agricultural University, Hyderabad, India

Bhimappa Honnappa Kittur Faculty of Technical Forestry, Indian Institute of Forest Management, Bhopal, MP, India

Barbara Klik Institute of Environmental Engineering, Warsaw University of Life Sciences, Warsaw, Poland

Richa Kothari Department of Environmental Sciences, Central University of Jammu, Rahya Suchani (Bagla) Samba, Jammu, Jammu and Kashmir, India

Dhiraj Kumar ICAR-Indian Institute of Soil Science, Bhopal, Madhya Pradesh, India

Kuldeep Kumar ICAR- Indian Institute of Soil and Water Conservation, RC, Kota, Rajasthan, India

Mahesh Kumar ICAR-Central Arid Zone Research Institute, Jodhpur, India

Laxman Kumawat Junagadh Agricultural University, Junagadh, India

Narayan Lal ICAR-Indian Institute of Soil Science, Bhopal, Madhya Pradesh, India

Monika Mahajan Waste Management, Resource Recovery and Ecotoxicology (WRE) Laboratory, Institute of Environment and Sustainable Development, Banaras Hindu University, Varanasi, Uttar Pradesh, India

Department of Chemical Engineering, Loughborough University, Loughborough, UK

Taher Mechergui Faculte des Sciences de Bizerte (FSB), Zarzouna, Tunisie

Bharat Prakash Meena ICAR-Indian Institute of Soil Science, Bhopal, Madhya Pradesh, India

H. M. Meena ICAR-Central Arid Zone Research Institute, Jodhpur, India

Arpan Mukherjee Plant-Microbe Interaction Lab, Institute of Environment and Sustainable Development, Banaras Hindu University, Varanasi, Uttar Pradesh, India

Arun Jyoti Nath Department of Ecology and Environmental Science, Assam University, Silchar, Assam, India

Vimal Chandra Pandey CSIR – National Botanical Research Institute, Lucknow, Uttar Pradesh, India

Society for Green Development, Lucknow, Uttar Pradesh, India

Nagaraja Tejo Prakash Department of Energy & Environment, Thapar Institute of Engineering & Technology, Patiala, Punjab, India

Maja Radziemska Institute of Environmental Engineering, Warsaw University of Life Sciences, Warsaw, Poland

Abhishek Raj Pt. Deendayal Upadhyay College of Horticulture and Forestry, Dr. Rajendra Prasad Central Agriculture University, Samastipur, Bihar, India

S. Rajendiran ICAR—Indian Institute of Horticultural Research, Bengaluru, India

Vikas Ramteke Indira Gandhi Krishi Vishwavidyalaya, Raipur, Chhattisgarh, India

Khushboo Rani ICAR-Indian Institute of Soil Science, Bhopal, Madhya Pradesh, India

Ch. Srinivasa Rao ICAR-Indian Agricultural Research Institute, New Delhi, India

Uttam Kumar Sahoo Department of Forestry, School of Earth Sciences & Natural Resource Management, Mizoram University, Aizawl, Mizoram, India

Nayan Sahu Department of Botany, Indira Gandhi National Tribal University, Amarkantak, Madhya Pradesh, India

Nisha Sahu ICAR-Indian Institute of Soil Science, Bhopal, Madhya Pradesh, India

R. C. Sanwal Department of Soil Science & Agricultural Chemistry, SKRAU, Bikaner, India

Bansh Narayan Singh Plant-Microbe Interaction Lab, Institute of Environment and Sustainable Development, Banaras Hindu University, Varanasi, Uttar Pradesh, India

Harpreet Singh Regional Research Station, PAU, Gurdaspur, Punjab, India

Rajeev Pratap Singh Waste Management, Resource Recovery and Ecotoxicology (WRE) Laboratory, Institute of Environment and Sustainable Development, Banaras Hindu University, Varanasi, Uttar Pradesh, India

Uttam Thangjam Department of Forestry, School of Earth Sciences & Natural Resource Management, Mizoram University, Aizawl, Mizoram, India

Pentile Thong Department of Forestry, School of Earth Sciences & Natural Resource Management, Mizoram University, Aizawl, Mizoram, India

Landstack Advisory Services Pvt Ltd, Kalinga Nagar, Bhubaneshwar, India

Prachi Upadhyay Indian Institute of Forest Management, Bhopal, Madhya Pradesh, India

Forest Research Institute Deemed to be University, Dehradun, India

Jay Prakash Verma Plant-Microbe Interaction Lab, Institute of Environment and Sustainable Development, Banaras Hindu University, Varanasi, Uttar Pradesh, India

Rajhans Verma Department of Soil Science & Agricultural Chemistry, SKN College of Agriculture, Jobner, Rajasthan, India

Santosh Kumar Verma Department of Civil Engineering, G H Raisoni University, Amravati, Maharastra, India

Vivek Verma Department of Energy & Environment, Thapar Institute of Engineering & Technology, Patiala, Punjab, India

Thapar School of Liberal Arts & Sciences, Thapar Institute of Engineering & Technology, Patiala, Punjab, India

Phytoremediation Strategies for Mitigating Global Concerns: A Comprehensive Review

1

Rajeev Pratap Singh, Diganta B. Das, Monika Mahajan, and Richa Kothari

Abstract

Contamination with heavy metals and trace elements has been a growing source of concern worldwide. Since the Industrial Revolution, natural and human activities have resulted in an increase in the release of toxic metals and metalloids into the environment, including cadmium, mercury, arsenic, chromium, nickel, copper, lead, zinc, and others. The health of people is seriously endangered by these toxins. Phytoremediation is a commonly used, affordable, and sustainable alternative to conventional procedures for cleaning up soils polluted with metals and metalloids. This chapter deals with the advantages and challenges of phytoremediation especially on phytostabilizations as well as handling requirements, correct disposal, and management of byproducts produced during phytoremediation operations. In the area of phytoremediation, phytoextraction, which employs hyperaccumulator plants to permanently remove heavy metals and metalloids from soils, has turned out to be the most successful technique. There are already

R. P. Singh (✉)
Waste Management, Resource Recovery and Ecotoxicology (WRE) Laboratory, Institute of Environment and Sustainable Development, Banaras Hindu University, Varanasi, Uttar Pradesh, India

D. B. Das
Department of Chemical Engineering, Loughborough University, Loughborough, UK

M. Mahajan
Waste Management, Resource Recovery and Ecotoxicology (WRE) Laboratory, Institute of Environment and Sustainable Development, Banaras Hindu University, Varanasi, Uttar Pradesh, India

Department of Chemical Engineering, Loughborough University, Loughborough, UK

R. Kothari
Department of Environmental Sciences, Central University of Jammu, Rahya Suchani (Bagla) Samba, Jammu, Jammu and Kashmir, India

V. C. Pandey (ed.), *Plantation-Based Land Restoration*,
https://doi.org/10.1007/978-981-95-2860-8_1

more than 400 hyperaccumulator plants known to be capable of collecting different heavy metals and metalloids from soils all over the globe.

Keywords

Phytoremediation · Heavy metal · Soil · Plant

1.1 Introduction

In today's world, the issue of coastal land and pollution poses significant challenges, as there is a growing concern about the increasing levels of heavy metal and trace element contamination on a global scale (Li et al. 2019; Liu et al. 2020). Throughout history, the release of harmful metals into the environment has been on the rise due to both natural and human factors (Dhaliwal et al. 2020). These metals include Cd, Hg, As, Cr, Ni, Cu, Pb, Zn, and various others (Qin et al. 2021). Thus, there is a need to understand and address the overall environmental consequences of the contamination of even metals and trace elements. In this context, this chapter discusses the perspective of pollution caused by heavy metals and alkaloids and the uniqueness of phytoremediation as a standard and reliable option for cleaning land using plants.

In addition, this chapter will discuss important technologies that have shown great success in eliminating heavy metals and trace elements from land. Along with that, it also discusses the laboratory and large-scale experiments and their advantages and challenges. One such technology is phytoextraction, which utilizes high-storage plants (Prasad et al. 2022), and another is phytostabilization, a technique effective in condensing metals and alkaloids around plant roots and reducing their movement and survivability in the soil (Kucaj et al. 2023). This chapter will explore the handling of requirements and the proper selection and management of byproducts generated during phytoremediation operations, and it will present a self-reliant alternative to the contamination of heavy metals with which we can move toward a healthy and balanced environment.

1.2 Global Concerns and Historical Context

Heavy metal contamination, a global issue due to industrialization and technological advancements, poses severe threats to ecosystems and human health. Vulnerable populations, like children and pregnant women, are at risk. Phytoremediation is a promising solution.

Potentially toxic metals have become a pressing global concern, marking a pivotal moment in industrialization, environmental degradation, and human health (Mishra 2020). This issue has deep historical roots, with its intensity escalating since the onset of the Industrial Revolution. The unprecedented technological advancements that fueled progress have also triggered an unintended consequence: the proliferation of toxic metals and metalloids in the environment (Padmavathiamma

Table 1.1 Increase in releases of toxic metals and metallic compounds

Global concerns and historical context	Description	References
Impact of the industrial revolution	Unprecedented technological advancements during the industrial revolution	Padmavathiamma and Li (2007), Srivastava and Giri (2021)
	Positive progress accompanied by unintended consequences, including environmental challenges	
	Accelerated release of pollutants due to industrial activities, setting the stage for global environmental concerns	
Increase in release of toxic metals and metalloids	Escalation in the release of toxic metals and metalloids into the environment	Mishra (2020), Mensah et al. (2021)
	Result of natural and human activities associated with industrialization	
	Cadmium (Cd), mercury (Hg), arsenic (As), chromium (Cr), nickel (Ni), copper (Cu), lead (Pb), zinc (Zn), and others contributing to environmental contamination	

and Li 2007). Since the Industrial Revolution, natural processes and human activities have significantly contributed to escalating poisonous substances released into the atmosphere (Srivastava and Giri 2021). The release of metals into the air, water, and soil has become ubiquitous (Martini et al. 2021) (Table 1.1). These elements, often byproducts of industrial processes, find their way into ecosystems, posing severe threats to the health of living organisms (Mensah et al. 2021).

The health implications of exposure to these toxic metals and metalloids are alarming. They can enter the food chain, affecting the environment and human populations (Mensah et al. 2021). The detrimental effects range from debilitating health conditions to long-term consequences, particularly for vulnerable populations such as children and pregnant women. As a result, the urgency to address this global predicament has led to exploring innovative and sustainable solutions, with phytoremediation emerging as a beacon of hope.

1.2.1 Impact of the Industrial Revolution

The Industrial Revolution has led to severe pollution and environmental deterioration, causing the proliferation of toxic metals like lead (Pb), mercury (Hg), arsenic (As), chromium (Cr), and nickel (Ni). These metals pose health risks, including neurological disorders, cancer, reproductive issues, immune system suppression, and organ damage. They also pose environmental threats, including ecosystem disruption, loss of biodiversity, and climate change due to energy-intensive extraction and processing techniques (Table 1.1).

1.2.1.1 Multifaceted Impact on People and Environment

The Industrial Revolution ushered in an era of unprecedented technological and economic progress that continues to shape our world today, but amidst the advancements in machinery and production came a dark side—a dramatic rise in pollution and environmental degradation with far-reaching consequences for human health and the planet itself (Garces-Pastor et al. 2023). One of the most concerning aspects of this environmental transformation is the significant increase in the concentration of toxic metals (Allan ct al. 2015). These metals pose a serious threat to human health in a multitude of ways. Exposure to them can occur through various pathways including inhalation of contaminated air, ingestion of polluted food or water, and direct skin contact (Sabir et al. 2022). Each metal presents its own unique set of risks, but some common adverse effects include:

(i) *Neurological damage:* Metals like lead, mercury, and arsenic can impair cognitive function, memory, and motor skills and have been linked to conditions like Alzheimer's disease and Parkinson's disease (Xu et al. 2018; Moon et al. 2022; Taux et al. 2022).
(ii) *Cancer:* Chronic exposure to certain metals like chromium and nickel is associated with an increased risk of developing various types of cancer (Moradmand and Harchegani 2011; Yang et al. 2021).
(iii) *Reproductive health issues:* Metals can disrupt hormonal function and fertility and have been linked to birth defects and developmental problems in children (Lauwerys and Lison 1994).
(iv) *Immune system suppression:* Exposure to metals can weaken the immune system making individuals more susceptible to infections and diseases (Okereafor et al. 2020).
(v) *Organ damage:* Depending on the specific metal and the route of exposure, various organs can be targeted including the kidneys, liver, lungs, and gastrointestinal tract (Meltzer et al. 2020; Kulathunga et al. 2021; Paithankar et al. 2021). The impact of these toxic metals extends beyond human health and poses significant threats to the environment as well.
(vi) *Ecosystem disruption:* The accumulation of metals in soil and water bodies can harm plant and animal life affecting the entire ecosystems and disrupting food chains (Moiseenko et al. 2018).
(vii) *Biodiversity loss:* Pollution can lead to habitat loss and species extinction directly through toxicity or indirectly by impacting food availability and reproductive success (Moiseenko et al. 2018).
(viii) *Climate change:* The mining, smelting, and processing of metals are energy intensive and contribute to greenhouse gas emissions further exacerbating the challenges of climate change (Bhattacharyya 2022).

Addressing issues of toxic metals and their widespread impact requires a multifaceted approach that considers both historical and contemporary factors.

1.2.1.2 Understanding the Legacy of the Industrial Revolution

Examining the historical context and identifying the specific industrial processes and technologies that contributed to the initial rise in metal pollution are crucial for developing effective mitigation strategies that include:

(i) *Implementing strict regulations and enforcement:* Strengthening environmental regulations and ensuring robust enforcement mechanisms are essential to prevent further pollution and hold polluters accountable (Teh et al. 2016).
(ii) *Investing in cleaner technologies:* Researching and developing sustainable and cleaner technologies for mining, smelting, and processing of metals can significantly reduce the environmental impact of these activities (Marques et al. 2009).
(iii) *Remediating contaminated sites:* Cleaning up existing sites contaminated with toxic metals through various remediation techniques like phytoremediation or soil washing can help restore ecosystems and reduce human exposure risks (Ponce-Hernández et al. 2020).
(iv) *Promoting public awareness and education:* Raising awareness about the danger of toxic metals and educating communities about the sources of exposure and ways to minimize risks are crucial for fostering collective action and responsible behavior (Cao et al. 2016).
(v) *International cooperation:* Addressing the challenges posed by toxic metals transcends national borders and requires global collaboration. Effective information exchange, technology transfer, and capacity building initiatives can empower developing countries to implement stricter environmental standards and manage metal pollution effectively (Murphy et al. 2009).

1.3 Environmental Impacts of Toxic Metal Pollution

Rapidly increasing urbanization and industrialization emit toxic metals and metal compounds into the environment due to mining, metal refining, industrial processes, and the use of consumer products. Elements such as lead, mercury, arsenic, cadmium, and chromium are highly toxic and can persist in the environment for long periods. These elements can enter the food chain and seriously affect the health of living organisms. Multiple routes of entry of toxic metals into the human body exist through contamination of drinking water sources, consumption of foods grown from contaminated soil, and air pollution (Sabir et al. 2022). Exposure to these elements can cause acute and long-term health problems. Some immediate effects include acute toxicity, brain and nervous system damage, kidney damage, cancer, and reproductive issues. Long-term effects may consist of organ damage, heart disease, disability, and neurodegenerative disorders (Das et al. 2023; Frazzoli et al. 2022; Sabir et al. 2022; van der Voet et al. 2007).

1.3.1 Toxic Metals and Nonmetals: Sources

Aftermath of past events, there were numerous reasons along with unprecedented technological advancement to disseminate metal pollution worldwide. However, this progress has also exposed its dark side, one of which is the increasing levels of toxic metals and nonmetals in the environment. These elements are highly poisonous, enter the food chain, and cause serious harm to the health of living beings (Singh and Prasad 2011). In this article, we will highlight the harmful health effects of some significant toxic metals and nonmetals:

Cadmium (Cd): Cadmium, found mainly in batteries, plastics, and fertilizers, is highly harmful to bones and kidneys. It increases the risk of high blood pressure, osteoporosis, and kidney failure. Additionally, cadmium is particularly detrimental to developing children's brains and can cause learning disabilities and other neurodegenerative problems (Al-Naemi and Das 2020; Li et al. 2023).

Mercury (Hg): Used in thermometers, batteries, and some chemical processes, mercury seriously affects the nerves and brain. Mercury exposure can increase the risk of tremors, memory loss, Alzheimer's disease, and congenital disabilities. Additionally, mercury accumulates in the environment and proves destructive to aquatic life (Bjorklund et al. 2017; Li et al. 2023).

Arsenic (As): Arsenic, used in pesticides and wood preservatives, can cause skin cancer, lung cancer, and many other health problems. Drinking groundwater contaminated with arsenic is particularly dangerous and increases the risk of diabetes, heart disease, and stroke (Dudka et al. 2014; Gupta et al. 2022).

Chromium (Cr): Chromium, found in stainless steel and other alloys, exists in two health-imposing forms: hexavalent chromium is highly toxic and can cause skin cancer, while trivalent chromium is an essential nutrient (Permenter et al. 2011). Unfortunately, industrial activities mainly emit hexavalent chromium, threatening the environment and human health.

Nickel (Ni): Nickel, found in batteries, stainless steel, and electronic devices, can cause allergic reactions in some individuals. However, research on its health effects is still ongoing (Permenter et al. 2011).

Copper (Cu): Copper is an essential nutrient found widely in nature, but too much can be harmful. Consuming excessive amounts of copper can damage the liver (Coelho et al. 2020).

Lead (Pb): Lead, used in paint, gasoline, and batteries, is especially harmful to children's developing brains. It can cause learning disabilities, behavioral problems, and low intelligence. In adults, lead may increase the risk of heart disease, high blood pressure, and kidney problems (Pappas et al. 2007; Bulka et al. 2019).

1.4 Possible Solutions for Dealing with Toxic Metals and Nonmetals

Toxic metals and nonmetals have wide-ranging negative impacts on the environment and human health. Addressing these threats requires a multi-pronged strategy to reduce the use and emissions of these elements and protect affected communities. Here are some possible solutions:

1.4.1 Regulation and Compliance

It is necessary to enforce strict environmental regulations and standards and ensure their compliance effectively. Strict laws should be made for mining, industrial activities, and metal disposal processes to reduce emissions into the environment. Additionally, there is a need to invest in cleaning up existing pollution.

1.4.2 Replacement and Purifying Technologies

Efforts should be made to phase out or limit the use of toxic substances with less harmful alternatives. It also encourages research and development of alternative materials and clean production techniques.

1.4.3 Awareness and Education

Educating the public about the dangers of toxic metals and encouraging their participation in keeping the environment healthy are essential. Awareness campaigns can be run through schools, community centers, and media, and consumers can be motivated to make consumption choices that support products with fewer toxins (Cao et al. 2016).

1.4.4 Risk Reduction Measures

It is necessary to develop strategies to reduce exposure to toxic metals for at-risk communities, especially children and pregnant women (Wu et al. 2023). For example, it is essential to identify contaminated water and food sources and ensure access to safe alternatives. Furthermore, regular monitoring of the health of affected individuals and timely medical intervention are necessary.

1.4.5 Collaborative Antidote

To effectively combat this global menace, international mutual aid is not merely desirable; it is indispensable. By fostering collective action and knowledge exchange, we can leverage the strengths of diverse nations and develop comprehensive strategies for managing and mitigating toxic metal pollution.

1.5 Phytoremediation: Successful Environmental Cleanup Technology

The accumulation of heavy metals and nonmetals in the soil has serious challenges for environment. These elements are highly toxic and can persist for long periods in the environment, where they can enter the food chain and seriously affect the health of living beings. Traditional treatment methods, such as excavation and thermal destruction, are often expensive, inefficient, and environmentally damaging. Phytoremediation technology has emerged as a potential and sustainable solution to this problem.

Phytoremediation is the process of using the natural ability of plants to remove heavy metals and nonmetals from contaminated soil and water. Some plants, called hyperaccumulators, can absorb and tolerate these elements in unusually high concentrations (Mani et al. 2015; Mengoni et al. 2010; Qiong et al. 2021). Phytoremediation uses hyperaccumulator plants to remove heavy metals and nonmetals from contaminated environments effectively. Phytoremediation technology has many benefits: (i) It is more cost-effective and environment-friendly than traditional methods and (ii) it uses less energy and does not harm the soil structure (Li et al. 2016; Yang et al. 2014).

1.5.1 Use of Hyperaccumulator Plants

The success of phytoremediation depends on the selection and management of hyperaccumulator plants. Various hyperaccumulator plants are efficient at absorbing multiple heavy metals and nonmetals. For example, willow is effective in removing cadmium, saltwort is suitable for chromium, and sesbania is efficient in the absorption of zinc and nickel (Tőzsér et al. 2017; Vandecasteele et al. 2002). Selecting the correct hyperaccumulator plants by considering soil properties, type, and concentration of contaminant compounds is essential.

During the phytoremediation process, hyperaccumulator plants are grown in the contaminated area. Plants absorb heavy metals through their roots and transport them into the stems and leaves (Garbisu and Alkorta 2001; Saha et al. 2022). The contaminated plants are then harvested and removed, permanently removing the heavy metals. Alternatively, plants can be burned, and heavy metals are vaporized (Olayiwola et al. 2023).

1.5.2 Stable Removal of Potential Toxic Metals from Soil

Phytoremediation processes can be designed for sustainable removal of heavy metals. Some hyperaccumulator plants convert heavy metals into harmless compounds in their tissues. This process is called phytostabilization (Madejon et al. 2009; Mataruga et al. 2020).

In phytostabilization, hyperaccumulator plants absorb heavy metals into their roots and store them in their tissues (Kucaj et al. 2023). However, they do not store these metals in toxic form (Eissa 2019; Kucaj et al. 2023). Instead, they convert them into less toxic or even harmless compounds (Testiati et al. 2013). The phytostabilization process is still poorly understood, but it is believed to involve several physiological and biochemical mechanisms. One possibility is that plants oxidize heavy metals into less toxic forms before accumulating them in their tissues (Madejon et al. 2009). Another possibility is that plants combine heavy metals into complex compounds that make them more stable and less harmful (Testiati et al. 2013). Phytostabilization is more cost-effective and environmentally friendly than traditional methods and does not harm the soil structure (Testiati et al. 2013). The success of phytostabilization depends on the selection and management of hyperaccumulator plants. Plants that can convert heavy metals into less toxic forms are considered most suitable for phytostabilization. Additionally, plants should be grown in contaminated areas with low concentrations of heavy metals.

(A) *Some Instances of Phytostabilization*

Phytostabilization has been used in many different applications to combat heavy metal pollution. For example, this technology has been used to purify contaminated soil and water in industrial areas, mining areas, and waste disposal sites (Sandhi and Greger 2012). Phytostabilization emerges as a powerful tool in our environmental arsenal, offering a sustainable and cost-effective approach to managing a diverse range of contaminants (Jung et al. 2020). Through careful selection of plant species, consideration of site-specific conditions, and ongoing research, this technology can play a key role in safeguarding environmental health and protecting human well-being. As we delve deeper into its intricate mechanisms and unlock its full potential, phytostabilization holds the promise of a cleaner and greener future for our planet.

(i) *Heavy Metal Immobilization*

(a) *Lead (Pb)* in Mine Tailings: Studies conducted in abandoned mine tailings contaminated with lead illustrate the effectiveness of phytostabilization. Species like Indian mustard (*Brassica juncea*) and vetiver grass (*Chrysopogon zizanioides*) have demonstrated exceptional Pb uptake and root exudate production (Chen et al. 2004; Lim et al. 2004; Otunola et al. 2022). These exudates chelate Pb, reducing its bioavailability and mobility in the soil. Research from the University of Salamanca, Spain, highlighted a significant decrease in Pb leaching from tailings amended

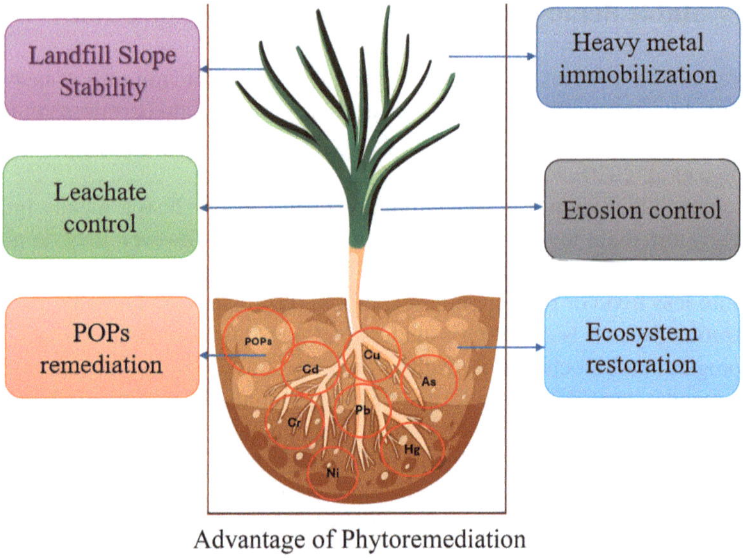

Fig. 1.1 Advantages of phytoremediation

with vetiver grass, proving its potential for long-term stabilization (Latif et al. 2023) (Fig. 1.1).

(b) *Arsenic (As)* in Contaminated Agricultural Lands: Arsenic contamination in agricultural lands poses a serious threat to food safety. Sunflower (*Helianthus annuus*) has emerged as a promising agent for As phytostabilization (Sharma et al. 2023). Its extensive root system and high biomass production enable efficient As uptake and accumulation in plant tissues. A published study demonstrated that sunflower cultivation reduced As bioavailability in soil effectively minimizing its uptake by food crops (Zhao et al. 2023). A total of 37 oilseed sunflower accessions were assessed in a natural field setting to determine their growth performance, ability to produce lignocellulosic biomass for bioenergy, and capacity to absorb metals while remediating soil contaminated with arsenic and mercury. Noticeable discrepancies were noted in the lignocellulosic constituents of plant stems upon reaching maturity, with a descending sequence of total sugars > cellulose > hemicellulose > lignin > ash. Furthermore, there was a notable disparity in metal absorption among the various accessions. The As and Hg concentrations in the above-ground sections reached maximum levels of 14.08 and 0.40 mg/kg, respectively. The concentrations of inorganic arsenic species (69.73 mg/kg for arsenate and 4.41 mg/kg for arsenite) were greater in the rhizosphere soil, compared to the concentrations of organic arsenic species (0.22 mg/kg for dimethyl arsenic and 0.02 mg/kg for monomethyl arsenic). The concentration of arsenate (3.72 mg/kg) was greater in the roots, while the concentration of arsenite (0.47 mg/kg) was greater in the shoots. Several accessions exhibited high oil production rates, ranging from 26% to 45.6%. The most abundant fatty acids were oleic acid, accounting for 75% of the total, and linoleic acid, accounting for 37.6%. These findings indicate that certain varieties of oilseed sunflower could be used as crops to efficiently clean up soils with low to moderate levels of arsenic (As) and mercury (Hg) contamination while also supporting agricultural productivity (Sahito et al. 2021).

(ii) *Organic Pollutant Containment*

Polycyclic Aromatic Hydrocarbons (PAHs) in Industrial Waste Sites: PAHs, common contaminants in industrial waste sites, pose significant environmental and human health risks. Certain poplar species, like *Salix matsudana*, are adept at PAH phytoremediation. Their deep root systems facilitate access to deeper contamination zones, while microbial activity in the rhizosphere degrades PAHs into less harmful compounds. A joint research project by the University of Alberta, Canada, and the National Research Council of Italy demonstrated significant reductions in PAH concentrations around *S. matsudana* roots, highlighting its potential for long-term containment and biodegradation (Abbaslou et al. 2018).

Persistent Organic Pollutants (POPs) in Agricultural Runoff: POPs, such as DDT and PCBs, are notorious for their persistence and bioaccumulation in food chains. Plants like hemp (*Cannabis sativa*) have been identified as effective POP phytorestabilizers. Their rapid growth and extensive root systems facilitate POP uptake and sequestration in plant tissues. Research from the University of Waterloo, Canada, revealed that hemp cultivation significantly reduced DDT and PCB mobility in agricultural soils, preventing their entry into waterways and potentially mitigating their impact on aquatic ecosystems (Sarkar et al. 2022).

(iii) *Erosion Control and Ecosystem Restoration*

Coastal Erosion and Habitat Degradation: Coastal erosion due to anthropogenic activities and climate change threatens biodiversity and infrastructure. Seagrasses, like *Spartina alterniflora*, play a crucial role in phytostabilization by binding sediments with their extensive root networks (Syed et al. 2021). This reduces wave erosion, promotes sediment accretion, and fosters habitat restoration for coastal organisms (Fig. 1.1).

(iv) *Landfill Slope Stability and Leachate Control*

Landfill slopes are susceptible to erosion and leachate migration, potentially contaminating surrounding environments. Plants like switchgrass (*Panicum virgatum*) with deep roots and robust root systems can be used for phytostabilization (Phouthavong-Murphy et al. 2020). These roots bind soil particles, enhancing slope stability and preventing erosion. Additionally, their rhizosphere microbial communities can degrade leachate pollutants, minimizing their environmental impact. A research has been conducted by Novak et al. (2019) on cadmium- and zinc-contaminated mine soil. The study explored phytostabilization by applying a biochar–compost mixture (derived from beef cattle manure and poultry litter) to cultivate switchgrass. Findings demonstrated that combining organic amendments with phytostabilization effectively remediated the mine site. The result concludes with successful mine site remediation with the combination of organic amendment and phytostabilization (Novak et al. 2019) (Fig. 1.1).

(B) *Phytoextraction*

The process of contaminants being absorbed and transported from plant roots to shoots is referred to as phytoextraction, also known as phytoaccumulation or phyto-sequestration. To eliminate soil pollutants, the roots and shoots have collected contaminants (Jaswal et al. 2023). Chandra and Kumar (2017) and Bhat et al. (2022) showed that 12 appropriate indigenous plants, which were growing on sludge, were observed for their capacity to deposit Fe. The study demonstrated that *A. spinosus* L. can accumulate high levels of zinc (Zn) and manganese (Mn) in its shoots. All plants, with the exception of *A. spinosus* L. and *Ricinus communis*, were identified as Fe, Mn, and Zn accumulators in leaf. The research conducted by Sharma et al. (2024) showed that ornamental plants have a high potential for removing heavy metals (HMs) from polluted environments through a process called phytoextraction. When compared to typical cleansing approaches, this strategy looks to be more ecologically friendly (Bhat et al. 2022).

Ugya et al.'s (2019) research examined the phytoextraction potential of specific vegetables and the potential health hazards associated with consuming vegetables cultivated in agricultural soils irrigated with refinery wastewater. The findings demonstrated a substantial accumulation of Cd and Hg in the vegetable. In research study, Makarova et al. (2021) investigated the potential for *Trifolium repens* to improve the phytoextraction of Cd, Ni, and Cu. The findings demonstrated the efficacy of K_2HEDP potassium salt, plant growth regulators, and iron chelate in facilitating the phytoextraction of Cd.

1.6 Considerations and Challenges in Phytostabilization

While the academic examples presented showcase the remarkable potential of phytostabilization, it's crucial to acknowledge its limitations and specific considerations (Kumar 2020; Mishra 2020).

Plant Selection

Choosing the appropriate plant species is vital for successful phytostabilization. Factors like contaminant type, soil conditions, and desired outcome must be carefully considered (Du et al. 2017; Prasad et al. 2022; Kumar et al. 2023).

Long-Term Monitoring and Maintenance
Phytostabilization is often a slow process requiring long-term monitoring and maintenance. Regular assessment of contaminant concentrations and plant health is essential (Odoh et al. 2019).

Contaminated Plant Biomass Management
The fate of the harvested contaminated plant biomass needs careful planning to avoid contaminant release and ensure safe disposal or recycling (Ahmad et al. 2022; Mahar et al. 2016; Yang et al. 2023).

Using phytostabilization, scientists have successfully removed lead, mercury, chromium, and other heavy metals from contaminated soils. These studies show that phytostabilization can be an effective and sustainable way to combat heavy metal pollution.

1.6.1 Phytostabilization and Hyperaccumulator Plants: A Promising Future for Combating Heavy Metal Pollution

The Industrial Revolution and rapid urbanization have left a legacy of environmental contamination, with heavy metals and metalloids posing a significant threat to human health and ecosystems. Technological advances like phytostabilization and nature's tools like hyperaccumulator plants offer promising solutions to combat this challenge (Mani et al. 2015; Qiong et al. 2021; Siyar et al. 2022).

1.6.2 Phytostabilization: A Green Hope for Contaminated Sites

Phytostabilization harnesses the power of plants to immobilize and reduce the bioavailability of heavy metals in soil and water. These plants act as nature's filters, absorbing and sequestering harmful contaminants within their tissues, preventing them from entering the food chain or leaching into groundwater (Sandhi and Greger 2012; Singh and Prasad 2011; Zhen et al. 2022). Compared to traditional, often harsh, remediation methods, phytostabilization offers significant advantages:

1.6.2.1 Cost-Effectiveness
This plant-based approach is typically cheaper than excavation or chemical treatment, making it a viable option for large-scale contaminated sites (Chen and Li 2018).

1.6.2.2 Environmental Friendliness
Phytostabilization avoids harmful chemicals and soil disturbance, promoting ecosystem health and biodiversity (Ayilara et al. 2023; Malaviya and Singh 2012; Zhou and Wang 2019).

1.6.2.3 Sustainability

Plants naturally renew themselves, providing long-term stability and reducing the need for repeated interventions. While phytostabilization holds immense promise, further research and development are crucial to maximize its effectiveness (Anand et al. 2023; Ganguly et al. 2023; Saxena et al. 2020). Understanding the mechanisms behind plant-metal interactions can guide scientists in selecting the most efficient hyperaccumulator species for specific contaminants. Optimizing planting strategies, tailoring soil amendments, and exploring synergistic combinations of plants could further boost the technology's potential (Ganguly et al. 2023).

1.7 Considerations for Optimizing Phytoremediation Strategies

Hyperaccumulator plants are nature's answer to metal-rich environments. Over thousands of hyperaccumulator species exist, each with unique metal affinities (van der Ent et al. 2013). Certain willows excel at cadmium extraction, while others, like *Alyssum*, capture nickel with impressive efficiency (Tőzsér et al. 2017; Vandecasteele et al. 2002). This diversity presents a powerful toolkit for targeted remediation, offering solutions for various contaminated sites. Harnessing the potential of hyperaccumulators in phytostabilization requires careful consideration. Choosing the suitable species for the contaminant of concern is critical. Understanding plant growth requirements and environmental factors allows for optimizing planting strategies and maximizing metal uptake. Additionally, exploring innovative applications, such as using harvested biomass for bio-ore production, could unlock further ecological and economic benefits (Ayilara et al. 2023; Malaviya and Singh 2012; Ponce-Hernández et al. 2020; Suman et al. 2018; Xu et al. 2021). The future of combating heavy metal pollution lies in harnessing the combined power of technology and nature. By advancing phytostabilization techniques and utilizing the natural prowess of hyperaccumulator plants, we can create a cleaner and healthier future for generations to come (Kumar et al. 2020). As both areas evolve, the potential for effective and sustainable remediation becomes increasingly tangible, offering hope for a world free from the harmful grip of heavy metal contamination.

Hyperaccumulator plants can absorb and tolerate heavy metals and nonmetals in unusually high concentrations. The effectiveness of hyperaccumulator plants is measured by their ability to remove heavy metals and their level of competition with cultivated crops (Koley 2022; Mishra 2020; Shamsi et al. 2023). Some hyperaccumulator plants slowly but steadily remove heavy metals, providing long-term sustainability (Saxena et al. 2020). An environmental cleanup technique called phytoremediation uses hyperaccumulator plants to remove heavy metals from contaminated soil and water (Dou et al. 2023; Jung et al. 2020; Perlein et al. 2021; Sandhi and Greger 2012; Saxena et al. 2020). Phytoremediation offers several advantages over traditional removal methods. It is more cost-effective and eco-friendly and does not harm the soil structure (Mench et al. 2010).

Additionally, phytoremediation also provides the possibility of permanent removal. Selection, management, and crop rotation are important factors in using hyperaccumulator plants in phytoremediation. The correct hyperaccumulator plant should be selected based on the contaminating metal, soil properties, and environmental conditions. Optimizing crop management techniques such as plant density, irrigation, and harvesting timing ensures that the phytoremediation process is efficient and sustainable (do Nascimento et al. 2021; Farid et al. 2020; Iqbal et al. 2022; Madejon et al. 2009; Mench et al. 2010).

1.8 Laboratory and Large-Scale Phytoremediation Experiments: Feasibility of Utilizing Plants for Soil Remediation

Environmental contamination is a global challenge, with soil pollution by heavy metals and organic pollutants being one of the severe concerns (Chirakadze et al. 2016; Demirel 2007). Traditional treatment methods are often expensive, ineffective, and damaging to the environment (Zamora-Ledezma et al. 2021). Various laboratory and large-scale studies have been conducted to assess the effectiveness of phytoremediation. These studies investigate the pollution removal capabilities of different plant species, the effects of other soil properties, and the importance of optimizing the treatment process.

1.8.1 Laboratory Studies: Preliminary Assessment and Clarification of Mechanisms

Laboratory studies are a valuable tool to assess the potential efficacy of phytoremediation and understand the underlying mechanisms. These studies often use soil and plant samples in controlled environments in small containers. Korzeniowska and Stanislawska-Glubiak (2023) evaluated the phytoremediation potential of five grass species in zinc-contaminated soil. They found that all five species absorbed zinc to some extent. The study's goal was to assess the phytoremediation capabilities of cultivated grasses with surrounding wild grass in soil contaminated with zinc. Two pot tests were conducted on soil purposely poisoned with zinc. Four species of cultivated grasses were employed as test plants—*Lolium perenne*, *Festuca pratensis*, *Festuca rubra*, and *Poa pratensis*—and one wild, native grass: *Deschampsia caespitosa*. Wild grass seedlings were taken from soil contaminated with heavy metals near a zinc smelter. The phytoremediation potential of grasses was assessed on the basis of the tolerance index (TI), bioaccumulation (BF), and translocation (TF) components. Differences were identified between the species in the reduction in the shoot and root biomass with increasing soil pollution with Zn. The investigated grasses' tolerance to excess zinc in the soil was in the following order: *D. caespitosa* > *L. perenne* > *F. rubra* > *F. pratensis* > *P. pratensis*. In addition, there were variations in the accumulation and distribution of Zn between the roots and shoots, which is connected to the various defensive mechanisms of the investigated grasses against zinc phytotoxicity. Of the five grasses studied, the maximum phytoremediation capacity was exhibited by *D. caespitosa.* This grass displayed a much higher tolerance to zinc and a reduced transfer of zinc from the roots to shoots than the other cultivated grasses studied. All four cultivated grasses can be beneficial for phytostabilization because they accumulate zinc primarily in the roots and inhibit its transfer to the shoots. Unlike wild grass seeds, cultured grass seeds are freely available commercially and can be employed for the phytoremediation of HM-contaminated areas (Korzeniowska and Stanislawska-Glubiak 2023). Laboratory studies also shed light on the influence of soil characteristics and other environmental elements. For example, another study indicated that variations in soil pH can impact plants' absorption of heavy metals (Adamczyk-Szabela and Wolf 2022). The researchers observed that cadmium absorption by plants was more significant in less acidic soils, suggesting that soil pH levels are necessary to enhance the phytoremediation process (Haider et al. 2021).

1.8.2 Large-Scale Phytoremediation: Testing Feasibility in the Real World

Large-scale phytoremediation studies are essential in translating laboratory findings to real-world situations. These studies evaluate the effectiveness of phytoremediation over areas of hundreds or thousands of square meters and consider the site's specific characteristics.

An experiment was conducted in the Sahl El Husseiniya region of Alsharqia government (Shehata et al. 2019). There were various cultivated regions confronted by water scarcity for agricultural activities, resulting in the consumption of low-quality waters from the drainage system. Unfortunately, these drainage fluids include high quantities of heavy metals, mostly concentrated in the soil's surface layer, eventually damaging plant and food systems. Specifically, the region relies significantly on wastewater from the Bahr El-Baqar drain for agriculture (Shehata et al. 2019).

To solve this environmental concern, a study was done with the purpose of deploying hyperaccumulator plants, especially kenaf (*Hibiscus cannabinus* L.) and flax (*Linum usitatissimum* L.), in the area. The growing of these plants was increased by the administration of humic acid and gibberellin as foliar sprayings to enhance their capacity to absorb heavy metal ions from the soil. Additionally, three soil additives—sulfur, vermiculite, and compost—were used to further enhance the remediation process (Shehata et al. 2019).

The experiment's findings revealed sulfur's significant influence on chromium (Cr), cobalt (Co), cadmium (Cd), and manganese (Mn) uptake by hyperaccumulator plants. Additionally, humic acid, when applied as a foliar treatment, enhanced Co and Cr absorption in flax and Cd and Mn uptake in kenaf. Gibberellin, as another foliar treatment, not only stimulated plant growth but also proved most effective for

Co and Cr uptake in kenaf-cultivated soils and Cd and Mn uptake in flax-cultivated soils (Shehata et al. 2019).

Generally, heavy metals exhibited higher concentrations in plant roots than in shoots. A comparative study on soil remediation efficacy between kenaf and flax revealed kenaf's superiority in removing Cr, Co, and Cd, whereas flax excelled in Mn removal. Kenaf achieved removal percentages of 50.71% for Cr, 38.27% for Co, 33.98% for Mn, and 14.43% for Cd, while flax attained clearance percentages of 54.36% for Mn, 36.95% for Cr, 28.72% for Co, and 11.37% for Cd. Across two seasons, the combined phytoremediation efficiency followed the order of Cr $\geq$ Mn > Co > Cd, achieving removal rates of 66.87%, 65.63%, 54.66%, and 23.40%, respectively (Shehata et al. 2019). Large-scale studies have shown that phytoremediation can be a viable method for removing pollutants. However, the success of this technique depends on several factors, including:

1.8.2.1 Plant Species Selection

Different plant species can remove different pollutants. Selecting the plant species suitable for the contaminated site is essential (Tőzsér et al. 2017).

1.8.2.2 Optimization of Soil Properties

Soil pH, salinity, and other properties can affect the efficiency of phytoremediation. Optimizing the soil properties of a contaminated site can make the phytoremediation activity more effective (Abukari et al. 2022).

1.8.2.3 Proper Management

Proper management of plants is essential to make the phytoremediation process successful. This may include plant irrigation, fertilization, and weed control.

Phytoremediation stands as a promising technology for the cleansing of soils tainted with heavy metals and organic pollutants. Both laboratory experiments and large-scale studies have showcased the feasibility of this approach. However, broader implementation of phytoremediation in practical settings necessitates further research and development efforts.

1.8.3 Challenges to the Feasibility: Phytoremediation

Phytoremediation, a plant-based method for cleaning contaminated environments, faces limitations like time-consuming biological processes. Strategies include selecting fast-growing plants, optimizing plant-microbe interactions, and implementing bioaugmentation techniques. Research in plant biotechnology and genetic engineering offers potential solutions for enhancing phytoremediation efficiency.

Limitations of Phytoremediation: A Deeper Dive

Phytoremediation, harnessing the power of plants to clean up contaminated environments, holds immense promise for a sustainable future. However, as with any technology, it comes with limitations that warrant careful consideration. Let's delve deeper into three key limitations:

1.8.3.1 Time: A Race Against the Clock

Time becomes a formidable opponent in phytoremediation (Kumar 2020). Unlike physical or chemical treatment methods, phytoremediation relies on the biological processes of plants, which inherently operate at a slower pace. Removing certain pollutants, particularly recalcitrant organics and tightly bound heavy metals, can be a marathon, taking years or even decades. This time constraint has several implications:

(i) *Long-term monitoring and maintenance:* Sites require extended monitoring to track progress and ensure remediation goals are met. This translates to ongoing cost commitments (Mishra 2020).
(ii) *Delayed land use recovery:* Contaminated land remains unusable for its intended purpose until remediation is complete, potentially hindering economic and social development (Prakash 2020).
(iii) *Public perception and impatience:* The slow pace can lead to frustration and skepticism, potentially hampering public support for phytoremediation projects (Mahar et al. 2016; Odoh et al. 2019; Prasad et al. 2022).
(iv) *Addressing these challenges requires:*

- Selection of fast-growing plants with high pollutant uptake capacities. Research in plant biotechnology and genetic engineering offers promising avenues for developing enhanced phytoremediation crops.
- Optimization of plant-microbe interactions. Symbiotic relationships between plants and rhizosphere microbes can significantly accelerate pollutant degradation (Kumar et al. 2023).

Implementation of bioaugmentation and biostimulation techniques. Introducing specific microbes or providing nutrients can enhance microbial activity and pollutant breakdown.

1.8.3.2 Land Use: A Balancing Act

Phytoremediation often demands sacrificing significant land area. Establishing dense vegetation for effective pollutant removal might not be feasible in densely populated areas or where land is already under pressure for agriculture or other purposes. This conflict raises serious considerations:

(i) *Competition for land resources:* Balancing the need for clean land with competing uses like food production or infrastructure development presents a complex challenge (Onyena et al. 2023).
(ii) *Impacts on biodiversity:* Large-scale planting for phytoremediation might displace existing ecosystems and affect native species (Pirzadah et al. 2014).

(iii) *Transportation and logistics:* Managing large biomass generated during phytoremediation can be logistically challenging and add to the overall cost (Ahmad et al. 2015; Juwarkar et al. 2012; Onyena et al. 2023; Pirzadah et al. 2014).

To navigate these challenges, the necessary requirements are as follows:

1. Exploration of vertical phytoremediation systems. Utilizing rooftops, walls, and other vertical spaces can maximize land use efficiency, especially in urban settings.
2. Development of high-density planting techniques. Optimizing plant spacing and selection can maximize pollutant removal on smaller land areas.
3. Integration of phytoremediation with other land uses. Agroforestry systems, where crops are grown alongside phytoremediation plants, can offer a multifunctional land use solution (Gori et al. 2019; Juwarkar et al. 2012; Kim and Owens 2011, 2019; Poria et al. 2022; Wu et al. 2021).

1.8.3.3 Safety: Walking a Tightrope

While plants act as nature's filters, accumulating pollutants in their tissues raises concerns about their safety after remediation. The potential concern are as follows:

(i) Entry into the food chain: If contaminated plants are used for grazing or compost, pollutants can enter the food chain, posing risks to human and animal health (Singh and Prasad 2011; Zhen et al. 2022).
(ii) Accidental exposure: Improper handling of contaminated plant biomass can expose workers and the public to harmful substances (Cao et al. 2015; Politis et al. 2022).
(iii) Leaching and dispersion: Over time, pollutants may leach from plant residues back into the soil or groundwater, jeopardizing the long-term success of remediation efforts (Krueger et al. 2013; Wu et al. 2004).
(iv) Ensuring safety demands: Selection of non-food, non-invasive plant species. Choosing plants unsuitable for human or animal consumption minimizes the risk of food chain contamination (Adel et al. 2018; Qin et al. 2021).
(v) Development of safe disposal and recycling methods: Techniques like incineration, pyrolysis, or biorefinery can safely dispose of contaminated biomass while potentially recovering valuable resources (Adel et al. 2016; Kim et al. 2020).
(vi) Long-term monitoring and risk assessment: Continuous monitoring of the remediated site ensures sustained safety and early detection of any potential leaching or dispersion issues.

1.9 Management of Phytoremediation Byproducts: Handling Requirements, Correct Disposal, and Overall Management

Phytoremediation is emerging as a promising technology to tackle the challenge of environmental pollution. Using plants' natural ability is a sustainable and cost-effective way to remove or immobilize heavy metals and organic pollutants from soil and water. However, safe and effective management of byproducts released during the phytoremediation process is essential.

The most common strategy for removing heavy-metal-contaminated plants is heat treatment, which includes processes such as liquefaction, gasification, pyrolysis, and incineration (Tran and Liu 2023; Wen et al. 2024). These approaches differ slightly in their technical characteristics. While hydrothermal carbonization and gasification require a pressured atmosphere, incineration, pyrolysis, and gasification often occur at higher temperatures (Cavali et al. 2023) (Fig. 1.2). Higher pressure is typically needed for hydrothermal processes than for gasification, pyrolysis, or incineration. While gasification and incineration are carried out in an oxidative environment, pyrolysis is carried out in an inert atmosphere (Cui et al. 2021) (Fig. 1.2). Supercritical water is used for hydrothermal gasification, while subcritical water is used for hydrothermal carbonization and liquefaction (Ojewumi and Chen 2024).

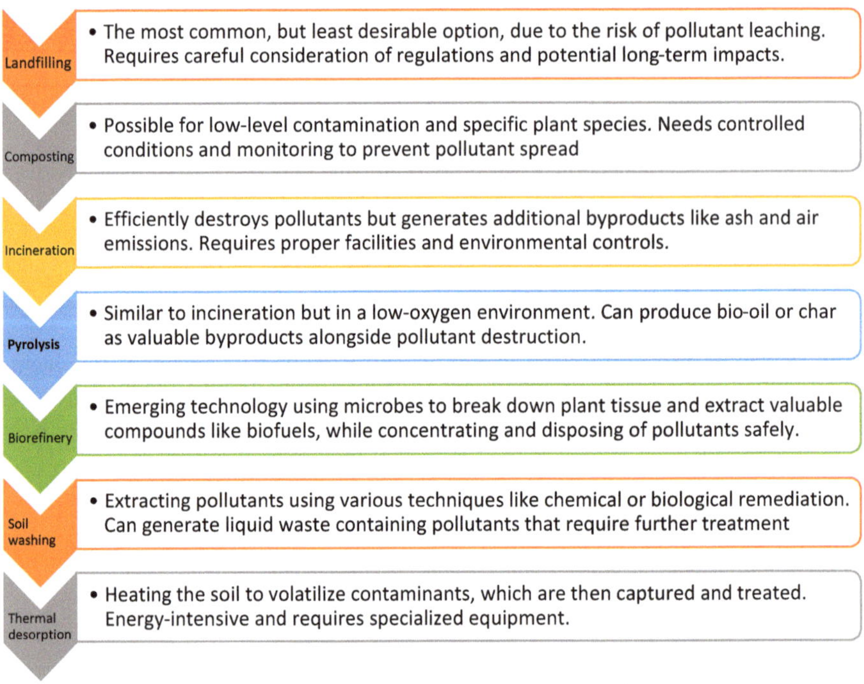

Fig. 1.2 Management option

Several publications have mentioned that one of the post-harvest biomass treatments is composting and compaction (Olson et al. 2024). Composting material leached to form soluble organic compounds, which increased metal solubility (Ameen and Dohuki 2024). Composting can dramatically reduce the amount of biomass removed from phytoremediation techniques, as demonstrated by Ameen and Dohuki (2024). Metal-contaminated plant biomass would still need to be treated before disposal. Compaction of contaminated plant biomass could result in a decrease in total dry weight (Madhav et al. 2024). This is useful because it may reduce the cost of transportation to a hazardous waste disposal site. Paniagua-López et al. (2024) proposed that compaction of collected plant material was important for processing metal-rich phytoextraction residue.

1.10 Operational Requirements Expanded

Identifying pollutants in phytoremediation byproducts is crucial for handling, storage, and disposal. Chemical analysis, specialized equipment, and labeling prevent cross-contamination. Worker safety is prioritized through training, protective equipment, and emergency protocols.

1.10.1 Identification and Classification

Importance Knowing the exact type and concentration of pollutants in the byproducts is crucial for determining the appropriate handling, storage, and disposal methods (Pathak et al. 2020).

Methods Chemical analysis, laboratory testing, and specialized equipment can be used to identify and quantify the contaminants.

Benefits Enables informed decision-making about safe and responsible management options, minimizing risks and optimizing costs (Abbaslou et al. 2018; Pathak et al. 2020).

1.10.2 Separation and Storage

Importance Separating and containing contaminated materials prevents cross-contamination and protects workers, the environment, and the general public (Gupta and Prakash 2013).

Methods Dedicated containers, designated storage areas, and proper labeling are essential. Depending on the byproducts, additional measures like leak-proof containers or temperature-controlled storage may be necessary (Ahmed et al. 2001; Gupta and Prakash 2013).

Benefits Mitigates potential risks, ensures traceability, and facilitates safe transportation and disposal when ready (Alshehri et al. 2023).

1.10.3 Labeling and Tracking

Importance Clear and accurate labeling on containers and documentation throughout the handling process ensure accountability and prevent accidental mishandling (Celin et al. 2020).

Methods Using standardized labels that specify the type of byproduct, date, origin, and potential hazards. Maintain records of transportation, storage, and disposal activities (Celin et al. 2020).

Benefits Promotes transparency, simplifies regulatory compliance, and enables tracing the material throughout its lifecycle for potential future audits or investigations (Latif et al. 2023).

1.10.4 Worker Safety

Importance Workers involved in handling phytoremediation byproducts should be aware of the potential risks and equipped with appropriate protective measures (Dessai et al. 2023).

Methods Training on safe handling procedures, personal protective equipment (PPE) suitable for the specific contaminants, emergency response protocols, and hygiene practices (Dessai et al. 2023).

Benefits Protects workers' health and safety, minimizes the risk of accidents and exposure, and fosters a culture of responsible environmental management (Adel et al. 2018). Adherence to these requirements helps reduce the risk of unsafe handling of byproducts and environmental harm.

Table 1.2 Management of phytoremediation byproduct

Management of phytoremediation byproducts	Handling requirements	Correct disposal	Overall management	References
Operational requirements	Correct identification and classification of byproducts for proper disposal guidance	Implementation of separation procedures to isolate contaminated material from clean material	Development of a comprehensive management plan, including labeling, tracking, and worker safety measures	Miller (1996)
Handling requirements	Proper training for workers on potential risks associated with byproducts	Workers utilize appropriate safety equipment to minimize exposure	Periodic monitoring and evaluation of handling procedures for compliance with safety standards	Liu and Tran (2021)
Correct disposal	Identification of suitable disposal methods based on the nature of the byproducts	Compliance with local and international environmental regulations for proper disposal	Collaboration with authorized waste disposal facilities for secure and legal disposal	Liu and Tran (2021)

Treaties and Agreements: Building a Legal Framework

International environmental treaties provide a crucial legal framework for coordinated action. The Minamata Convention on Mercury, for instance, aims to control and ultimately eliminate mercury use, a significant source of global metal pollution. Similarly, the Stockholm Convention on Persistent Organic Pollutants addresses the production and use of persistent organic pollutants, many of which contain toxic metals (Selin et al. 2018). These treaties, coupled with regional agreements like the Espoo Convention on Environmental Impact Assessment in a Transboundary Context, form the bedrock of international cooperation in tackling metal pollution (Environmental Assessment, Espoo Convention on Environmental Impact Assessment in a Transboundary, 2003) (Table 1.2).

Knowledge Exchange and Best Practices: A Pillar of International Cooperation

Exchanging knowledge and best practices is another critical pillar of international cooperation. Developed countries, with their advanced technologies and expertise, can play a vital role in assisting developing nations grappling with metal pollution. This can be achieved through capacity-building programs, technical assistance, and knowledge dissemination initiatives. For instance, the Global Environment Facility (GEF), an international partnership for environmental action, has supported numerous projects in developing countries aimed at reducing mercury use and promoting cleaner technologies.

1.11 Conclusion

Environmental pollution, particularly soil contamination with heavy metals and organic pollutants, poses a global challenge. Traditional treatment methods often fall short, being expensive, ineffective, and environmentally damaging. Enter phytoremediation, a shining ray of hope offering a promising, greener solution. Harnessing nature's power, this technology utilizes plants' innate ability to extract or stabilize pollutants from soil, making it cost-effective and sustainable. Unlike its harsh counterparts, phytoremediation permanently sequesters pollutants near plant roots, leaving the soil structure intact and avoiding harmful chemicals.

However, this green marvel isn't without its limitations. Time can be a foe, especially on heavily contaminated sites. Additionally, some plants may accumulate metals, risking their entry into the food chain. Careful safety evaluation is crucial, precluding the use of such plants as food sources.

Looking ahead, several avenues can pave the way for enhanced efficiency and broader application of phytoremediation:

(i) Phytochemical research: Unveiling and isolating the powerful agents within plants responsible for tackling specific pollutants, paving the way for more targeted solutions.
(ii) Gene modification: Exploring the potential of genetic engineering to empower plants with enhanced pollutant-removal capabilities, allowing them to combat a wider range of contaminants.
(iii) Byproduct management: Developing sustainable and safe strategies for handling the byproducts of phytoremediation, such as recycling or reuse, minimizing environmental impacts and resource waste.
(iv) Social acceptance: Raising public awareness and garnering social acceptance for this green technology by highlighting its diverse applications and environmental benefits.
(v) Phytoremediation, with its unique potential for environmentally friendly and sustainable soil remediation, stands poised to make a significant contribution to our collective future. By harnessing the combined efforts of scientists, policymakers, and communities, we can unlock the full potential of this technology and ensure a cleaner, healthier planet for generations to come.

Future Horizon

The battle against toxic metal pollution is far from over. However, by harnessing the power of international cooperation, knowledge sharing, and unwavering commitment, we can build a future where communities across the globe breathe easier, knowing they are no longer at the mercy of a borderless threat. By working together, we can ensure that the specter of toxic metals fades into a distant memory, replaced by a world where clean air, clean water, and healthy ecosystems are not privileges but fundamental rights enjoyed by all. Combating the dangers of toxic metals and nonmetals is an essential challenge. This challenge requires sustainable solutions that balance environmental protection, public health protection, and economic growth. Only through a concerted effort of scientific research, technological innovation, responsible industry practices, and public awareness can we create a healthy and sustainable future free from the harmful effects of toxic metals.

References

Abbaslou H, Bakhtiari S, Hashemi SS (2018) Rehabilitation of iron ore mine soil contaminated with heavy metals using rosemary phytoremediation-assisted Mycorrhizal Arbuscular fungi bioaugmentation and fibrous clay mineral immobilization. Iran J SciTechnol Trans A Sci 42:431–441. https://doi.org/10.1007/s40995-018-0543-7

Abukari A, Kaba JS, Dawoe E, Abunyewa AA (2022) A comprehensive review of the effects of biochar on soil physicochemical properties and crop productivity. Waste Dispos Sustain Energy 4:343–359. https://doi.org/10.1007/s42768-022-00114-2

Adamczyk-Szabela D, Wolf WM (2022) The impact of soil pH on heavy metals uptake and photosynthesis efficiency in Melissa officinalis, Taraxacum officinalis, Ocimum basilicum. Molecules 27:4671. https://doi.org/10.3390/molecules27154671

Adel M, Conti GO, Dadar M, Masoud M, Copat C, Ferrante M (2016) Heavy metal concentrations in edible muscle of whitecheek shark, *Carcharhinus dussumieri* (elasmobranchii, chondrichthyes) from the Persian Gulf: a food safety issue. Food Chem Toxicol 97:135–140. https://doi.org/10.1016/j.fct.2016.09.002

Adel M, Copat C, Asl MRS, Conti GO, Babazadeh M, Ferrante M (2018) Bioaccumulation of trace metals in banded Persian bamboo shark (*Chiloscyllium arabicum*) from the Persian Gulf: a food safety issue. Food Chem Toxicol 113:198–203. https://doi.org/10.1016/j.fct.2018.01.027

Ahmad W, Najeeb U, Zia MH (2015) Soil contamination with metals: sources, types and implications. In: Sources, types and implications., soil remediation and plants: prospects and challenges. Elsevier Inc. https://doi.org/10.1016/B978-0-12-799937-1.00002-4

Ahmad A, Tariq S, Zaman JU, Perales AIM, Mubashir M, Luque R (2022) Recent trends and challenges with the synthesis of membranes: industrial opportunities towards environmental remediation. Chemosphere 306. https://doi.org/10.1016/j.chemosphere.2022.135634

Ahmed KS, Panwar BS, Gupta SP (2001) Phytoremediation of cadmium-contaminated soil by brassica species. Acta Agronomica Hungarica 49:351–360. https://doi.org/10.1556/AAgr.49.2001.4.6

Allan M, Fagel N, Van Rampelbergh M, Baldini J, Riotte J, Cheng H, Lawrence ER, Gillikin D, Quinif Y, Verheyden S (2015) Lead concentrations and isotope ratios in speleothems as proxies for atmospheric metal pollution since the industrial revolution. Chem Geol 401:140–150. https://doi.org/10.1016/j.chemgeo.2015.02.035

Al-Naemi HA, Das SC (2020) Cadmium-induced endothelial dysfunction mediated by asymmetric dimethylarginine. Environ Sci Pollut Res 27:16246–16253. https://doi.org/10.1007/s11356-020-08116-5

Alshehri K, Gao Z, Harbottle M, Sapsford D, Cleall P (2023) Life cycle assessment and cost-benefit analysis of nature-based solutions for contaminated land remediation: a mini-review. Heliyon 9. https://doi.org/10.1016/j.heliyon.2023.e20632

Ameen HA, Dohuki MSS (2024) The effect of leaching and bulking agents on the quality of municipal solid waste compost. Environ Technol 45(1):144–157

Anand U, Pal T, Yadav N, Singh VK, Tripathi V, Choudhary KK, Shukla AK, Sunita K, Kumar A, Bontempi E, Ma Y, Kolton M, Singh AK (2023) Current scenario and future prospects of Endophytic microbes: promising candidates for abiotic and biotic stress management for agricultural and environmental sustainability. Microb Ecol 86:1455–1486. https://doi.org/10.1007/s00248-023-02190-1

Ayilara MS, Adeleke BS, Adebajo MT, Akinola SA, Fayose CA, Adeyemi UT, Gbadegesin LA, Omole RK, Johnson RM, Edhemuino M, Ogundolie FA, Babalola OO (2023) Remediation by enhanced natural attenuation; an environment-friendly remediation approach. Front Environ Sci 11. https://doi.org/10.3389/fenvs.2023.1182586

Bhat SA, Bashir O, Haq SAU, Amin T, Rafiq A, Ali M et al (2022) Phytoremediation of heavy metals in soil and water: an eco-friendly, sustainable and multidisciplinary approach. Chemosphere 303:134788

Bhattacharyya N (2022) Naturally growing native plants of wastelands: their stress management strategies and prospects in changing climate. In: Advances in science, technology and innovation, pp 149–168. https://doi.org/10.1007/978-3-030-95365-2_10

Bjorklund G, Dadar M, Mutter J, Aaseth J (2017) The toxicology of mercury: current research and emerging trends. Environ Res 159:545–554. https://doi.org/10.1016/j.envres.2017.08.051

Bulka CM, Bryan MS, Persky VW, Daviglus ML, Durazo-Arvizu RA, Faruque P, Slavkovich V, Graziano JH, Islam T, Baron JA, Ahsan H, Argos M (2019) Changes in blood pressure associated with lead, manganese, and selenium in a Bangladeshi cohort. Environ Pollut 248:28–35. https://doi.org/10.1016/j.envpol.2019.01.129

Cao S, Duan X, Zhao X, Wang B, Ma J, Fan D, Sun C, He B, Wei F, Jiang G (2015) Levels and source apportionment of children's lead exposure: could urinary lead be used to identify the levels and sources of children's lead pollution? Environ Pollut 199:18–25. https://doi.org/10.1016/j.envpol.2014.12.035

Cao J, Lu B, Chen Y, Zhang X, Zhai G, Zhou G, Jiang B, Schnoor JL (2016) Extended producer responsibility system in China improves e-waste recycling: government policies, enterprise, and public awareness. Renew Sustain Energy Rev 62:882–894. https://doi.org/10.1016/j.rser.2016.04.078

Cavali M, Soccol CR, Tavares D, Zevallos Torres LA, Oliveira de Andrade Tanobe V, Zandona Filho A, Woiciechowski AL (2023) Valorization of lignin from pine (Pinus spp.) residual sawdust: antioxidant activity and application in the green synthesis of silver nanoparticles for antibacterial purpose. Biomass Conversion and Biorefinery 13(11):10051–10063

Celin SM, Sahai S, Kalsi A, Bhanot P (2020) Environmental monitoring approaches used during bioremediation of soils contaminated with hazardous explosive chemicals. Trends Environ Anal Chem 26:e00088. https://doi.org/10.1016/j.teac.2020.e00088

Chandra R, Kumar V (2017) Phytoextraction of heavy metals by potential native plants and their microscopic observation of root growing on stabilised distillery sludge as a prospective tool for in situ phytoremediation of industrial waste. Environ Sci Pollut Res Int 24(3):2605–2619

Chen W, Li H (2018) Cost-effectiveness analysis for soil heavy metal contamination treatments. Water Air Soil Pollut 229. https://doi.org/10.1007/s11270-018-3784-3

Chen Y, Shen Z, Li X (2004) The use of vetiver grass (Vetiveria zizanioides) in the phytoremediation of soils contaminated with heavy metals. Appl Geochem 19:1553–1565. https://doi.org/10.1016/j.apgeochem.2004.02.003

Chirakadze A, Buachidze Z, Khomeriki IO, Gvakharia V, Stamateli M, Chakvetadze N, Chokheli M, Toscano WA, Chigogidze K, Gvertsiteli L, Bagrationi N (2016) Arsenic pollution of soils and morbidity prevalence in Racha-lower Svaneti district of Georgia. Int J Glob Warming 10:92–114. https://doi.org/10.1504/IJGW.2016.077908

Coelho FC, Squitti R, Ventriglia M, Cerchiaro G, Daher JP, Rocha JG, Rongioletti MCA, Moonen A-C (2020) Agricultural use of copper and its link to Alzheimer's disease. Biomolecules 10. https://doi.org/10.3390/biom10060897

Cui X, Mao P, Sun S, Huang R, Fan Y, Li Y, Li Z (2021) Phytoremediation of cadmium contaminated soils by Amaranthus Hypochondriacus L.: The effects of soil properties highlighting cation exchange capacity. Chemosphere 283:131067

Das S, Sultana KW, Ndhlala AR, Mondal M, Chandra I (2023) Heavy metal pollution in the environment and its impact on health: exploring green technology for remediation. Environ Health Insights 17. https://doi.org/10.1177/11786302231201259

Demirel Z (2007) Monitoring of heavy metal pollution of groundwater in a phreatic aquifer in Mersin-Turkey. Environ Monit Assess 132:15–23. https://doi.org/10.1007/s10661-006-9498-9

Dessai NSP, Juvekar VS, Nasnodkar MR (2023) Assessment of metal bioavailability in sediments and bioaccumulation in edible bivalves, and phyto-remediation potential of mangrove plants in the tropical (Kali) estuary, India. Mar Pollut Bull 194:115419. https://doi.org/10.1016/j.marpolbul.2023.115419

Dhaliwal SS, Singh J, Taneja PK, Mandal A (2020) Remediation techniques for removal of heavy metals from the soil contaminated through different sources: a review. Environ Sci Pollut Res 27:1319–1333. https://doi.org/10.1007/s11356-019-06967-1

do Nascimento CW, Miranda BC, Vieira da Silva FB, Vieira Lima LH (2021) Using plants to remediate or manage metal-polluted soils: an overview on the current state of phytotechnologies. Acta Sci Agron 43. https://doi.org/10.4025/actasciagron.v43i1.58283

Dou C, Cui H, Zhang W, Yu W, Sheng X, Zheng X (2023) Copper and cadmium accumulation and Phytorextraction potential of native and cultivated plants growing around a copper smelter. Agronomy (Basel) 13. https://doi.org/10.3390/agronomy13122874

Du Z, Zhu H, Zhou Q, Wong YD (2017) Challenges and solutions for ship recycling in China. Ocean Eng 137:429–439. https://doi.org/10.1016/j.oceaneng.2017.04.004

Dudka I, Kossowska B, Senhadri H, Latajka R, Hajek J, Andrzejak R, Antonowicz-Juchniewicz J, Gancarz R (2014) Metabonomic analysis of serum of workers occupationally exposed to arsenic, cadmium and lead for biomarker research: a preliminary study. Environ Int 68:71–81. https://doi.org/10.1016/j.envint.2014.03.015

Eissa MA (2019) Effect of compost and biochar on heavy metals Phytostabilization by the halophytic plant old man saltbush [*Atriplex Nummularia* Lindl]. Soil Sediment Contam 28:135–147. https://doi.org/10.1080/15320383.2018.1551325

Farid M, Farid S, Zubair M, Ghani MA, Rizwan M, Ishaq HK, Alkahtani S, Abdel-Daim MM, Ali S (2020) Glutamic acid-assisted Phytomanagement of chromium contaminated soil by sunflower (Helianthus annuus L.): morphophysiological and biochemical alterations. Front Plant Sci 11. https://doi.org/10.3389/fpls.2020.01297

Frazzoli C, Ruggieri F, Battistini B, Orisakwe OE, Igbo JK, Bocca B (2022) E-Waste threatens health: the scientific solution adopts the one health strategy. Environ Res 212. https://doi.org/10.1016/j.envres.2022.113227

Ganguly S, Basuli D, Chatterjee M, Seal S, Sinha D, Chowdhury S (2023) Sustainability of aromatic plants in remediation of cadmium from contaminated sites. In: Phytoremediation potential of medicinal and aromatic plants: a bioeconomical approach. CRC Press. https://doi.org/10.1201/9781003290780-10

Garbisu C, Alkorta I (2001) Phytoextraction: a cost-effective plant-based technology for the removal of metals from the environment. Bioresour Technol 77:229–236. https://doi.org/10.1016/S0960-8524(00)00108-5

Garces-Pastor S, Fletcher WJ, Ryan PA (2023) Ecological impacts of the industrial revolution in a lowland raised peat bog near Manchester, NW England. Ecol Evol 13. https://doi.org/10.1002/ece3.9807

Gori A, Ferrini F, Fini A (2019) Growing healthy food under heavy metal pollution load: overview and major challenges of tree based edible landscapes. Urban For Urban Green 38:403–406. https://doi.org/10.1016/j.ufug.2019.01.010

Gupta C, Prakash D (2013) Duckweed: an effective tool for phyto-remediation. Toxicol Environ Chem 95:1256–1266. https://doi.org/10.1080/02772248.2013.879309

Gupta A, Dubey P, Kumar M, Roy A, Sharma D, Khan MM, Bajpai AB, Shukla RP, Pathak N, Hasanuzzaman M (2022) Consequences of arsenic contamination on plants and Mycoremediation-mediated arsenic stress tolerance for sustainable agriculture. Plants (Basel) 11. https://doi.org/10.3390/plants11233220

Haider FU, Liqun C, Coulter JA, Cheema SA, Wu J, Zhang R, Wenjun M, Farooq M (2021) Cadmium toxicity in plants: impacts and remediation strategies. Ecotoxicol Environ Saf 211:111887. https://doi.org/10.1016/j.ecoenv.2020.111887

Iqbal S, Farid M, Zubair M, Asam ZUZ, Ali S, Abubakar M, Farid S, Rizwan M (2022) Efficacy of various amendments for the phytomanagement of heavy metal contaminated sites and sus-

tainable agriculture. A review. In: Managing plant production under changing environment. Springer Nature. https://doi.org/10.1007/978-981-16-5059-8_9
Jaswal A, Singh A, Sarkar S, Singh M (2023) Optimized phytoremediation process for the sustainable management of radionuclides in the food chain. J Food Chem Nanotechnol 9(S1):S291–S299
Jung MH, Ji WH, Lee JS, Yang IJ (2020) Soil neutralizer selection for Phytostabilzation using *Miscanthus sinensis* Anderss. In heavy metal contaminated soil of abandoned metal mine. Econ Environ Geol 53:517–528. https://doi.org/10.9719/EEG.2020.53.5.517
Juwarkar AA, Yadav SK, Kumar GP (2012) Jatropha and phytoremediation of metal contaminated land. In: Jatropha, challenges for a new energy crop: Volume 1: farming, economics and biofuel. Springer, New York. https://doi.org/10.1007/978-1-4614-4806-8_23
Kim K-R, Owens G (2011) Potential for enhanced phytoremediation of landfills using biosolids – a review. In: Comprehensive biotechnology, 2nd edn. Elsevier Inc. https://doi.org/10.1016/B978-0-08-088504-9.00373-1
Kim K-R, Owens G (2019) Potential for enhanced phytoremediation of landfills using biosolids – a review. In: Comprehensive biotechnology. Elsevier. https://doi.org/10.1016/B978-0-444-64046-8.00354-2
Kim YT, Kim WJ, Choi JE, Bae M, Jang H, Lee CJ, Lee H-J, Im DJ, Ye BS, Kim M-J, Jeong Y, Oh SS, Jung Y-C, Kang ES, Park S, Koo LS, Park KS, Koh SB, Kim C (2020) Cohort profile: firefighter research on the enhancement of safety and health (FRESH), a prospective cohort study on Korean firefighters. Yonsei Med J 61:103–109. https://doi.org/10.3349/ymj.2020.61.1.103
Koley S (2022) Future perspectives and mitigation strategies towards groundwater arsenic contamination in West Bengal, India. Environ Qual Manag 31:75–97. https://doi.org/10.1002/tqem.21784
Korzeniowska J, Stanislawska-Glubiak E (2023) The phytoremediation potential of local wild grass versus cultivated grass species for zinc-contaminated soil. Agronomy 13:160. https://doi.org/10.3390/agronomy13010160
Krueger E, Darland J, Goldyn S, Swanson R, Lehmann R, Shepardson S, Karpovich D (2013) Water leaching of chelated Pb complexes from post-phytoremediation biomass. Water Air Soil Pollut 224. https://doi.org/10.1007/s11270-013-1615-0
Kucaj E, Osmani M, Gjoni A, Bardhi A, Kucaj B (2023) Phytostabilization potential of Salix alba as biomonitors of heavy metals in soil pollution near the Erzeni Riverbank. Int J Environ Sci Dev 14:30–36. https://doi.org/10.18178/ijesd.2023.14.1.1411
Kulathunga MRDL, Wijayawardena MAA, Naidu R (2021) Heavy metal(loid)s and health risk assessment of Dambulla vegetable market in Sri Lanka. Environ Monit Assess 193. https://doi.org/10.1007/s10661-021-09020-2
Kumar V (2020) Phytoremediation of distillery effluent: current progress, challenges, and future opportunities. In: Bioremediation for environmental sustainability: toxicity, mechanisms of contaminants degradation, detoxification and challenges. Elsevier. https://doi.org/10.1016/B978-0-12-820524-2.00014-6
Kumar V, Kaushal A, Shah MP, Singh K (2020) Phytoaugmentation technology for phytoremediation of environmental pollutants: current scenario and future prospects. In: Bioremediation for environmental sustainability: approaches to tackle pollution for cleaner and greener society. Elsevier Applied Science. https://doi.org/10.1016/B978-0-12-820318-7.00016-2
Kumar K, Shinde A, Aeron V, Verma A, Arif NS (2023) Genetic engineering of plants for phytoremediation: advances and challenges. J Plant Biochem Biotechnol 32:12–30. https://doi.org/10.1007/s13562-022-00776-3
Latif A, Abbas A, Iqbal J, Azeem M, Asghar W, Ullah R, Bilal M, Arsalan M, Khan M, Latif R, Ehsan M, Abbas A, Bashir S, Bashir S, Saifullah Khan K, Sun K, Kang W, Bashir F, Chen Z (2023) Remediation of environmental contaminants through Phytotechnology. Water Air Soil Pollut 234. https://doi.org/10.1007/s11270-023-06112-2
Lauwerys R, Lison D (1994) Health risks associated with cobalt exposure – an overview. Sci Total Environ 150:1–6. https://doi.org/10.1016/0048-9697(94)90125-2

Li Z, Jia M, Wu L, Christie P, Luo Y (2016) Changes in metal availability, desorption kinetics and speciation in contaminated soils during repeated phytoextraction with the Zn/Cd hyperaccumulator *Sedum plumbizincicola*. Environ Pollut 209:123–131. https://doi.org/10.1016/j.envpol.2015.11.015

Li C, Ji X, Luo X (2019) Phytoremediation of heavy metal pollution: a bibliometric and Scientometric analysis from 1989 to 2018. Int J Environ Res Public Health 16. https://doi.org/10.3390/ijerph16234755

Li X, Zhao L, Teng Y, Luo Y, Zhao Q (2023) Effects of cadmium on mercury accumulation and transformation by *Arundo donax* L. Environ Sci Pollut Res 30:62461–62469. https://doi.org/10.1007/s11356-023-26516-1

Lim J-M, Salido AL, Butcher DJ (2004) Phytoremediation of lead using Indian mustard (Brassica juncea) with EDTA and electrodics. Microchem J 76:3–9. https://doi.org/10.1016/j.microc.2003.10.002

Liu Z, Tran KQ (2021) A review on disposal and utilization of phytoremediation plants containing heavy metals. Ecotoxicol Environ Saf 226:112821. https://doi.org/10.1016/j.ecoenv.2021.112821

Liu S, Yang B, Liang Y, Xiao Y, Fang J (2020) Prospect of phytoremediation combined with other approaches for remediation of heavy metal-polluted soils. Environ Sci Pollut Res 27:16069–16085. https://doi.org/10.1007/s11356-020-08282-6

Madejon P, Burgos P, Cabrera F, Madejon E (2009) Phytostabilization of amended soils polluted with trace elements using the Mediterranean shrub: Rosmarinus Officinalis. Int J Phytoremediation 11:542–557. https://doi.org/10.1080/15226510902717572

Madhav S, Mishra R, Kumari A, Srivastav AL, Ahamad A, Singh P et al (2024) A review on sources identification of heavy metals in soil and remediation measures by phytoremediation-induced methods. Int J Environ Sci Technol 21(1):1099–1120

Mahar A, Wang P, Ali A, Kumar AM, Lahori AH, Wang Q, Li R, Zhang Z (2016) Challenges and opportunities in the phytoremediation of heavy metals contaminated soils: a review. Ecotoxicol Environ Saf 126:111–121. https://doi.org/10.1016/j.ecoenv.2015.12.023

Malaviya P, Singh A (2012) Phytoremediation strategies for remediation of uranium-contaminated environments: a review. Crit Rev Environ Sci Technol 42:2575–2647. https://doi.org/10.1080/10643389.2011.592761

Mani D, Kumar C, Patel NK (2015) Hyperaccumulator oilcake manure as an alternative for chelate-induced phytoremediation of heavy metals contaminated alluvial soils. Int J Phytoremediation 17:256–263. https://doi.org/10.1080/15226514.2014.883497

Marques APGC, Rangel AOSS, Castro PML (2009) Remediation of heavy metal contaminated soils: phytoremediation as a potentially promising clean-up technology. Crit Rev Environ Sci Technol 39:622–654. https://doi.org/10.1080/10643380701798272

Makarova A, Nikulina E, Avdeenkova T, Pishaeva K (2021) The improved phytoextraction of heavy metals and the growth of Trifolium repens L.: The role of K2HEDP and plant growth regulators alone and in combination. Sustainability 13(5):2432

Martini S, Afroze S, Roni KA, Setiawati M, Kharismadewi D (2021) A review of fruit waste-derived sorbents for dyes and metals removal from contaminated water and wastewater. Desalin Water Treat 235:300–323. https://doi.org/10.5004/dwt.2021.27658

Mataruga Z, Jaric S, Kostic O, Markovic M, Jakovljevic K, Mitrovic M, Pavlovic P (2020) The potential of elm trees (*Ulmus glabra* Huds.) for the phytostabilisation of potentially toxic elements in the riparian zone of the Sava River. Environ Sci Pollut Res 27:4309–4324. https://doi.org/10.1007/s11356-019-07173-9

Meltzer GY, Watkins B-X, Dorice V, Zelikoff JT, Boden-Albala B (2020) A systematic review of environmental health outcomes in selected American Indian and Alaska native populations. J Racial Ethn Health Disparities 7:698–739. https://doi.org/10.1007/s40615-020-00700-2

Mench M, Lepp N, Bert V, Schwitzguebel J-P, Gawronski SW, Schroeder P, Vangronsveld J (2010) Successes and limitations of phytotechnologies at field scale: outcomes, assessment and outlook from COST Action 859. J Soils Sediments 10:1039–1070. https://doi.org/10.1007/s11368-010-0190-x

Mengoni A, Schat H, Vangronsveld J (2010) Plants as extreme environments? Ni-resistant bacteria and Ni-hyperaccumulators of serpentine flora. Plant Soil 331:5–16. https://doi.org/10.1007/s11104-009-0242-4

Mensah AK, Marschner B, Antoniadis V, Stemn E, Shaheen SM, Rinklebe J (2021) Human health risk via soil ingestion of potentially toxic elements and remediation potential of native plants near an abandoned mine spoil in Ghana. Sci Total Environ 798. https://doi.org/10.1016/j.scitotenv.2021.149272

Miller RR (1996) Phytoremediation, ground water remediation technologies analysis center. https://clu-in.org/download/toolkit/phyto_o.pdf

Mishra A (2020) Phytoremediation of heavy metal-contaminated soils: recent advances, challenges, and future prospects. In: Bioremediation for environmental sustainability: toxicity, mechanisms of contaminants degradation, detoxification and challenges. Elsevier. https://doi.org/10.1016/B978-0-12-820524-2.00002-X

Moiseenko TI, Morgunov BA, Gashkina NA, Megorskiy VV, Pesiakova AA (2018) Ecosystem and human health assessment in relation to aquatic environment pollution by heavy metals: case study of the Murmansk region, northwest of the Kola Peninsula, Russia. Environ Res Lett 13. https://doi.org/10.1088/1748-9326/aab5d2

Moon MK, Lee I, Lee A, Park H, Kim MJ, Kim S, Cho YH, Hong S, Yoo J, Cheon GJ, Choi K, Park YJ, Park J (2022) Lead, mercury, and cadmium exposures are associated with obesity but not with diabetes mellitus: Korean National Environmental Health Survey (KoNEHS) 2015–2017. Environ Res 204. https://doi.org/10.1016/j.envres.2021.111888

Moradmand M, Harchegani HB (2011) Treated municipal wastewater irrigation effect on lead content and health risks of nickel in soil and pepper in Shahrekord, Iran. Desalin Water Treat 28:42–45. https://doi.org/10.5004/dwt.2011.2198

Murphy HM, McBean EA, Farahbakhsh K (2009) Appropriate technology—a comprehensive approach for water and sanitation in the developing world. Technol Soc 31:158–167. https://doi.org/10.1016/j.techsoc.2009.03.010

Novak JM, Ippolito JA, Watts DW, Sigua GC, Ducey TF, Johnson MG (2019) Biochar compost blends facilitate switchgrass growth in mine soils by reducing Cd and Zn bioavailability. Biochar 1:97–114. https://doi.org/10.1007/s42773-019-00004-7

Odoh CK, Zabbey N, Sam K, Eze CN (2019) Status, progress and challenges of phytoremediation – an African scenario. J Environ Manag 237:365–378. https://doi.org/10.1016/j.jenvman.2019.02.090

Ojewumi ME, Chen G (2024) Hydrochar production by hydrothermal carbonization: microwave versus supercritical water treatment. Biomass 4(2):574–598

Okereafor U, Makhatha M, Mekuto L, Uche-Okereafor N, Sebola T, Mavumengwana V (2020) Toxic metal implications on agricultural soils, plants, animals, aquatic life and human health. Int J Environ Res Public Health 17. https://doi.org/10.3390/ijerph17072204

Olayiwola VA, Mesele SA, Ajayi EO (2023) Towards a clean environment: to what extent can trees serve as phytoremediators for chemically polluted soils? Environ Chall 13. https://doi.org/10.1016/j.envc.2023.100764

Olson NE, Neher DA, Holden VI (2024) On-farm conversion of cannabis sativa waste biomass into an organic fertilizer by microbial digestion. Compost Sci Util 31(1–2):38–54

Onyena AP, Folorunso OM, Nwanganga N, Udom GJ, Ekhator OC, Frazzoli C, Ruggieri F, Bocca B, Orisakwe OE (2023) Engaging one health in heavy metal pollution in some selected Nigerian Niger Delta cities. A systematic review of pervasiveness, bioaccumulation and subduing environmental health challenges. Biol Trace Elem Res. https://doi.org/10.1007/s12011-023-03762-5

Otunola BO, Aghoghovwia MP, Thwala M, Ololade OO (2022) Heavy metal phytoremediation potential of Vetiver grass and Indian mustard update on enhancements and research opportunities. Water Air Soil Pollut 233. https://doi.org/10.1007/s11270-022-05620-x

Padmavathiamma PK, Li LY (2007) Phytoremediation technology: hyper-accumulation metals in plants. Water Air Soil Pollut 184:105–126. https://doi.org/10.1007/s11270-007-9401-5

Paithankar JG, Saini S, Dwivedi S, Sharma A, Chowdhuri DK (2021) Heavy metal associated health hazards: an interplay of oxidative stress and signal transduction. Chemosphere 262. https://doi.org/10.1016/j.chemosphere.2020.128350

Paniagua-López M, Silva-Castro GA, Romero-Freire A, Martín-Peinado FJ, Sierra-Aragón M, García-Romera I (2024) Integrating waste valorization and symbiotic microorganisms for sustainable bioremediation of metal (loid)-polluted soils. Sci Total Environ 945:174030

Pappas RS, Polzin GM, Watson CH, Ashley DL (2007) Cadmium, lead, and thallium in smoke particulate from counterfeit cigarettes compared to authentic US brands. Food Chem Toxicol 45:202–209. https://doi.org/10.1016/j.fct.2006.08.001

Pathak S, Agarwal AV, Pandey VC (2020) Phytoremediation—a holistic approach for remediation of heavy metals and metalloids. In: Bioremediation of pollutants: from genetic engineering to genome engineering. Elsevier. https://doi.org/10.1016/B978-0-12-819025-8.00001-6

Perlein A, Bert V, Desannaux O, Fernandes de Souza M, Papin A, Gaucher R, Zdanevitch I, Meers E (2021) The use of *sorghum* in a phytoattenuation strategy: a field experiment on a TE-contaminated site. Appl Sci (Basel) 11. https://doi.org/10.3390/app11083471

Permenter MG, Lewis JA, Jackson DA (2011) Exposure to nickel, chromium, or cadmium causes distinct changes in the gene expression patterns of a rat liver derived cell line. PLoS One 6. https://doi.org/10.1371/journal.pone.0027730

Phouthavong-Murphy JC, Merrill AK, Zamule S, Giacherio D, Brown B, Roote C, Das P (2020) Phytoremediation potential of switchgrass (Panicum virgatum), two United States native varieties, to remove bisphenol-A (BPA) from aqueous media. Sci Rep 10:835. https://doi.org/10.1038/s41598-019-56655-w

Pirzadah TB, Malik B, Tahir I, Kumar M, Varma A, Rehman RU (2014) Phytoremediation: an eco-friendly green Technology for Pollution Prevention, control and remediation. An eco-friendly green technology for pollution prevention, control and remediation. In: Soil remediation and plants: prospects and challenges. Elsevier Inc. https://doi.org/10.1016/B978-0-12-799937-1.00005-X

Politis MD, Yao M, Gennings C, Tamayo-Ortiz M, Valvi D, Kim-Schulze S, Jingjing Q, Amarasiriwardena C, Pantic I, Cruz TM, Estrada-Gutierrez G, Greenberg JH, Tellez-Rojo MM, Wright RO, Sanders AP, Rosa MJ (2022) Prenatal metal exposures and associations with kidney injury biomarkers in children. Toxics 10. https://doi.org/10.3390/toxics10110692

Ponce-Hernández A, Maldonado-Miranda JJ, Medellin-Castillo NA, Alonso-Castro AJ, Carranza-Alvarez C (2020) Phytoremediation technology: sustainable solution for cleaning up of recalcitrant pollutants from disturbed environs. In: Bioremediation and biotechnology, Vol 3: persistent and recalcitrant toxic substances. Springer International Publishing. https://doi.org/10.1007/978-3-030-46075-4_11

Poria V, Dębiec-Andrzejewska K, Fiodor A, Lyzohub M, Ajijah N, Singh S, Pranaw K (2022) Plant growth-promoting bacteria (PGPB) integrated phytotechnology: a sustainable approach for remediation of marginal lands. Front Plant Sci 13. https://doi.org/10.3389/fpls.2022.999866

Prakash J (2020) Potential application of endophytes in bioremediation of heavy metals and organic pollutants and growth promotion: mechanism, challenges, and future prospects. In: Bioremediation for environmental sustainability: toxicity, mechanisms of contaminants degradation, detoxification and challenges. Elsevier. https://doi.org/10.1016/B978-0-12-820524-2.00005-5

Prasad J, Tiwari S, Singh BK, Dubey NK (2022) Phytoextraction of heavy metals: challenges and opportunities. In: Phytoremediation technology for the removal of heavy metals and other contaminants from soil and water. Elsevier. https://doi.org/10.1016/B978-0-323-85763-5.00011-8

Qin G, Niu Z, Yu J, Li Z, Ma J, Xiang P (2021) Soil heavy metal pollution and food safety in China: effects, sources and removing technology. Chemosphere 267. https://doi.org/10.1016/j.chemosphere.2020.129205

Qiong W, Fengshan P, Xiaomeng X, Tariq RM, Xiao'e Y, Bao C, Ying F (2021) Cadmium level and soil type played a selective role in the endophytic bacterial community of hyperaccumulator *Sedum alfredii* Hance. Chemosphere 263. https://doi.org/10.1016/j.chemosphere.2020.127986

Sabir M, Baltrėnaitė-Gedienė E, Ditta A, Ullah H, Kanwal A, Ullah S, Faraj TK (2022) Bioaccumulation of heavy metals in a soil–plant system from an open dumpsite and the associated health risks through multiple routes. Sustainability (Switzerland) 14. https://doi.org/10.3390/su142013223

Saha A, Mukherjee P, Roy K, Sen K, Sanyal T (2022) A review on phyto-remediation by aquatic macrophytes: a natural promising tool for sustainable management of ecosystem. Int J Exp Res Rev:27, 9–31. https://doi.org/10.52756/ijerr.2022.v27.002

Sahito ZA, Zehra A, Tang L, Ali Z, Laeeq ur Rehman Hashmi M, Bano N, Ullah MA, He Z, Yang X (2021) Arsenic and mercury uptake and accumulation in oilseed sunflower accessions selected to mitigate co-contaminated soil coupled with oil and bioenergy production. J Clean Prod 291:125226. https://doi.org/10.1016/j.jclepro.2020.125226

Sandhi A, Greger M (2012) Phytostablization of arsenic: is it possible by growing *Salix spp.* in the contaminated sites? In: Ng JC, Noller BN, Naidu R, Bundschuh J, Bhattacharya P (eds) Understanding the geological and medical interface of arsenic, as 2012, Arsenic in the environment. CRC Press/Taylor & Francis Group, Boca Raton, pp 328–329

Sarkar S, Basu S, Prasad R, Kumar G (2022) Phytoremediation: mechanistic approach for eliminating heavy metal toxicity from environment. In: Phytoremediation for environmental sustainability. Springer Singapore. https://doi.org/10.1007/978-981-16-5621-7_19

Saxena G, Purchase D, Mulla SI, Saratale GD, Bharagava RN (2020) Phytoremediation of heavy metal-contaminated sites: eco-environmental concerns, field studies, sustainability issues, and future prospects. Rev Environ Contam Toxicol 249:71–131. https://doi.org/10.1007/398_2019_24

Selin H, Keane SE, Wang S, Selin NE, Davis K, Bally D (2018) Linking science and policy to support the implementation of the Minamata convention on mercury. Ambio 47:198–215. https://doi.org/10.1007/s13280-017-1003-x

Shamsi IH, Jin X, Zhang X, Feng Q, Ibrahim Z, Adil MF, Karim MF, Ullah N (2023) The dynamics of phytoremediation of heavy metals: recent progress and future perspective. In: Heavy metal toxicity and tolerance in plants: a biological, omics, and genetic engineering approach. Wiley. https://doi.org/10.1002/9781119906506.ch25

Sharma M, Mathur J, Goswami P (2023) Ultrastructural analysis and biochemical evaluation provide evidence for arsenic reclamation from contaminated soil through Helianthus annuus L. Acta Physiol Plant 45. https://doi.org/10.1007/s11738-023-03561-4

Sharma A, Sharma S, Sharma S, Kumar A, Sharma V (2024) Phytoremediation: a clean and green approach for heavy metal remediation. In: Microbial applications for environmental sustainability, pp 257–276

Shehata SM, Badawy RK, Aboulsoud YIE (2019) Phytoremediation of some heavy metals in contaminated soil. Bull Natl Res Cent 43:189. https://doi.org/10.1186/s42269-019-0214-7

Singh A, Prasad SM (2011) Reduction of heavy metal load in food chain: technology assessment. Rev Environ Sci Biotechnol 10:199–214. https://doi.org/10.1007/s11157-011-9241-z

Siyar R, Ardejani FD, Norouzi P, Maghsoudy S, Yavarzadeh M, Taherdangkoo R, Butscher C (2022) Phytoremediation potential of native Hyperaccumulator plants growing on heavy metal-contaminated soil of Khatunabad copper smelter and refinery, Iran. Water (Basel) 14. https://doi.org/10.3390/w14223597

Srivastava P, Giri N (2021) Role of plants in phytoremediation of industrial waste. In: Bioprospecting of plant biodiversity for industrial molecules. Wiley. https://doi.org/10.1002/9781119718017.ch4

Suman J, Uhlik O, Viktorova J, Macek T (2018) Phytoextraction of heavy metals: a promising tool for clean-up of polluted environment? Front Plant Sci 9. https://doi.org/10.3389/fpls.2018.01476

Syed S, Xu M, Wang Z, Yu C, Lian B (2021) Invasive Spartina alterniflora in controlled cultivation: environmental implications of converging future technologies. Ecol Indic 130:108027. https://doi.org/10.1016/j.ecolind.2021.108027

Taux K, Kraus T, Kaifie A (2022) Mercury exposure and its health effects in Workers in the Artisanal and Small-Scale Gold Mining (ASGM) Sector-A systematic review. Int J Environ Res Public Health 19. https://doi.org/10.3390/ijerph19042081

Teh T, Norulaini NARN, Mohammad S, Wong Y, Omar AKM (2016) Risk assessment of metal contamination in soil and groundwater in Asia: a review of recent trends as well as existing environmental laws and regulations. Pedosphere 26:431–450. https://doi.org/10.1016/S1002-0160(15)60055-8

Testiati E, Parinet J, Massiani C, Laffont-Schwob I, Rabier J, Hans-Rudolf P, Lenoble V, Masotti V, Prudent P (2013) Trace metal and metalloid contamination levels in soils and in two native plant species of a former industrial site: evaluation of the phytostabilization potential. J Hazard Mater 248:131–141. https://doi.org/10.1016/j.jhazmat.2012.12.039

Tőzsér D, Magura T, Simon E (2017) Heavy metal uptake by plant parts of willow species: a meta-analysis. J Hazard Mater 336:101–109. https://doi.org/10.1016/j.jhazmat.2017.03.068

Tran KQ, Liu Z (2023) Thermochemical conversions of contaminated biomass for sustainable phytoremediation. In: Biofuel Technologies for a Sustainable Future: India and beyond. River Publishers, pp 193–223

van der Ent A, Baker AJM, Reeves RD, Pollard AJ, Schat H (2013) Hyperaccumulators of metal and metalloid trace elements: facts and fiction. Plant Soil 362:319–334. https://doi.org/10.1007/s11104-012-1287-3

van der Voet GB, Todorov TI, Centeno JA, Jonas W, Ives J, Mullick FG (2007) Metals and health: a clinical toxicological perspective on tungsten and review of the literature. Mil Med 172:1002–1005. https://doi.org/10.7205/MILMED.172.9.1002

Vandecasteele B, De Vos B, Tack FMG (2002) Cadmium and zinc uptake by volunteer willow species and elder rooting in polluted dredged sediment disposal sites. Sci Total Environ 299:191–205. https://doi.org/10.1016/S0048-9697(02)00275-9

Wen J, Zhou J, Zhang R, Ren W, Zhao J, Cai D (2024) Current advances of the valorization technologies for heavy metal containing hyperaccumulators. Ind Crop Prod 209:118051

Wu LH, Luo YM, Xing XR, Christie P (2004) EDTA-enhanced phytoremediation of heavy metal contaminated soil with Indian mustard and associated potential leaching risk. Agric Ecosyst Environ 102:307–318. https://doi.org/10.1016/j.agee.2003.09.002

Wu D, Yu X, Lai M, Feng J, Dong X, Peng W, Su S, Zhang X, Wan L, Jacobs DF, Zeng S (2021) Diversified effects of co-planting landscape plants on heavy metals pollution remediation in urban soil amended with sewage sludge. J Hazard Mater 403. https://doi.org/10.1016/j.jhazmat.2020.123855

Wu X, Zhang D, Wang F, Luo L, Yining C, Lu S (2023) Risk assessment of metal(loid)s in tea from seven producing provinces in China. Sci Total Environ 856. https://doi.org/10.1016/j.scitotenv.2022.159140

Xu X, Liao W, Lin Y, Dai Y, Shi Z, Huo X (2018) Blood concentrations of lead, cadmium, mercury and their association with biomarkers of DNA oxidative damage in preschool children living in an e-waste recycling area. Environ Geochem Health 40:1481–1494. https://doi.org/10.1007/s10653-017-9997-3

Xu Y, Mu W, Li J, Ba Q, Wang H (2021) Chronic cadmium exposure at environmental-relevant level accelerates the development of hepatotoxicity to hepatocarcinogenesis. Sci Total Environ 783. https://doi.org/10.1016/j.scitotenv.2021.146958

Yang W, Zhang T, Li S, Ni W (2014) Metal removal from and microbial property improvement of a multiple heavy metals contaminated soil by phytoextraction with a cadmium hyperaccumulator *Sedum alfredii* H. J Soils Sediments 14:1385–1396. https://doi.org/10.1007/s11368-014-0875-7

Yang J, Xie Q, Wang Y, Wang J, Zhang Y, Zhang C, Wang D (2021) Exposure of the residents around the Three Gorges Reservoir, China to chromium, lead and arsenic and their health risk via food consumption. Ecotoxicol Environ Saf 228. https://doi.org/10.1016/j.ecoenv.2021.112997

Yang J, Hu R, Zhao C, Wang L, Mei L, Guo G, Shi H, Liao X, Chen T (2023) Challenges and opportunities for improving the environmental quality of cadmium-contaminated soil in China. J Hazard Mater 445. https://doi.org/10.1016/j.jhazmat.2022.130560

Zamora-Ledezma C, Negrete-Bolagay D, Figueroa F, Zamora-Ledezma E, Ni M, Alexis F, Guerrero VH (2021) Heavy metal water pollution: a fresh look about hazards, novel and conventional remediation methods. Environ Technol Innov 22:101504. https://doi.org/10.1016/j.eti.2021.101504

Zhao Y, Cao J, Chen P (2023) Enhanced effect of Phytoextraction on arsenic-contaminated soil by microbial reduction. Appl Sci (Basel) 13. https://doi.org/10.3390/app131910921

Zhen Y, Ge L, Chen Q, Xu J, Duan Z, Loor JJ, Wang M (2022) Latent benefits and toxicity risks transmission chain of high dietary copper along the livestock-environment-plant-human health Axis and microbial homeostasis: a review. J Agric Food Chem 70:6943–6962. https://doi.org/10.1021/acs.jafc.2c01367

Zhou X-Y, Wang X (2019) Cd contamination status and cost-benefits analysis in agriculture soils of Yangtze River basin. Environ Pollut 254. https://doi.org/10.1016/j.envpol.2019.112962

Biochar-Assisted Phytostabilization of Heavy Metal-Polluted Land

2

Maja Radziemska and Barbara Klik

Abstract

The contamination of soils with heavy metals is strictly connected with the intensive development of industry, progressing urbanization, agriculture, and mining. The search for effective and safe methods of cultivation in areas degraded with heavy metals plays a key role in the current times. The technique of phytostabilization has found its application in immobilizing excessive heavy metals in soil. It is often supplemented with biochar, which is considered a promising and effective material to enhance the process. This approach has gained considerable attention due to its potential to reduce the negative impact of heavy metals on soil and the environment. The present chapter contains the role of biochar in aided phytostabilization and the accompanying mechanisms that take place throughout the process in connection with the physicochemical properties of biochar. Moreover, the most recent examples of applying this material in phytostabilization are described.

Keywords

Immobilization · Remediation methods · Soil amendments · Soil contamination

2.1 Introduction

The contamination of soil with heavy metals poses a significant challenge to the natural environment's ongoing development. In light of progressing industrialization and mining sprawl deposition, increasing agricultural intensification, and

M. Radziemska (✉) · B. Klik
Institute of Environmental Engineering, Warsaw University of Life Sciences, Warsaw, Poland
e-mail: maja_radziemska@sggw.edu.pl

V. C. Pandey (ed.), *Plantation-Based Land Restoration*,
https://doi.org/10.1007/978-981-95-2860-8_2

increased use of chemical substances, heavy metals accumulate in soils, leading to the degradation of the ecosystem (Ali et al. 2019; Jadaa and Mohammed 2023; Wuana and Okieimen 2011). Heavy metals such as Pb, Cd, Cu, or Zn not only degrade the soil quality but also have the potential to make their way into the food chain, posing a significant threat to the human health of ecosystems (Wu et al. 2022; Klik et al. 2021).

The consequences of the presence of heavy metals in soils are multifaceted that influence physical, chemical, as well as biological properties of soil (Dhaliwal et al. 2020; Aponte et al. 2020). The degradation of soil that results from the processes of contamination with heavy metals declasses its structure and has a negative influence on the ecological functions of soil as well as its fertility (Ayub et al. 2021). Plants subjected to the toxic effects of heavy metals reveal a limited growth and development capability, which in turn influences the quality and quantity of the supplied nutrients in the food chain (Jayakumar et al. 2021; Klik et al. 2024).

When considering the long-term implications of the presence of heavy metals, the threat to biodiversity, as well as to the balance of soil ecosystems, should be stressed (Ali et al. 2019; Ayub et al. 2021). In response to these challenges, it becomes necessary not only to minimize the emission of heavy metals but also to develop innovative and effective strategies for remediating the soil, which is an integral part of environmental protection as well as social health (Lacalle et al. 2023; Wu et al. 2022). The necessity for an interdisciplinary approach that takes into account ecological, health, and social aspects directs our attention to the multifaceted nature of the problem of heavy metal contamination.

Among the innovative technological strategies, aided phytostabilization has emerged as a promising method for the remediation of soils, which deserves special attention among researchers and practitioners. This technology, based on the synergy between the capabilities of plants and additives, has been drawing attention due to its efficiency in reducing the toxic chemical substances available in the soil (Radziemska et al. 2020). As an effect of aided phytostabilization, plants not only absorb contaminants but also they cooperate with additives, such as compost, biochar, marble waste, charcoal, iron grit, dolomite, bentonite, diatomite, or chelating substances to effectively limit the mobility of heavy metals (Garau et al. 2021; Giagnoni et al. 2020; Kiran and Prasad 2021; Klik et al. 2022; Lebrun et al. 2021; Radziemska 2018; Radziemska et al. 2013, 2020; Ramírez-Zamora et al. 2022; Scattolin et al. 2021). Such additives play a key role in increasing the abilities of plants to adsorb heavy metals, at the same time influencing the chemical speciation of these substances in the soil (Alsafran et al. 2023; Yan et al. 2020). Among these additives, the potential of biochar, playing a key role in the soil remediation process, is particularly noteworthy (Radziemska et al. 2021a, b; Dwibedi et al. 2022). Its large adsorption surface effectively limits the availability of heavy metals for plants, at the same time having a positive influence on the soil structure (Guo et al. 2020). The stability of biochar in the soil ensures long-term effects, and its influence on soil microorganisms may lead to the transformation of heavy metals into less toxic forms (Mierzwa-Hersztek et al. 2020; Xu 2022). The application of biochar in phytostabilization not only increases the adsorption capabilities of plants but also influences the complex effect of heavy metals in soils, making it a promising tool in the context of long-term management of soil contamination (Eissa 2019; Radziemska et al. 2021a).

The present chapter focuses on analyzing the technique of aided phytostabilization of soils contaminated by heavy metals, placing particular emphasis on the role of biochar as a soil amendment. The mechanisms of phytostabilization will be presented, focusing on the ability of plants to limit the availability of heavy metals. The evolution of scientific thought in this field will be discussed, accounting for the current trends and practical challenges. The main focus of the chapter will be dedicated to the role of biochar, analyzing its physicochemical properties and interactions with heavy metals. Specific examples of applying phytostabilization with biochar will be traced, identifying key results and the effectiveness of this technology under different soil conditions. The chapter aims to provide not only an overview of the current knowledge but also a way forward for further studies on phytostabilization, especially the use of biochar as an effective method for improving soil quality.

2.2 Techniques of Aided Phytostabilization of Soils Contaminated with Heavy Metals

The technique of aided phytostabilization falls into safe phyto-technologies and relies on the application of adequately chosen plants as well as soil additives aiding the immobilization of heavy metals present in a contaminated area (Lacalle et al. 2023). Various non-organic as well as organic materials, which lead to increasing the effect of binding the mobile fractions of heavy metals, are used individually as well as in combination with soil additives (Caravaca et al. 2023; Trippe et al. 2021). The phytostabilization process relies on the immobilization of contaminants in the soil via adsorption and accumulation in plant roots, adsorption on root surfaces, and transformation of soluble compounds into poorly soluble forms within the rhizosphere (Zhang et al. 2024).

A very significant factor in ensuring the best results when applying the technique of aided phytostabilization is the appropriate choice of plants, which ought to be characterized by the lowest possible ability to allow heavy metal ions to migrate to above-ground parts (Bakshe and Jugade 2023). Both in the process of phytostabilization and aided phytostabilization, plant species such as *Agrostis tenuis* L., *Festuca rubra* L., *Hyparrhenia hirta* L., *Typha latifolia* L., *Trifolium repens* L., or *Lolium perenne* L. are most frequently used (Garau et al. 2021; Radziemska et al. 2020). These plants protect the soil against erosion through stabilization with a developed root system. Through their root systems, the plants release organic compounds as well as CO_2, which influence a change in the pH of the soil as well as the redox potential (Zine et al. 2020). They also can transform heavy metal ions into less available forms. Plants grown under phytostabilization should be growing with a high biomass-producing ability, canopy cover, and tolerance to high concentration of heavy metals and widely adaptable to different habitats and a dense root system.

Moreover, the roots restrict the permeation of rainwater into the soil, limiting the leaching of contaminants deeper into the soil (Berti and Cunningham 2000). Furthermore, the below-ground parts of plants can serve as an area for the sorption or precipitation of metallic elements, playing a key role in the protection of groundwaters and the reduction of metal dispersion (Radziemska et al. 2013).

2.3 Significance of Biochar as an Additive in Aided Phytostabilization

The preferred soil additives in the technique of aided phytostabilization over a long-term perspective are those characterized by a high potential for practical application. Therefore, those characterized by high efficiency also in terms of costs, environmental safety, producing minimal production waste, and not requiring complicated and costly treatment are most frequently chosen. Moreover, it is important to introduce such additives that, on the one hand, aid heavy metal immobilization processes in the soil and, at the same time, have a positive influence on plant growth, which tends to be scarce in areas affected by contamination. The immobilization of heavy metals in soils subjected to phytostabilization has been the focus of many researchers over recent years (Anbuganesan et al. 2024; Qiu et al. 2024; Visconti et al. 2020; Radziemska et al. 2021b, 2023; Trippe et al. 2021), which indicates that this field of research opens a new direction for utilizing biochars that may form as a byproduct of biomass pyrolysis processes.

Physical and chemical properties of biochar are connected, above all, with the type of biomass used to produce the product as well as the conditions under which production takes place, and it is they which, for the most part, determine whether biochars will prove to be useful for a given type of application. The production process at a temperature of 600 °C leads to the formation of a stable structure of biochar, whereas biochar produced at 400 °C retains volatile matter as well as unstable compounds (Buss et al. 2016). The structure of biochar derived at a high temperature is characterized by a high specific surface area as well as a high concentration of aromatic carboxylic acids, which can lead to increasing sorption capacity and thus can be potentially used for the remediation of soil contaminated with heavy metals (Radziemska et al. 2023). Due to the above fact that the physicochemical properties of biochar are a function of the type of raw material and pyrolysis conditions (e.g., the temperature, residential period, and rate at which the temperature is increased), it is necessary to determine the appropriate raw material for the production of biochar, which exhibits potential to be used in the aided phytostabilization of heavy metals.

In the technique of aided phytostabilization, the pH parameter plays an important role in both the soils subjected to the remediation process and the actual applied additive. The alkalinity of biochar depends, to a large degree, on the type of raw materials from which it was created as well as the temperature of the pyrolysis that was carried out. It is assumed that the alkalinity of biochar increases along with an increase in the temperature of pyrolysis, during which functional groups (carboxyl and hydroxyl molecules) are formed on the surface of biochar, which largely contribute to the alkalinity of biochar (Fidel et al. 2017). In addition to soil pH, the content of organic matter also significantly influences the adsorption and desorption of heavy metals in soil. In this case, the addition of biochar to soil can have an influence not only on the pH but also on the dissolved organic matter and, as a consequence, shape the mobility of metals.

In the soil environment, metal ions can occur both in the cation and anion forms, which can interact with positive and negative charges on the surface of biochar. Thus, upon being introduced into the soil, biochar with negative charges can strictly adsorb cation forms of metals, such as Cd^{2+} (Beesley and Marmiroli 2011), Pb^{2+} (Herath et al. 2015), and Fe^{2+} and Fe^{+3} (Mitzia et al. 2023), whereas biochar with positive charges can retain anion forms of metals, such as H_2AsO^{-3} and $H_2AsO_3{}^{-4}$ (Zhang et al. 2015). Many scientific studies have found that biochar can decrease the uptake of heavy metals by plants (Chen et al. 2018; Lei et al. 2019; Odinga et al. 2020). It can create conditions in the rhizosphere that increase the supply and absorption of nutrients, immobilize phytotoxic substances, and stimulate the growth and development of plants (Joseph et al. 2021; Radziemska et al. 2021a). Moreover, by increasing pH, biochar mitigates the toxicity of Al and heavy metals, which stunt the growth of roots (Lauricella et al. 2021).

2.4 Biochar Application for Phytostabilization: Practical Examples

Biochar has become the subject of intense studies due to its unique physicochemical properties, as described above. In this section, an attempt has been made to synthesize the novel scientific findings related to the phytostabilization of soils contaminated with heavy metals, with special regard to the application of biochar. The salient findings of the authors are presented in detail in Table 2.1.

Biochar, which is a product of processing various organic raw materials, turns out to be an effective tool in the context of the phytostabilization of soils contaminated with heavy metals. The application of biochar under various environmental conditions and on different types of soils indicates its potential for reducing the mobility of heavy metals and limiting their accumulation in plants (Nkoh et al. 2022). When analyzing different sources of biochar remediating various types of contaminants, we can observe their effectiveness in diverse soil conditions.

Studies on biochar derived from plant sources suggest its wide-ranging application in heavy metal remediation. Rice straw biochar, applied for the phytostabilization of soils contaminated with Pb, Zn, and Cr, turned out to be an effective tool in facilitating the growth of lavender (*L. dentata*) (Ye et al. 2020). Similar advantages for plants were also observed in the case of wheat straw and rice husk biochars, which effectively improved soil conditions, especially in the case of *Lolium perenne* L. in areas contaminated with Pb and Zn, derived from mine tips (Alhar et al. 2021). In another context, wheat straw biochar revealed abilities to reduce Cr under laboratory conditions (Naveed et al. 2021). Wheat-derived biochar, on the other hand,

Table 2.1 Overview of worldwide research on the biochar-assisted phytostabilization of soils contaminated with heavy metals

Biochar used	Soil origin	Contaminants	Plants	References
Banana pith-derived biochar	Industrial soil	Cd, Zn	*Sorghum bicolor*	Anbuganesan et al. (2024)
Pinus halepensis	Mine-impacted soil	Cd, Pb, Zn	*Portulaca oleracea*	Thalassinos et al. (2023)
Hibiscus cannabinus core biochar and sewage sludge biochar	Mine tailings	Cd, Cu, Pb, Zn	*Cassia alata*, *Boehmeria nivea*	Li et al. (2023)
Willow wood chips	Steel and metal waste disposal soil	Cd, Cu, Ni, Pb, Zn	*Lolium perenne*	Radziemska et al. (2023)
Neem seed biochar	Mine-impacted soil	Cd, As, Zn, Pb, Hg	*Manihot esculenta*, *Jatropha curcas*	Mensah et al. (2023)
Prosopis laevigata biochar	Mine tailings	Cu, Pb, Cd	*Prosopis laevigata*	Ramirez-Zamora et al. (2022)
Pine shoots	Mine-impacted agricultural soil	Cd, Pb, Zn	*Sweet sorghum*	Wang et al. (2022)
Wheat straw-derived biochar	Coal mine-contaminated soil	Pb, Cd, Cr, Ni, Mn, Cu	*Tomato plants*	Zaman et al. (2022)
Pine chips	Tailings	Zn, Cu, Cd	*Poplar, willow*, *Miscanthus*	Al-Lami et al. (2022)
Tree wood (mixture of pine, oak, and eucalyptus)	Mine tailings	Cd, Mn, Zn	*Zygophyllum fabago*	Conesa and Parraga-Aguado (2022)
Willow chips biochar	Steel disposal dump soil	As, Cu, Pb, Zn	*Lolium perenne*	Radziemska et al. (2021a)
Wood chip biochar	Sandy soil near shooting range	Cd, Cr, Cu, Ni, Pb, Zn	*Trifolium alexandrinum*	Pescatore et al. (2022)
Manure-based biochar	Mine-impacted soils	Cd, Zn	*Panicum virgatum*, *Bouteloua dactyloides*	Ducey et al. (2021)
Wheat straw biochar	Artificially contaminated	Cr	*Brassica*	Naveed et al. (2021)
P-enriched sludge biochar	Mine-impacted soils	Cd	*Robinia pseudoacacia*	Zhang et al. (2021)
Willow chip biochar	Steel disposal dump	Cu, Cr, Ni, Zn, Pb	*Lupinus luteus*	Radziemska et al. (2021b)
Wheat and rice biochars	Mine spoil heaps	Pb, Zn	*Lolium perenne*	Alhar et al. (2021)

(continued)

Table 2.1 (continued)

Biochar used	Soil origin	Contaminants	Plants	References
Acacia arabica wood-derived biochar	Artificially contaminated	Cd, Pb, Ni	*L. esculentum*	Zeeshan et al. (2020)
Acacia wood-derived biochar	Mine tailings	As, Mn	*Pennisetum purpureum*	Kowitwiwat and Sampanpanish (2020)
Rice straw biochar	Contaminated slag	Cr	*L. dentata*	Ye et al. (2020)
Willow chip biochar	Mine-impacted soils	Pb, As	*Salix viminalis*	Lebrun et al. (2020)
Coconut shell biochar	Mine-impacted soils	Cd, Pb	*B. nivea*	Lan et al. (2020)
Apple wood biochar	Mine tailings	Pb, Zn, Cd, Cu	*Amorpha fruticosa*	Sikdar et al. (2020)
Peat-based biochar	Mine tailings	Pb, Zn, Cd, As	*P. purpureum, P. thyphoideum*	Gao et al. (2020)

applied for phytostabilization of soil derived from mining activities (contaminated with Pb, Cd, Cr, Ni, Mn, and Cu) turned out to be an effective tool in improving the soil conditions for tomatoes (Rio Grande—a tomato variety) (Zaman et al. 2022). Neem-seed-derived biochar has been shown to effectively immobilize Cd, As, Zn, Pb, and Hg in soils affected by mining activities, thereby promoting the growth of *Manihot esculenta* and *Jatropha curcas* (Mensah et al. 2023), while banana-peel-derived biochar applied to industrial soils contaminated with Cd and Zn enhanced the production of *Sorghum bicolor* L. (Anbuganesan et al. 2024).

Research on biochar from woody species indicates its effective application in various areas of soil remediation. Willow wood chip biochar, applied in soils contaminated with Cd, Cu, Ni, Pb, and Zn in steel and metal waste dumping yards, was found to be effective in improving soil conditions, especially for *Lolium perenne* L. (Radziemska et al. 2023). Similar advantages of pine chip biochar application, a mixture of pine, oak, and eucalyptus, were observed in soils contaminated with Zn, Cu, and Cd (Al-Lami et al. 2022). Moreover, tree wood biochar turned out to be an effective additive in the phytostabilization of Cd, Mn, and Zn in soils from mining regions (Conesa and Párraga-Aguado 2022). Willow chip biochar, applied in soils contaminated with As, Cu, Pb, and Zn derived from dumping yards of steel waste, revealed a positive influence on *Lolium perenne* L. (Radziemska et al. 2021a). Similar benefits for soils contaminated with Cd, Cr, Cu, Ni, Pb, and Zn in areas of shooting ranges were also noted in the case of chip biochar by *Trifolium alexandrinum* L. (Pescatore et al. 2022).

Recently, there has been increasing interest in the application of biochar derived from organic materials. Sewage sludge-derived biochar or directly applied in the case of Cd, Cu, Pb, and Zn contamination turned out to be an effective phytostabilization tool in the context of growing *Cassia alata* L. and *Boehmeria nivea* L. (Li et al. 2023). Sewage sludge enriched with phosphorus also proved to be effective in the phytostabilization of soil contaminated with Cd, especially when growing *Robinia pseudoacacia* L. (Zhang et al. 2021). Similar benefits to soil from mining areas contaminated with Cd and Zn were found when using manure-based biochar when growing *Panicum virgatum* and *Bouteloua dactyloides* (Ducey et al. 2021). On the other hand, biochar derived from coconut shells revealed phytostabilization capabilities in soils contaminated with Cd and Pb in mining areas, especially when growing ramie (*B. nivea* L.) (Lan et al. 2020). The effective use of organic waste for the production of biochar can lead to minimizing negative effects on the environment, which is a key aspect in striving for more sustainable practices, especially in the context of managing a closed system.

2.5 Conclusions

Phytostabilization assisted with biochar is included among remediation techniques, which are characterized by the high potential of immobilizing heavy metals from soils. Biochar can limit the bioavailability and leachability of heavy metals in the soil, as well as facilitate the growth of plants in contaminated areas by improving soil fertility. The key to this is the production of biochar under appropriate conditions to optimize the effectiveness of heavy metal immobilization. The efficiency of biochar over a longer timeframe, however, is yet to be well understood, and thus future research should focus on this aspect.

References

Alhar MAM, Thompson DF, Oliver IW (2021) Mine spoil remediation via biochar addition to immobilise potentially toxic elements and promote plant growth for phytostabilisation. J Environ Manag 277:111500

Ali H, Khan E, Ilahi I (2019) Environmental chemistry and ecotoxicology of hazardous heavy metals: environmental persistence, toxicity, and bioaccumulation. J Chem 2019:6730305

Al-Lami MK, Oustriere N, Gonzales E, Burken JG (2022) Phytomanagement of Pb/Zn/Cu tailings using biosolids-biochar or -humus combinations: enhancement of bioenergy crop production, substrate functionality, and ecosystem services. Sci Total Environ 836:155676

Alsafran M, Saleem MH, Al Jabri H, Rizwan M, Usman K (2023) Principles and applicability of integrated remediation strategies for heavy metal removal/recovery from contaminated environments. J Plant Growth Regul 42:3419–3440

Anbuganesan V, Vishnupradeep R, Mehnaz N, Kumar A, Freitas H, Rajkumar M (2024) Synergistic effect of biochar and plant growth promoting bacteria improve the growth and phytostabilization potential of Sorghum bicolor in cd and Zn contaminated soils. Rhizosphere 29:100844

Aponte H, Medina J, Butler B, Meier S, Cornejo P, Kuzyakov Y (2020) Soil quality indices for metal(loid) contamination: an enzymatic perspective. Land Degrad Dev 31:2700–2719

Ayub MA, Zia ur Rehman M, Umar W, Jamil A, Aslam MZ, Ahmad HR, Ahmad Z, Siddique A, Quinones JE (2021) Uptake and transformation of heavy metals/metalloids in plants. In: Heavy metal toxicity in plants: physiological and molecular perspectives, pp 111–125

Bakshe P, Jugade R (2023) Phytostabilization and rhizofiltration of toxic heavy metals by heavy metal accumulator plants for sustainable management of contaminated industrial sites: a comprehensive review. J Hazard Mat Adv 10:100293

Beesley L, Marmiroli M (2011) The immobilisation and retention of soluble arsenic, cadmium and zinc by biochar. Environ Poll 159:474–480
Berti WR, Cunningham SD (2000) Phytostabilization of metals. In: Raskin I, Ensley BD (eds) Phytoremediation of toxic metals: using plants to clean up the environment, pp 71–88
Buss W, Graham MC, Shepherd JG, Mašek O (2016) Suitability of marginal biomass-derived biochars for soil amendment. Sci Total Environ 547:314–322
Caravaca F, Díaz G, Torres P, Campoy M, Roldán A (2023) Synergistic enhancement of the phytostabilization of a semiarid mine tailing by a combination of organic amendment and native microorganisms (*Funneliformis mosseae* and *Bacillus cereus*). Chemosphere 312(1):137106
Chen D, Liu X, Bian R, Cheng K, Zhang X, Zheng J, Joseph S, Crowley D, Pan G, Li L (2018) Effects of biochar on availability and plant uptake of heavy metals—a meta-analysis. J Environ Manag 222:76–85
Conesa HM, Párraga-Aguado I (2022) Effects on metal availability of the application of tree biochar and municipal waste biosolid in a metalliferous mine tailings substrate. Environ Geochem Health 44:1317–1327
Dhaliwal SS, Singh J, Taneja PK, Mandal A (2020) Remediation techniques for removal of heavy metals from the soil contaminated through different sources: a review. Environ Sci Poll Res 27:1319–1333
Ducey TF, Johnson MG, Sigua GC, Novak JM, Ippolito JA, Spokas KA (2021) Microbial response to phytostabilization in mining impacted soils using maize in conjunction with biochar and compost. Microorganisms 9(12):2545
Dwibedi SK, Pandey VC, Divyasree D, Bajpai O (2022) Biochar-based land development. Land Degrad Dev 33:1139–1158. https://doi.org/10.1002/ldr.4185
Eissa MA (2019) Effect of compost and biochar on heavy metals Phytostabilization by the halophytic plant old man saltbush [*Atriplex Nummularia* Lindl]. Soil Sedim Contam Int J 28:135–147
Fidel RB, Laird DA, Thompson ML, Lawrinenko M (2017) Characterization and quantification of biochar alkalinity. Chemosphere 167:367–373
Gao B, Zhang X, Tian C, Zhang X, Liu J (2020) Effects of amendments and aided phytostabilization of an energy crop on the metal availability and leaching in mine tailings using a pot test. Environ Sci Pollut Res 27:2745–2759
Garau M, Castaldi P, Diquattro S, Pinna MV, Senette C, Roggero PP, Garau G (2021) Combining grass and legume species with compost for assisted phytostabilization of contaminated soils. Environ Technol Innov 22:101387
Giagnoni L, dos Anjos Borges LG, Giongo A, de Oliveira Silveira A, Ardissone AN, Triplett EW, Mench M, Renella G (2020) Dolomite and compost amendments enhance Cu phytostabilization and increase microbiota of the leachates from a Cu-contaminated soil. Agronomy 10:719
Guo M, Song W, Tian J (2020) Biochar-facilitated soil remediation: mechanisms and efficacy variations. Front Environ Sci 8:521512
Herath I, Kumarathilaka P, Navaratne A, Rajakaruna N, Vithanage M (2015) Immobilization and phytotoxicity reduction of heavy metals in serpentine soil using biochar. J Soils Sediments 15:126–138
Jadaa W, Mohammed H (2023) Heavy metals—definition, natural and anthropogenic sources of releasing into ecosystems, toxicity, and removal methods—an overview study. J Ecol Eng 24:249–271
Jayakumar M, Surendran U, Raja P, Kumar A, Senapathi V (2021) A review of heavy metals accumulation pathways, sources and management in soils. Arab J Geosci 14:20
Joseph S, Cowie AL, Van Zwieten L, Bolan N, Budai A, Buss W, Cayuela ML, Graber ER, Ippolito JA, Kuzyakov Y, Luo Y, Ok YS, Palansooriya KN, Shepherd J, Stephens S, Weng Z, Lehmann J (2021) How biochar works, and when it doesn't: a review of mechanisms controlling soil and plant responses to biochar. GCB Bioenergy 13:1731–1764
Kiran BR, Prasad MNV (2021) Assisted phytostabilization of Pb-spiked soils amended with charcoal and banana compost and vegetated with *Ricinus communis* L. (Castor bean). Environ Geochem Health 43:1507–1521

Klik B, Gusiatin ZM, Kulikowska D (2021) Quality of heavy metal-contaminated soil before and after column flushing with washing agents derived from municipal sewage sludge. Sci Rep 11:1

Klik B, Holatko J, Jaskulska I, Gusiatin MZ, Hammerschmiedt T, Brtnicky M, Liniauskienė E, Baltazar T, Jaskulski D, Kintl A, Radziemska M (2022) Bentonite as a functional material enhancing Phytostabilization of post-industrial contaminated soils with heavy metals. Materials 15:8331

Klik B, Jachimowicz P, Liniauskiene E, Gusiatin M, Radzevičius A, Brtnicky M, Šadzevičius R, Bęś A, Mazur Z, Dapkienė M, Radziemska M (2024) Ash from gasification of poultry feathers for heavy metals immobilization under assisted phytostabilization in soil. Sci Rev Eng Environ Sci 33:2

Kowitwiwat A, Sampanpanish P (2020) Phytostabilization of arsenic and manganese in mine tailings using Pennisetum purpureum cv. Mott supplemented with cow manure and acacia wood-derived biochar. Heliyon 6(7):e04552

Lacalle RG, Bernal MP, Álvarez-Robles MJ, Clemente R (2023) Phytostabilization of soils contaminated with As, Cd, Cu, Pb and Zn: physicochemical, toxicological and biological evaluations. Soil Environ Health 1(2):100014

Lan MM, Liu C, Liu SJ, Qiu RL, Tang YT (2020) Phytostabilization of cd and pb in highly polluted farmland soils using ramie and amendments. Int J Environ Res Pub Health 17(5):1661

Lauricella D, Weng Z, Clark GJ, Butterly CR, Li G, Gazey C, Sale PWG, Tang C (2021) Biochars and their feedstocks differ in their short-term effects in ameliorating acid soils grown with aluminium-sensitive wheat. J Soils Sediments 21(8):2805–2816

Lebrun M, De Zio E, Miard F, Scippa GS, Renzone G, Scaloni A, Bourgerie S, Morabito D, Trupiano D (2020) Amending an As/Pb contaminated soil with biochar, compost and iron grit: effect on *Salix viminalis* growth, root proteome profiles and metal(loid) accumulation indexes. Chemosphere 244:125397

Lebrun M, Miard F, Bucci A, Fougère L, Nandillon R, Naclerio G, Scippa GS, Destandeau E, Morabito D, Bourgerie S (2021) The rhizosphere of *Salix viminalis* plants after a phytostabilization process assisted by biochar, compost, and iron grit: chemical and (micro)-biological analyses. Environ Sci Pollut Res 28:47447–47462

Lei S, Shi Y, Qiu Y, Che L, Xue C (2019) Performance and mechanisms of emerging animal-derived biochars for immobilization of heavy metals. Sci Total Environ 646:1281–1289

Li M, Chen X, Chen C, Huang L, Chi H, Zhao N, Yan B, Chao Y, Tang Y, Qiu R, Qiu R, Wang S (2023) The effectiveness of sewage sludge biochar amendment with Boehmeria nivea L. in improving physicochemical properties and rehabilitating microbial communities in mine tailings. J Environ Manag 345:118552

Mensah MK, Drebenstedt C, Ola IM, Hoth N, Damptey FG, Wiafe ED (2023) Immobilization effects of co-pyrolyzed neem seed mixed with poultry manure on potentially toxic elements in soil and the phytoremediation potentials of native *Manihot esculenta* and *Jatropha curcas* in ensuring sustainable land use. Environ Monit Asses 195:793

Mierzwa-Hersztek M, Wolny-Koładka K, Gondek K, Gałązka A, Gawryjołek K (2020) Effect of Coapplication of biochar and nutrients on Microbiocenotic composition, dehydrogenase activity index and chemical properties of Sandy soil. Waste Bioms Valor 11:3911–3923

Mitzia A, Vítková M, Ratié G, Chotěborský R, Vantelon D, Neaman A, Komárek M (2023) Revealing the long-term behaviour of nZVI and biochar in metal (loid)-contaminated soil: focus on Fe transformations. Environ Sci Nano 10(10):2861–2879

Naveed M, Tanvir B, Xiukang W, Brtnicky M, Ditta A, Kucerik J, Subhani Z, Nazir MZ, Radziemska M, Saeed Q, Saeed Q, Mustafa A (2021) Co-composted biochar enhances growth, physiological, and phytostabilization efficiency of Brassica napus and reduces associated health risks under chromium stress. Front Plant Sci 12:775785

Nkoh JN, Ajibade FO, Atakpa EO, Baquy MAA, Mia S, Odii EC, Xu R (2022) Reduction of heavy metal uptake from polluted soils and associated health risks through biochar amendment: a critical synthesis. J Hazard Mater Adv 6:100086

Odinga ES, Waigi MG, Gudda FO, Wang J, Yang B, Hu X, Li S, Gao Y (2020) Occurrence, formation, environmental fate and risks of environmentally persistent free radicals in biochars. Environ Int 134:105172

Pescatore A, Grassi C, Rizzo AM, Orlandini S, Napoli M (2022) Effects of biochar on berseem clover (*Trifolium alexandrinum*, L.) growth and heavy metal (Cd, Cr, Cu, Ni, Pb, and Zn) accumulation. Chemosphere 287:131986

Qiu J, Souza MF, Wang X, Ok YS, Meers E (2024) Influence of biochar addition and plant management (cutting and time) on ryegrass growth and migration of As and Pb during phytostabilization. Sci Total Environ 913:169771

Radziemska M (2018) Study of applying naturally occurring mineral sorbents of Poland (dolomite, halloysite, chalcedonite) for aided phytostabilization of soil polluted with heavy metals. Catena 163:123–129. https://doi.org/10.1016/j.catena.2017.12.015

Radziemska M, Mazur Z, Jeznach J (2013) Influence of applying halloysite and zeolite to soil contaminated with nickel on the content of selected elements in Maize (*Zea mays* L.). Chem Eng Trans 32:301–306

Radziemska M, Bęś A, Gusiatin ZM, Cerda A, Jeznach J, Mazur Z, Brtnický M (2020) Assisted phytostabilization of soil from a former military area with mineral amendments. Ecotoxicol Environ Saf 188:109934

Radziemska M, Gusiatin ZM, Kumar V, Brtnicky M (2021a) Co-application of nanosized halloysite and biochar as soil amendments in aided phytostabilization of metal(-oid)s-contaminated soil under different temperature conditions. Chemosphere 288:132452

Radziemska M, Gusiatin ZM, Cydzik-Kwiatkowska A, Cerdà A, Pecina V, Bęś A, Datta R, Majewski G, Mazur Z, Dzięcioł J, Danish S, Brtnický M (2021b) Insight into metal immobilization and microbial community structure in soil from a steel disposal dump phytostabilized with composted, pyrolyzed or gasified wastes. Chemosphere 272:129576

Radziemska M, Gusiatin MZ, Cydzik-Kwiatkowska A, Blazejczyk A, Holatko J, Brtnicky M (2023) Does biochar in combination with compost effectively promote phytostabilization of heavy metals in soil under different temperature regimes? Sci Total Environ 882:163634

Ramírez-Zamora J, Mussali-Galante P, Rodríguez A, Castrejón-Godínez ML, Valencia-Cuevas L, Tovar-Sánchez E (2022) Assisted Phytostabilization of Mine-Tailings with Prosopis laevigata (Fabaceae) and Biochar. Plants 11:3441

Scattolin M, Peuble S, Pereira F, Paran F, Moutte J, Menad N, Faure O (2021) Aided-phytostabilization of steel slag dumps: the key-role of pH adjustment in decreasing chromium toxicity and improving manganese, phosphorus and zinc phytoavailability. J Hazard Mater 405:124225

Sikdar A, Wang J, Hasanuzzaman M, Liu X, Feng S, Roy R, Sial TA, Lahori AH, Arockiam Jeyasundar PGS, Wang X (2020) Phytostabilization of Pb-Zn mine tailings with Amorpha fruticosa aided by organic amendments and triple superphosphate. Molecules 25(7):1617

Thalassinos G, Levizou E, Antoniadis V (2023) Can soil improvers (biochar, compost, insect Frass, lime, and zeolite) achieve Phytostabilization of potentially toxic elements in heavily contaminated soil with the use of purslane (*Portulaca oleracea*)? Agronomy 13(11):2827

Trippe KM, Manning VA, Reardon CL, Klein AM, Weidman C, Ducey TF, Novak JM, Watts DW, Rushmiller H, Spokas KA, Ippolito JA, Johnson MG (2021) Phytostabilization of acidic mine tailings with biochar, biosolids, lime, and locally-sourced microbial inoculum: do amendment mixtures influence plant growth, tailing chemistry, and microbial composition? Appl Soil Ecol 165:103962

Visconti D, Álvarez-Robles MJ, Fiorentino N, Fagnano M, Clemente R (2020) Use of Brassica juncea and Dactylis glomerata for the phytostabilization of mine soils amended with compost or biochar. Chemosphere 260:127661

Wang F, Cheng P, Zhang S, Zhang S, Sun Y (2022) Contribution of arbuscular mycorrhizal fungi and soil amendments to remediation of a heavy metal-contaminated soil using sweet sorghum. Pedosphere 32:844–855

Wu G, Wang L, Yang R, Hou W, Zhang S, Guo X, Zhao W (2022) Pollution characteristics and risk assessment of heavy metals in the soil of a construction waste landfill site. Ecol Inf 70:101700

Wuana RA, Okieimen FE (2011) Heavy metals in contaminated soils: a review of sources, chemistry, risks and best available strategies for remediation. ISRN Ecol:1–20
Xu XK (2022) Effect of freeze-thaw disturbance on soil C and N dynamics and GHG fluxes of East Asia forests: review and future perspectives. J Soil Sci Plant Nutr 68:15–26
Yan A, Wang Y, Tan SN, Mohd Yusof ML, Ghosh S, Chen Z (2020) Phytoremediation: a promising approach for revegetation of heavy metal-polluted land. Front Plant Sci 11:513099
Zaman A, Irfan M, Khan AM, Ali H, Iqbal N, Ahmad I, Fawad M, Muhammad F (2022) Toxicity assessment and phytostabilization of contaminated soil by using wheat straw-derived biochar in tomato plants. Gesunde Pflanz 74(3):705–713
Zhang W, Zheng J, Zheng P, Tsang DC, Qiu R (2015) Sludge-derived biochar for arsenic (III) immobilization: effects of solution chemistry on sorption behavior. J Environ Qual 44:1119–1126
Zhang Y, Zhan J, Ma C, Liu W, Huang H, Yu H, Christie P, Li T, Wu L (2024) Root-associated bacterial microbiome shaped by root selective effects benefits phytostabilization by *Athyrium wardii* (Hook.). Ecotoxicol Environ Saf 269:115739
Zine H, Midhat L, Hakkou R, Adnani ME, Ouhammou A (2020) Guidelines for a phytomanagement plan by the phytostabilization of mining wastes. Sci Afr 10:e00654

Plantation-Based Land Restoration in Rainfed Ecosystems in India

3

Ch. Srinivasa Rao, V. Mani Chandana, Rayapati Karthik, and M. Jagadesh

Abstract

The pivotal role of plantation-based approaches in restoring degraded lands within India's rainfed ecosystems is underscored by the significant challenges these regions face, including soil erosion, nutrient depletion, and water scarcity, all of which are further intensified by climate change. These challenges pose a threat to both agricultural productivity and the overall integrity of the ecosystems. The chapter assesses the current state of rainfed ecosystems in India, shedding light on the extent and underlying causes of land degradation. It introduces plantation-based land restoration as a promising strategy, examining its potential to address crucial aspects such as soil health, water conservation, biodiversity enhancement, and the improvement of rural livelihoods. Various plantation models, including agroforestry systems, silvopasture, and mixed species plantations, are analyzed, with insightful case studies from diverse agroecological zones of India demonstrating successful implementation and evaluating impacts on soil quality, water retention, biodiversity, and agricultural productivity. The chapter also explores the challenges and opportunities associated with scaling up these approaches, encompassing discussions on policy frameworks, community

Ch. S. Rao (✉)
ICAR-Indian Agricultural Research Institute, New Delhi, India

V. M. Chandana
ICAR-Indian Agricultural Research Institute, New Delhi, India

ICAR-National Academy of Agricultural Research Management, Hyderabad, India

R. Karthik
ICAR-National Academy of Agricultural Research Management, Hyderabad, India

Professor Jayashankar Telangana Agricultural University, Hyderabad, India

M. Jagadesh
ICAR-National Academy of Agricultural Research Management, Hyderabad, India

V. C. Pandey (ed.), *Plantation-Based Land Restoration*,
https://doi.org/10.1007/978-981-95-2860-8_3

engagement, and technological innovations. Ultimately, the analysis underscores the significant potential of plantation-based approaches to contribute to land restoration, climate change mitigation, and sustainable development in India's vulnerable rainfed areas.

Keywords

Rainfed agriculture · Soil erosion · Nutrient depletion · Water scarcity · Climate change · Land restoration · Climate change mitigation · Sustainable development · Biodiversity enhancement

3.1 Introduction

Rainfed ecosystems in India are agricultural areas that primarily depend on natural rainfall for water supply, as opposed to irrigated lands that rely on artificial water sources such as canals, wells, or tube wells (Nanwal and Rajanna 2022). These ecosystems cover approximately 60% of India's net sown area, spanning about 85 million hectares (Wani et al. 2009a, b). Rainfed regions are characterized by their reliance on monsoon rains, which are often erratic and unevenly distributed both spatially and temporally. This variability in rainfall makes rainfed agriculture particularly challenging and vulnerable to climatic fluctuations (Venkateswarlu and Shanker 2009).

These ecosystems are predominantly found in the central, eastern, and western parts of India, including states such as Maharashtra, Madhya Pradesh, Gujarat, Rajasthan, Andhra Pradesh, and Karnataka (Venkateswarlu and Prasad 2012). They encompass a diverse range of agroecological zones, from arid and semi-arid regions to sub-humid and humid areas, each with its unique set of challenges and potentials (Singh 1996). The diversity of these regions means that rainfed agriculture practices must be tailored to local conditions to be effective.

Significance of Land Restoration in Rainfed Areas

Land restoration in rainfed ecosystems is of paramount importance for several reasons:

- *Food Security*: Rainfed areas contribute significantly to India's food production, accounting for 44% of rice, 87% of coarse cereals, 85% of food legumes, 72% of oilseeds, 65% of cotton, and 90% of minor millets (Wani et al. 2009a, b). However, the productivity of these lands is often compromised due to soil degradation and water scarcity. Restoring degraded lands can enhance productivity, ensuring a stable food supply to meet the needs of India's growing population (Sharma 2011).
- *Livelihood Support:* These regions are home to a large proportion of India's rural poor, including small and marginal farmers who are heavily dependent on agriculture for their livelihoods (Garg et al. 2020). Improved soil health and water availability through land restoration can lead to better agricultural yields, diversi-

fied income sources, and improved quality of life for these communities (Haddad et al. 2011). This is crucial for reducing poverty and promoting rural development. Practicing integrated farming and crop diversification could mitigate the risk of crop failure and provide the livelihood support to small and marginal farmers.

- *Environmental Sustainability:* Restoration efforts can help combat desertification, mitigate climate change impacts, and preserve biodiversity (Lal 2020). This is crucial for maintaining ecological balance and ensuring long-term environmental health. Healthy ecosystems provide a range of services, including carbon sequestration, water filtration, and habitat for wildlife, all of which are vital for sustainable development (Parikh et al. 2021).
- *Water Resource Management:* Improved land management practices, such as contour plowing, terracing, and agroforestry, can enhance water retention, reduce runoff, and increase groundwater recharge (Mandal et al. 2020). Water harvesting techniques should be promoted across rainfed areas not only to conserve water but also for enhancing the water use efficiency as well as crop productivity. This is particularly important in rainfed areas, which are often prone to drought and water scarcity. Efficient water management ensures that crops have sufficient water during critical growth periods, thus stabilizing agricultural output (Bhattacharya et al. 2020).
- *Climate Resilience:* Restored landscapes are better equipped to withstand extreme weather events and adapt to changing climatic conditions, thus reducing vulnerability to climate-related risks (Mirzabaev et al. 2019). Practices such as crop diversification, mulching, cover cropping, and integrated pest management can improve soil health and increase the resilience of agricultural systems to droughts, floods, and temperature extremes (Dubey et al. 2020).

3.2 Challenges and Impacts of Land Degradation in Rainfed Regions

Challenges

- *Soil Erosion:* Rainfed areas are particularly susceptible to soil erosion due to intense rainfall events, especially in regions with poor vegetation cover and improper land management practices (Mandal et al. 2020). Erosion removes the fertile topsoil, reducing the land's productivity and its ability to support healthy crop growth (Bhattacharyya et al. 2015).
- *Nutrient Depletion:* Continuous cropping without adequate nutrient replenishment leads to soil fertility decline, affecting crop productivity (Lal 2015). Traditional farming practices often fail to replace the nutrients that crops extract from the soil, leading to long-term degradation of soil health (Sharma et al. 2021).
- *Water Scarcity:* Erratic rainfall patterns, coupled with poor water management, result in frequent droughts and water stress for crops (Garg et al. 2020).

Inadequate water availability during critical crop growth stages can lead to reduced yields and crop failure (Singh et al. 2014).

- *Low Crop Productivity:* A combination of poor soil health, water scarcity, and limited access to improved agricultural technologies contributes to low and unstable crop yields (Pingali et al. 2019). Farmers in rainfed areas often lack access to high-yielding seed varieties, fertilizers, and modern farming techniques that could improve productivity (Raju and Wani 2016).
- *Limited Access to Resources:* Many farmers in rainfed regions lack access to credit, quality inputs, and modern agricultural knowledge, hindering their ability to implement sustainable land management practices (Kerr 1996). This lack of resources constrains their ability to invest in land improvements and adopt new technologies (Harris and Orr 2014).
- *Climate Change:* Increasing temperatures and changing precipitation patterns exacerbate existing challenges, making rainfed agriculture more vulnerable (Mirzabaev et al. 2019). Climate change can lead to more frequent and severe droughts, unpredictable rainfall, and increased incidence of pests and diseases, all of which threaten agricultural productivity (Dubey et al. 2020).

Impacts

- *Reduced Agricultural Productivity:* Land degradation leads to declining crop yields, threatening food security and farmer incomes (Nkonya et al. 2016). Lower productivity can result in higher food prices and increased dependence on food imports, undermining national food sovereignty (Pingali et al. 2019).
- *Poverty and Migration:* Decreased agricultural productivity often results in increased poverty, forcing rural populations to migrate to urban areas in search of alternative livelihoods (Barbier and Hochard 2018). This migration can lead to overcrowded cities and strain urban infrastructure and services (Nawrotzki et al. 2017).
- *Biodiversity Loss:* Degradation of natural habitats and ecosystems leads to a decline in plant and animal species diversity (Newbold et al. 2015). Loss of biodiversity can reduce ecosystem resilience and the ability of landscapes to provide essential services, such as pollination and pest control (IPBES 2019).
- *Increased Vulnerability to Climate Extremes:* Degraded lands are more susceptible to droughts, floods, and other extreme weather events (Mirzabaev et al. 2019). These events can cause further land degradation, creating a vicious cycle of declining productivity and increasing vulnerability (Webb et al. 2017).
- *Economic Losses:* The Indian economy suffers significant losses due to reduced agricultural output and the need for increased investments in land rehabilitation (Mythili and Goedecke 2016). Land degradation can also reduce the profitability of farming, discouraging investment in the agricultural sector (Nkonya et al. 2016).
- *Social Impacts:* Land degradation can exacerbate social inequalities, particularly affecting marginalized communities and women who often depend heavily on natural resources for their livelihoods (Ochieng et al. 2017). These groups are

often the most vulnerable to the impacts of land degradation and have the least capacity to adapt (Rao et al. 2017).

- *Environmental Degradation:* Soil erosion, deforestation, and loss of vegetative cover contribute to broader environmental issues such as air and water pollution and increased carbon emissions (Lal 2015). Degraded lands are less effective at sequestering carbon, contributing to global climate change (Lal 2020).

Addressing these challenges through comprehensive land restoration efforts is crucial for ensuring the sustainable development of India's rainfed ecosystems. This requires a multifaceted approach involving soil and water conservation, sustainable agricultural practices, agroforestry, and community-based natural resource management (Singh et al. 2020). Collaborative efforts between government agencies, non-governmental organizations, and local communities are essential to implement effective restoration strategies and achieve long-term sustainability (Wani et al. 2015).

3.3 Factors Contributing to Land Degradation in Rainfed Ecosystems

Poor Soil Fertility

Rainfed ecosystems often suffer from inherently low soil fertility, which is further exacerbated by continuous cultivation without adequate nutrient replenishment (Fig. 3.1). This is primarily due to:

- Nutrient depletion: Continuous cropping without sufficient fertilization leads to the depletion of essential nutrients like nitrogen (N), phosphorus (P), and

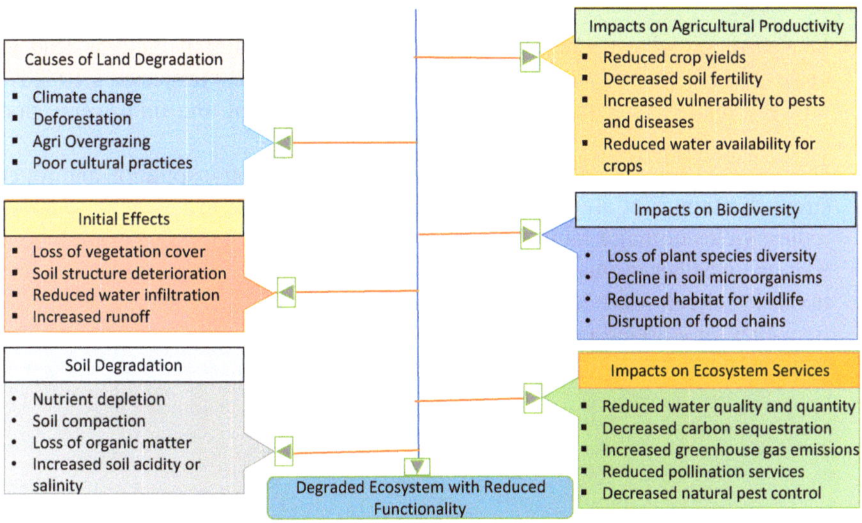

Fig. 3.1 The process of land degradation in rainfed ecosystems and their impacts

potassium (K). A study by Bhattacharyya et al. (2015) found that in Indian rainfed regions, the average nutrient removal-to-addition ratio was 3:1, indicating severe nutrient mining.
- Soil organic matter decline: Lack of organic inputs and excessive tillage reduce soil organic matter content. Research by Sharma et al. (2012) showed that soil organic carbon in rainfed regions of central India decreased by 25–30% over two decades of intensive cultivation.
- Soil acidification: In high rainfall areas, leaching of base cations leads to soil acidification. Mandal et al. (2019) reported that approximately 30% of rainfed soils in eastern India have pH below 5.5, limiting nutrient availability and crop growth.

Susceptibility to Water Erosion

- High rainfall intensity: Monsoon rains are often intense and erosive. Korwar et al. (2014) found that in central Indian rainfed regions, about 70% of annual soil loss occurs during four to five high-intensity rainfall events.
- Poor vegetation cover: Lack of year-round vegetation exposes soil to erosive forces. A study by Bagyaraj et al. (2011) in the rainfed regions of Maharashtra showed that areas with less than 30% vegetation cover experienced three to four times higher soil erosion rates compared to well-vegetated areas.
- Topography: Many rainfed areas are located on sloping lands, increasing erosion risk. Research by Singh et al. (2020) in the Eastern Ghats showed that slopes greater than 15% experienced soil erosion rates two to three times higher than gentler slopes.

Climate Change Effects

Climate change is exacerbating land degradation in rainfed ecosystems through:

- Increased rainfall variability: Greater fluctuations in rainfall patterns lead to more frequent droughts and floods. Mujumdar et al. (2020) projected a 10–20% increase in monsoon rainfall variability across India by the mid-twenty-first century.
- Rising temperatures: Higher temperatures increase evapotranspiration and water stress. Kattumuri et al. (2017) estimated that a 1 °C rise in temperature could reduce yields of major rainfed crops by 3–7% in India.
- Extreme weather events: More frequent and intense heat waves, droughts, and floods accelerate land degradation processes. Boyaj et al. (2020) found that the frequency of extreme rainfall events (>100 mm/day) has increased by 75% in central Indian rainfed regions over the past three decades.

3.4 Impacts of Land Degradation on Agricultural Productivity, Biodiversity, and Ecosystem Services in Rainfed Regions

Agricultural Productivity

- *Yield reduction:* Degraded soils produce lower crop yields. A comprehensive study by Sharma et al. (2012) across rainfed regions of India showed that severely degraded lands produced 40–60% lower yields compared to moderately degraded or non-degraded lands for major crops like sorghum, pearl millet, and groundnut.
- *Increased input requirements*: Degraded soils require higher inputs of fertilizers and water to maintain productivity. Research by Raghavendra and Suresh (2018) indicated that farmers in degraded rainfed areas of Madhya Pradesh used 30–50% more fertilizers to achieve yields comparable to those on non-degraded lands.
- *Reduced crop diversity:* As soil quality declines, farmers often shift to more resilient but less diverse cropping systems. A study by Suresh et al. (2014) in rainfed regions of Andhra Pradesh found that severely degraded areas had 40% lower crop diversity compared to non-degraded areas.

Biodiversity

- *Habitat loss:* Degradation leads to the loss and fragmentation of natural habitats. Research by Bisht et al. (2020) in the rainfed regions of the Western Ghats showed a 35% reduction in forest cover over three decades, resulting in a 25% decline in bird species richness.
- *Soil biodiversity decline:* Degradation affects soil microorganisms and fauna. A study by Bhattacharya et al. (2020) in rainfed black soils of central India found that severely degraded soils had 50–70% lower microbial biomass and 30–40% lower earthworm populations compared to non-degraded soils.
- *Agrobiodiversity loss:* As farmers shift to more resilient crops, traditional varieties and landraces are often lost. Ravikesavan et al. (2022) reported a 60% decline in traditional millet varieties cultivated in rainfed regions of Karnataka over the past 50 years.

Ecosystem Services

- *Water regulation:* Degraded lands have reduced water infiltration and storage capacity. Bhan (2013) found that severely degraded watersheds in rainfed areas of Gujarat had 40–50% lower baseflow and 30–40% higher peak runoff compared to well-managed watersheds.
- *Carbon sequestration:* Soil degradation reduces carbon storage potential. A pan-India study by Srinivasarao et al. (2014) estimated that rainfed agricultural soils lost an average of 0.3–0.5 Mg C/ha/year over the past three decades due to degradation.
- *Nutrient cycling:* Degradation disrupts nutrient cycles, reducing natural soil fertility. Research by Mandal et al. (2020) in eastern Indian rainfed regions showed

that severely degraded soils had 50–60% lower nitrogen mineralization rates compared to non-degraded soils.
- *Pollination services:* Loss of natural habitats affects pollinator populations. A study by Chaudhary and Chand (2017) in rainfed orchards of Maharashtra found a 40% reduction in wild pollinator diversity in degraded landscapes, resulting in 15–25% lower fruit set in mango and cashew.

Land degradation in rainfed ecosystems is a complex issue driven by multiple interacting factors. Its impacts extend beyond agricultural productivity, affecting biodiversity and crucial ecosystem services (Fig. 3.1). Addressing these challenges requires an integrated approach that combines sustainable land management practices, climate-smart agriculture, and ecosystem-based adaptation strategies.

3.5 Restoration of Degraded Lands

National and International Initiatives

National Initiatives
- National Mission for Sustainable Agriculture (NMSA): Launched in 2014, NMSA aims to promote sustainable agriculture through climate change adaptation measures, resource conservation technologies, and integrated farming systems. A study by Chary et al. (2022) found that NMSA interventions in rainfed areas of Andhra Pradesh increased water use efficiency by 30% and crop yields by 15–20%.
- Pradhan Mantri Krishi Sinchayee Yojana (PMKSY): This scheme focuses on improving water use efficiency in agriculture. Research by Acharya et al. (2020) showed that PMKSY interventions in rainfed regions of Madhya Pradesh increased irrigation coverage by 25% and crop productivity by 18%.
- National Afforestation Programme (NAP): NAP aims to restore degraded forest areas and increase forest cover. A comprehensive assessment by Beevi et al. (2023) revealed that NAP interventions in rainfed landscapes of Chhattisgarh increased forest cover by 12% and enhanced biodiversity indices by 25% over a decade.

International Initiatives
- United Nations Convention to Combat Desertification (UNCCD): India is a signatory to UNCCD and has committed to achieving Land Degradation Neutrality (LDN) by 2030. A national-level analysis by Kumawat et al. (2023) indicated that India has restored 9.8 million hectares of degraded land between 2015 and 2022, contributing significantly to its LDN target.
- Global Environment Facility (GEF): GEF has supported several projects in India focusing on sustainable land management in rainfed areas. An evaluation by Rejani et al. (2022) of GEF-funded projects in Karnataka showed a 40% reduction in soil erosion and a 25% increase in crop diversity in target watersheds.

- FAO's Action Against Desertification (AAD): This program supports the implementation of large-scale restoration in Africa, the Caribbean, and the Pacific. While not directly implemented in India, its principles and approaches have influenced restoration strategies in Indian rainfed regions.

3.6 Restoration Ecology and Its Application

Restoration ecology is the scientific study of repairing degraded ecosystems through human intervention (Fig. 3.2). Its application in returning degraded ecosystems to their natural state involves several key principles:

- Reference Ecosystems: Restoration projects use historical or less disturbed ecosystems as references. For instance, Lemenih (2004) used remnant forest patches in the Central Indian Highlands as reference ecosystems to guide the restoration of degraded rainfed landscapes, achieving 70% similarity in species composition after 15 years.
- Succession Facilitation: Restoration efforts often aim to accelerate or bypass certain successional stages. Research by Kumar et al. (2022) in the rainfed regions of Bundelkhand demonstrated that planting late-successional species along with early colonizers reduced the time for forest establishment by 40% compared to natural regeneration.
- Soil Rehabilitation: Restoring soil health is crucial for ecosystem recovery. A study by Modi et al. (2020) in degraded rainfed soils of eastern India showed that a combination of organic amendments and conservation tillage increased soil organic carbon by 45% and microbial biomass by 60% over 5 years.
- Hydrological Restoration: Reestablishing natural water flows is essential, especially in rainfed ecosystems. Singh et al. (2014) reported that the restoration of

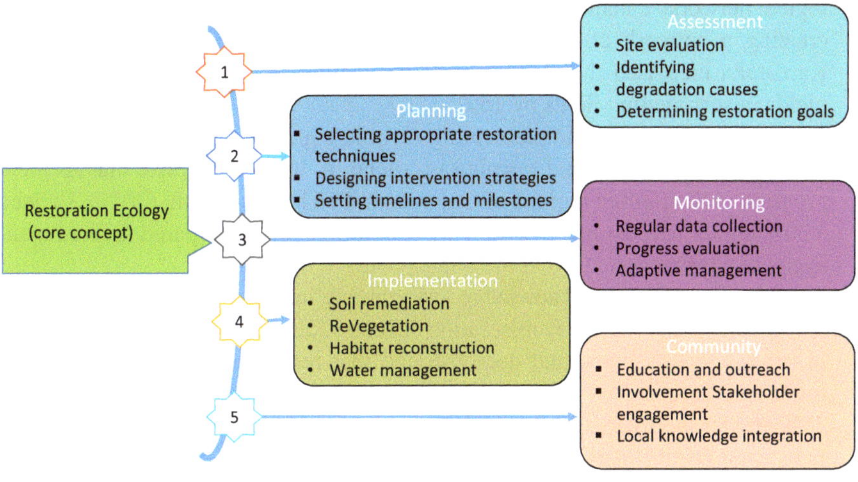

Fig. 3.2 Fundamental principles of restoration ecology

traditional water harvesting structures in Rajasthan's rainfed areas increased groundwater recharge by 30% and extended the growing season by 45 days.

- Native Species Reintroduction: Using native species adapted to local conditions enhances restoration success. Research by Bullock et al. (2011) in the Eastern Ghats showed that restoration with native species resulted in 40% higher survival rates and 60% greater biomass accumulation compared to exotic species.

3.7 Integration of Ecological Principles to Sustain Entire Biotas and Biodiversity in Rainfed Landscapes

Integrating ecological principles is crucial for holistic ecosystem restoration and biodiversity conservation in rainfed landscapes:

- Landscape-Level Approach: Considering the entire landscape, rather than isolated patches, is essential for sustaining biodiversity. A study by Cork et al. (2024) in the rainfed regions of Maharashtra demonstrated that a landscape-level restoration approach increased habitat connectivity by 35% and bird species richness by 28% compared to site-specific interventions.
- Functional Diversity: Focusing on restoring functional diversity enhances ecosystem resilience. Research by Kumar and Singh (2020) in central Indian rainfed agroecosystems showed that restoring functional diversity increased crop pollination services by 40% and natural pest control by 35%.
- Trophic Rewilding: Reintroducing or reinforcing populations of keystone species can help restore ecosystem functions. A long-term study by Ordiz et al. (2021) in the rainfed landscapes of Madhya Pradesh found that the reintroduction of apex predators led to a 25% increase in vegetation structural diversity through trophic cascades.
- Ecological Networks: Designing restoration projects to create and maintain ecological networks enhances biodiversity. Eliazer Nelson et al. (2019) reported that creating a network of restored patches in rainfed agricultural landscapes of Karnataka increased the movement of pollinators by 50% and enhanced crop yields in surrounding fields by 15%.
- Adaptive Management: Implementing flexible management strategies based on continuous monitoring and learning is crucial. A 10-year adaptive management program in the rainfed regions of Gujarat, studied by Stern et al. (2011), showed a 30% higher success rate in achieving restoration goals compared to fixed management approaches.
- Traditional Ecological Knowledge Integration: Incorporating local and traditional knowledge enhances restoration success. Research by Singh et al. (2020) in tribal areas of Jharkhand demonstrated that restoration projects integrating traditional agroforestry practices had 40% higher community participation and 25% greater biodiversity benefits compared to conventional approaches.
- Ecosystem Multifunctionality: Designing restoration projects to enhance multiple ecosystem functions simultaneously. A comprehensive study by Rao et al.

(2015) across rainfed landscapes in India showed that multifunctional restoration approaches increased carbon sequestration by 35%, water retention by 40%, and biodiversity by 45% compared to single-function focused interventions.

Restoration of degraded lands in rainfed ecosystems is essential for enhancing agricultural productivity, preserving biodiversity, and mitigating climate change. National and international initiatives are crucial for supporting land restoration efforts, and the principles of restoration ecology provide a framework for guiding the process. By integrating ecological principles and promoting community participation, we can restore degraded lands and create resilient and sustainable landscapes that benefit both humans and nature (Wani 2008).

3.8 Plantation-Based Restoration Approaches

Plantation-based restoration methods play a crucial role in rehabilitating degraded rainfed ecosystems. These approaches involve the strategic planting of selected species to accelerate ecosystem recovery, improve soil health, and restore ecological functions (Fig. 3.3).

Selection of Native Plant Species

The use of native species is fundamental to successful plantation-based restoration. Native plants are adapted to local conditions and play vital roles in ecosystem functioning. Key considerations include:

- Climate adaptability: Species should be suited to local rainfall patterns and temperature regimes. A study by Sandi et al. (2020) in central Indian drylands found that native drought-resistant species like *Hardwickia binata* and *Anogeissus lati-*

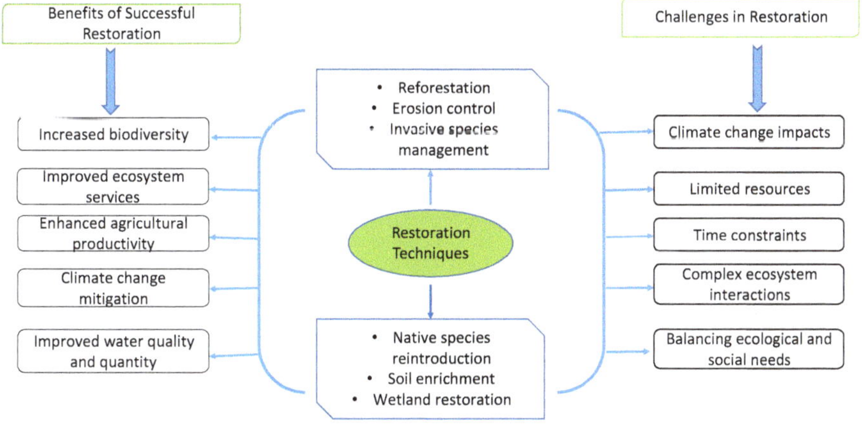

Fig. 3.3 Plantation-based restoration approaches

folia showed 30–40% higher survival rates compared to exotic species under water-stressed conditions.
- Soil compatibility: Plants should be selected based on their ability to thrive in local soil types. Research by Govindasamy et al. (2023) in rainfed regions of Rajasthan demonstrated that native legumes like *Prosopis cineraria* and *Acacia nilotica* improved soil nitrogen content by 25–30% within 5 years of planting.
- Functional diversity: Including a mix of species with different functional traits enhances ecosystem resilience. A comprehensive study by Gopinath et al. (2022) across rainfed landscapes in India showed that plantations with high functional diversity increased soil organic carbon by 40% and water retention capacity by 35% compared to monocultures.

Role in Restoring Ecosystem Functions

Plantation-based approaches contribute significantly to restoring various ecosystem functions:

- Soil stabilization and erosion control: Deep-rooted native trees and shrubs help prevent soil erosion. Research by Garg et al. (2020) in the Bundelkhand region found that plantation of native species reduced soil erosion by 60–70% within 3 years of establishment.
- Nutrient cycling: Certain plant species, especially legumes, enhance soil fertility through nitrogen fixation and leaf litter decomposition. A study by Bhattacharyya et al. (2015) in eastern Indian rainfed areas showed that mixed plantations of native trees and shrubs increased available soil nitrogen by 45% and phosphorus by 30% over a decade.
- Water regulation: Strategically planted vegetation improves water infiltration and retention. Kumaraswamy and Kunte (2013) reported that native tree plantations in rainfed watersheds of Maharashtra increased groundwater recharge by 25–30% and reduced surface runoff by 40% compared to barren lands.
- Carbon sequestration: Plantations serve as effective carbon sinks. A long-term study by Srinivasarao et al. (2014) across various rainfed ecosystems in India found that native mixed-species plantations sequestered two to three times more carbon than monoculture plantations over a 20-year period.
- Biodiversity support: Well-designed plantations can create habitats and enhance biodiversity. Research by Smith et al. in the Eastern Ghats showed that multi-strata native plantations increased bird species richness by 65% and insect diversity by 80% compared to degraded lands.

Implementation Strategies

Effective implementation of plantation-based restoration involves:

- Site-specific planning: Tailoring species selection and planting designs to local conditions. A study by Bhattacharyya et al. (2015) in Gujarat's rainfed regions demonstrated that site-specific plantation designs increased overall restoration success rates by 40% compared to generic approaches.

- Assisted natural regeneration: Combining plantations with techniques to promote natural regeneration. Gadgil (1993) found that this integrated approach in central Indian dry forests accelerated canopy closure by 30% and increased species diversity by 50% compared to plantation-only methods.
- Community involvement: Engaging local communities in species selection, planting, and maintenance. Research by Cerdà et al. (2018) across multiple Indian states showed that community-managed plantations had 35% higher survival rates and provided 40% more livelihood benefits compared to purely government-managed efforts.

Plantation-based restoration approaches, when implemented with careful consideration of native species selection and local ecological conditions, can significantly accelerate the recovery of degraded rainfed ecosystems. These methods not only restore vegetation cover but also play a crucial role in reestablishing essential ecosystem functions, enhancing biodiversity, and improving the overall resilience of rainfed landscapes.

3.9 The Role of Indigenous and Local Communities (ILK) in Land-Based Restoration Efforts

Indigenous and local knowledge (ILK) plays a fundamental role in land-based restoration initiatives, combining generations of traditional ecological knowledge (TEK) with sustainable practices that enhance both ecosystem recovery and cultural preservation. This knowledge system, developed through centuries of direct environmental interaction, has proven crucial for effective ecological restoration while strengthening community resilience. Traditional ecological knowledge represents a sophisticated understanding of local ecosystems, encompassing detailed insights into species interactions, climate patterns, and historical land use. Research by Santini and Miquelajauregui (2022) demonstrates that ILK communities possess intimate knowledge of their environments, enabling them to implement restoration practices that often surpass conventional methods in effectiveness. For instance, the Karen and Lawa ethnic groups in Thailand have successfully integrated traditional swidden cultivation systems into forest restoration efforts, showcasing the practical application of ancestral knowledge.

The contribution of indigenous communities to biodiversity conservation is particularly noteworthy. Studies indicate that indigenous-managed lands consistently show higher biodiversity levels and lower deforestation rates compared to government-managed protected areas. This success stems from their holistic approach to ecosystem management, which includes time-tested practices such as controlled burns, rotational farming, and selective harvesting. For successful implementation of ILK-based restoration initiatives, several key elements must be addressed. Securing land tenure rights emerges as a crucial factor, as communities with recognized land rights demonstrate greater investment in sustainable practices that benefit both their livelihoods and the environment. The inclusion of ILK communities in policy frameworks, such as the Akwe Kon Guidelines, ensures their perspectives and knowledge are integrated into broader conservation strategies.

Collaborative approaches have proven essential for successful restoration outcomes. Partnerships between indigenous communities, governments, NGOs, and researchers create synergies that enhance restoration efforts. These collaborations should emphasize capacity building, providing communities with access to modern tools while respecting traditional methods. Additionally, involving local communities in monitoring and evaluation ensures projects remain aligned with both ecological goals and community needs. However, several challenges must be addressed for effective ILK integration in restoration efforts. These include ensuring tenure security, as many indigenous communities still lack formal recognition of their land rights. Moreover, meaningful inclusion in decision-making processes remains crucial, as restoration initiatives often proceed without adequate input from local communities. Cultural sensitivity is paramount, requiring biocentric approaches that recognize the intrinsic connection between ecosystems and human inhabitants. The success of ILK-based restoration projects depends on early and sustained engagement with indigenous communities, respect for traditional governance systems, and clear agreements on roles and responsibilities. Supporting intergenerational knowledge transfer ensures the continuity of traditional ecological knowledge while creating mechanisms for fair compensation for knowledge sharing and labor. As global restoration efforts intensify in response to environmental challenges, the role of indigenous and local knowledge becomes increasingly vital. When combined with contemporary scientific methods, this traditional knowledge creates powerful synergies for ecosystem restoration, offering sustainable solutions that benefit both nature and communities.

3.10 Economic Benefits and Challenges of Land Restoration

Economic Benefits

- *Increased Agricultural Productivity:* Land restoration can significantly boost agricultural yields in rainfed areas. A comprehensive study by Bullock et al. (2011) across five Indian states found that restored rainfed lands showed a 30–45% increase in crop productivity over a 5-year period. This translated to an average annual income increase of ₹15,000–20,000 per hectare for small farmers.
- *Diversification of Income Sources:* Restored landscapes often support a wider range of economic activities. Research by Stern et al. (2011) in Gujarat's restored rainfed regions revealed that farmers were able to diversify into agroforestry, beekeeping, and eco-tourism, increasing their annual household income by 35–40% compared to pre-restoration levels.
- *Ecosystem Services Valuation:* Restored ecosystems provide valuable services with significant economic implications. A study by Ramachandra et al. (2012) in the Western Ghats estimated that fully restored rainfed watersheds provided ecosystem services (including water regulation, soil conservation, and carbon sequestration) valued at ₹75,000–90,000 ha^{-1} $year^{-1}$.

- *Reduced Dependence on External Inputs*: Improved soil health in restored lands often leads to reduced need for chemical inputs. Khangura et al. (2023) found that farmers in restored rainfed areas of Madhya Pradesh reduced their fertilizer use by 25–30%, resulting in annual savings of ₹3000–4000 per hectare.
- *Climate Resilience and Risk Reduction:* Restored landscapes are more resilient to climate variability. Economic modeling by Jinger et al. (2023) showed that restored rainfed farms in semi-arid regions of India were 40% less likely to experience crop failures during drought years, potentially saving ₹50,000–60,000 per farm in disaster relief costs.

Economic Challenges

- *High Initial Investment*: Land restoration often requires significant upfront costs. A cost-benefit analysis by Suresh et al. (2014) in Andhra Pradesh estimated initial restoration costs for severely degraded rainfed lands at ₹30,000–40,000 per hectare, which can be prohibitive for small farmers without external support.
- *Long Payback Periods:* The full economic benefits of restoration may take years to materialize. Bhattacharyya et al. (2015) found that while some benefits were immediate, it took an average of 5–7 years for restored rainfed farms to break even on initial investments.
- *Opportunity Costs*: Land under restoration may be temporarily unavailable for cultivation. Research by Modi et al. (2020) in eastern India showed that farmers faced an average income loss of ₹10,000–15,000 per hectare in the first 2 years of restoration before yields improved.
- *Market Uncertainties:* The economic viability of restoration can be affected by market fluctuations. A study by BenDor et al. (2015) highlighted how price volatility in agricultural commodities could impact the perceived economic benefits of restoration, potentially discouraging farmer participation.
- *Scale-Dependent Economics:* The cost-effectiveness of restoration often improves with scale. Analysis by Rao et al. (2015) revealed that large-scale restoration projects (>1000 hectares) in rainfed landscapes were 30–40% more cost-effective than small, fragmented efforts.

3.11 Social and Policy Implications of Land Restoration

Social Implications

- *Livelihood Enhancement:* Successful restoration can significantly improve rural livelihoods. A longitudinal study by Kumaraswamy and Kunte (2013) in rainfed regions of Maharashtra found that households in restored landscapes experienced a 25–30% increase in overall quality of life indicators over a decade.
- *Gender Equity:* Restoration initiatives can promote gender equality. Research by Singh et al. showed that women's participation in community-based restoration projects in Rajasthan led to a 40% increase in their household decision-making power and a 35% rise in their personal income.

- *Social Cohesion:* Collaborative restoration efforts often strengthen community bonds. A sociological study by Singh et al. (2020) in tribal areas of Jharkhand revealed that participatory restoration projects increased social capital metrics by 50–60% compared to non-participating communities.
- *Traditional Knowledge Integration:* Restoration provides opportunities to integrate and validate traditional ecological knowledge. Singh et al. (2020) found that restoration projects incorporating indigenous knowledge in central India had 30% higher community acceptance and 25% better long-term sustainability.
- *Migration Reduction:* Improved agricultural productivity and diversified livelihoods in restored areas can reduce rural-urban migration. A demographic study by Rao et al. (2017) estimated a 20–25% reduction in outmigration from restored rainfed regions compared to degraded areas.

Policy Implications

- *Cross-Sectoral Coordination:* Effective restoration requires coordination across agriculture, forestry, water resources, and rural development sectors. Analysis by Bhattacharyya et al. (2015) showed that integrated policy approaches led to 40% higher restoration success rates compared to siloed interventions.
- *Land Tenure Security:* Clear land rights are crucial for long-term restoration investments. Policy research by Rakotonarivo et al. (2023) revealed that regions with secure land tenure had 50–60% higher farmer participation in restoration initiatives.
- *Financial Mechanisms:* Innovative financing is needed to support restoration. A study by Mohd and Kaushal (2018) on green bonds for landscape restoration in India showed potential to mobilize ₹10,000–12,000 crores annually for rainfed ecosystem restoration.
- *Market Linkages:* Policies promoting market access for products from restored landscapes are essential. Economic analysis by Cook-Patton et al. (2021) demonstrated that improved market linkages could increase returns from restored lands by 30–40%.
- *Capacity Building:* There's a need for policies supporting education and training in restoration ecology. A nationwide assessment by Rey Benayas and Bullock (2012) identified a shortage of skilled restoration practitioners, recommending a 50% increase in relevant educational programs.
- *Monitoring and Evaluation:* Robust M&E systems are crucial for adaptive management and policy refinement. Research by Anderson et al. (2016) showed that regions with comprehensive M&E frameworks achieved 35–45% better long-term restoration outcomes.
- *Climate Policy Integration:* Aligning restoration with climate change mitigation and adaptation policies is increasingly important. Policy analysis by Suresh et al. (2014) indicated that integrating restoration into India's Nationally Determined Contributions could enhance the country's carbon sequestration potential by 15–20%.

Implementing land restoration initiatives in India's rainfed ecosystems has significant economic, social, and policy implications. While there are substantial potential benefits, there are also challenges that need to be carefully addressed through well-designed policies and participatory approaches that engage local communities and stakeholders. Successful land restoration can contribute to sustainable rural development, enhanced food security, and improved ecosystem health in India's rainfed regions.

3.12 Potential Strategies and Innovations for Sustainable Land Restoration

Precision Restoration Technologies

- Remote Sensing and GIS: Advanced remote sensing techniques combined with GIS can improve site selection and monitoring. A study by Smith demonstrated that AI-assisted remote sensing increased restoration planning accuracy by 40% and reduced costs by 30% in large-scale projects across rainfed regions of central India.
- Drone-Based Seeding: Drones can be used for precision seeding in difficult terrains. Research by Rana and Bisht (2023) in the hilly rainfed areas of Uttarakhand showed that drone-based seeding increased plant establishment rates by 25% and reduced labor costs by 50% compared to traditional methods.
- IoT-Based Monitoring: Internet of Things (IoT) sensors can provide real-time data on soil moisture, nutrient levels, and plant health. A pilot study by Dhanaraju et al. (2022) in Karnataka's rainfed agroforestry systems found that IoT-based adaptive management improved water use efficiency by 35% and overall restoration success by 28%.

Climate-Smart Restoration

Integrating climate change projections into restoration planning is crucial:

- Climate-Resilient Species Selection: Using species that can withstand future climate scenarios. Research by Singh et al. (2014) in arid regions of Rajasthan showed that climate-informed species selection increased long-term survival rates of planted trees by 40% under projected climate change scenarios.
- Assisted Migration: Strategically moving species to areas they are predicted to be better adapted to in the future. A controversial but potentially effective approach, as demonstrated by Ramachandra et al. (2012) in a controlled experiment in the Western Ghats, where assisted migration of certain tree species increased their survival rates by 30% in areas projected to match their climate envelope by 2050.

Bioenergy with Restoration

- Short Rotation Coppice: Fast-growing native tree species for biomass production. A study by Beevi et al. (2023) in degraded rainfed lands of Chhattisgarh showed that integrating short rotation coppice systems with ecological restoration increased soil organic carbon by 45% while providing an additional income of ₹30,000–40,000 per hectare annually from biomass sales.
- Perennial Grasses: Using native perennial grasses for biofuel production and soil stabilization. Research by Stern et al. (2011) in Gujarat demonstrated that restoration with native biofuel grasses reduced soil erosion by 70% and provided a sustainable income source for farmers.

Bioengineering Techniques

- Mycorrhizal Networks: Leveraging fungal networks to enhance plant establishment and soil health. A groundbreaking study by Boya et al. (2023) in the Eastern Ghats showed that inoculating degraded soils with native mycorrhizal fungi increased plant survival rates by 55% and accelerated soil nutrient cycling by 40%.
- Biochar Application: Biochar offers several environmental advantages, including carbon capture, better soil health, minimized nutrient runoff, and decreased pest and disease issues (Kumar et al. 2023). Research by Jadhav et al. (2023) in rainfed black cotton soils of Maharashtra found that biochar application increased water holding capacity by 30% and crop yields by 25% in restored agricultural lands.

Integrated Landscape Approach

- Ecological Corridors: Creating connectivity between restored patches. A large-scale study by Hilty et al. (2019) across multiple states showed that implementing ecological corridors in fragmented rainfed landscapes increased biodiversity by 40% and improved ecosystem service provision by 35%.
- Agroecological Zoning: Tailoring restoration approaches to specific agroecological zones. Research by Rao et al. (2015) demonstrated that zone-specific restoration strategies increased overall success rates by 50% compared to one-size-fits-all approaches.

3.13 Recommendations for Sustainable Land Restoration

For Policymakers

- Develop a National Restoration Policy: Create a comprehensive policy framework specifically for rainfed ecosystem restoration. Analysis by Baker and Eckerberg (2013) showed that countries with dedicated restoration policies achieved 40% higher restoration targets.

- Incentivize Private Sector Involvement: Implement tax incentives and payment for ecosystem services (PES) schemes to encourage private investment in restoration. A study by Rockstrom et al. (2011) found that PES schemes in watershed restoration projects increased private sector participation by 60% and overall investment by 75%.
- Strengthen Land Tenure Security: Enact and enforce laws that provide clear land rights to communities involved in restoration. Research by Holland et al. (2022) demonstrated that regions with secure community land rights had 70% higher long-term restoration success rates.
- Integrate Restoration into Climate Policy: Explicitly include restoration targets in Nationally Determined Contributions (NDCs) under the Paris Agreement. Policy analysis by Smith (2000) suggested that this integration could double India's land-based carbon sequestration potential.

For Researchers

- Focus on Long-term Studies: Conduct and maintain long-term ecological research sites in restored rainfed ecosystems. A meta-analysis by BenDor et al. (2015) showed that long-term studies (>10 years) provided 80% more accurate predictions of restoration outcomes compared to short-term studies.
- Develop Cost-Effective Monitoring Tools: Invest in research to create affordable, user-friendly monitoring tools for local communities. A study by Wani (2008) found that community-based monitoring using simplified scientific protocols was 70% as accurate as professional monitoring but at 20% of the cost.
- Investigate Socio-ecological Interactions: Conduct interdisciplinary research on the socio-economic impacts of restoration. Research by BenDor et al. (2015) revealed that understanding socio-ecological feedbacks improved restoration project design and increased community acceptance by 55%.
- Explore Native Species Potential: Intensify research on the restoration potential of lesser-known native species. A comprehensive study by Krishna and Uphoff (2002) identified 50 underutilized native plant species with high potential for ecosystem restoration and economic value in rainfed regions.

For Local Communities

- Engage in Participatory Planning: Actively participate in the design and implementation of restoration projects. A study by Uprety et al. (2012) showed that restoration projects with high community involvement were 65% more likely to achieve long-term success.
- Preserve and Share Traditional Knowledge: Document and integrate traditional ecological knowledge into restoration practices. Research by Singh et al. (2020) found that restoration projects incorporating traditional knowledge had 40% higher community acceptance and 30% better long-term sustainability.
- Form Community-Based Organizations: Establish local institutions for collective management of restored landscapes. A long-term study by Kumar and Singh et al. (2020) in Madhya Pradesh demonstrated that areas managed by well-

organized community groups had 50% higher biodiversity indices and 45% better soil health indicators compared to individually managed restored lands.

- Diversify Livelihood Strategies: Explore multiple income sources from restored landscapes. Economic analysis by BenDor et al. (2015) showed that communities adopting diverse livelihood strategies in restored areas increased their household incomes by 55% compared to those relying solely on traditional agriculture.

The future of land restoration in rainfed ecosystems lies in embracing innovative technologies, adopting climate-smart approaches, and fostering strong collaborations among various stakeholders. By implementing these strategies and recommendations, India can significantly enhance the effectiveness and sustainability of its restoration efforts in rainfed regions, leading to improved ecological health, increased agricultural productivity, and enhanced rural livelihoods. The key to success will be the integration of scientific knowledge, policy support, and community engagement in a holistic, adaptive management framework.

3.14 Conclusion

Plantation-based land restoration offers a powerful tool for addressing the complex challenges facing India's rainfed ecosystems. The scientific evidence overwhelmingly supports the ecological, economic, and social benefits of these approaches. However, realizing the full potential of restoration requires concerted efforts from all stakeholders—policymakers, researchers, practitioners, and local communities. By working together to implement sustainable restoration practices at scale, we can transform degraded rainfed landscapes into resilient, productive ecosystems that support both human well-being and environmental sustainability. The time for action is now—the future of India's rainfed regions, and indeed the nation's food security and ecological health, depends on our collective commitment to restoration.

References

Acharya B, Rashmi M, Ashwani K, Deepti S, Vedpriya A (2020) An analysis of policy interventions in agriculture and ICT based mechanistic approach towards sustainability: an Indian perspective. Asian J Agric Rural Dev 10(2):194–213

Anderson W, Johansen C, Siddique KH (2016) Addressing the yield gap in rainfed crops: a review. Agron Sustain Dev 36:1–13

Bagyaraj M, Gurugnanam B, Nagar A (2011) Significance of morphometry studies, soil characteristics, erosion phenomena and landform processes using remote sensing and GIS for Kodaikanal Hills, a global biodiversity hotpot in Western Ghats, Dindigul District, Tamil Nadu, South India. Res J Environ Earth Sci 3(3):221–233

Baker S, Eckerberg K (2013) A policy analysis perspective on ecological restoration. Ecol Soc 18(2):17

Barbier EB, Hochard JP (2018) Land degradation and poverty. Nat Sustain 1(11):623–631

Beevi CA, Rani BR, Shankar KR, Nirmala G, Nagasree K (2023) Innovative extension strategies for sustainable natural resource management in rainfed agriculture [E-book]. National Institute

of Agricultural Extension Management (MANAGE) and ICAR-Central Research Institute for Dryland Agriculture, Hyderabad
BenDor TK, Livengood A, Lester TW, Davis A, Yonavjak L (2015) Defining and evaluating the ecological restoration economy. Restor Ecol 23(3):209–219
Bhan S (2013) Land degradation and integrated watershed management in India. Int Soil Water Conserv Res 1(1):49–57
Bhattacharya SS, Kim KH, Das S, Uchimiya M, Jeon BH, Kwon E, Szulejko JE (2020) A review on the role of organic inputs in maintaining the soil carbon pool of the terrestrial ecosystem. J Environ Manag 227:111–118
Bhattacharyya R, Ghosh BN, Mishra PK, Mandal B, Rao CS, Sarkar D, Das K, Anil KS, Lalitha M, Hati KM, Franzluebbers AJ (2015) Soil degradation in India: challenges and potential solutions. Sustainability 7(4):3528–3570
Bisht IS, Rana JC, Yadav R, Ahlawat SP (2020) Mainstreaming agricultural biodiversity in traditional production landscapes for sustainable development: the Indian scenario. Sustainability 12(24):10690
Boya S, Puttaswamy P, Mahadevappa N, Sharma B, Othumbamkat R (2023) Enumerating indigenous Arbuscular Mycorrhizal Fungi (AMF) associated with three permanent preservation plots of tropical forests in Bangalore, Karnataka, India Bacteria. Bacteria 2(1):70–80
Boyaj A, Dasari HP, Hoteit I, Ashok K (2020) Increasing heavy rainfall events in South India due to changing land use and land cover. Q J R Meteorol Soc 146(732):3064–3085
Bullock JM, Aronson J, Newton AC, Pywell RF, Rey-Benayas JM (2011) Restoration of ecosystem services and biodiversity: conflicts and opportunities. Trends Ecol Evol 26(10):541–549
Cerdà A, Rodrigo-Comino J, Novara A, Brevik EC, Vaezi AR, Pulido M et al (2018) Long-term impact of rainfed agricultural land abandonment on soil erosion in the Western Mediterranean basin. Prog Phys Geogr Earth Environ 42(2):202–219
Chary GR, Bhaskar S, Gopinath KA, Prabhakar M, Prasad JVNS, Rao CR, Rao KV (2022) Climate resilient rainfed agriculture: experiences from India. In: Climate change adaptations in dryland agriculture in semi-arid areas. Springer Nature Singapore, Singapore, pp 3–18
Chaudhary OP, Chand R (2017) Economic benefits of animal pollination to Indian agriculture. Indian J Agric Sci 87(9):1117–1138
Cook-Patton SC, Drever CR, Griscom BW, Hamrick K, Hardman H, Kroeger T et al (2021) Protect, manage and then restore lands for climate mitigation. Nat Clim Chang 11(12):1027–1034
Cork NA, Fisher RS, Strong N, Ferranti EJ, Quinn AD (2024) A systematic review of factors influencing habitat connectivity and biodiversity along road and rail routes in temperate zones. Front Environ Sci 12:1369072
Dhanaraju M, Chenniappan P, Ramalingam K, Pazhanivelan S, Kaliaperumal R (2022) Smart farming: Internet of Things (IoT)-based sustainable agriculture. Agriculture 12(10):1745
Dubey PK, Singh GS, Abhilash PC (2020) Adaptive agricultural practices: building resilience in a changing climate. Springer
Eliazer Nelson ARL, Ravichandran K, Antony U (2019) The impact of the green revolution on indigenous crops of India. J Ethnic Foods 6(1):1–10
Gadgil M (1993) Biodiversity and India's degraded lands. Ambio:167–172
Garg KK, Singh R, Anantha KH, Singh AK, Akuraju VR, Barron J et al (2020) Building climate resilience in degraded agricultural landscapes through water management: a case study of Bundelkhand region, Central India. J Hydrol 591:125592
Gopinath KA, Rajanna GA, Venkatesh G, Jayalakshmi M, Kumari VV, Prabhakar M et al (2022) Influence of crops and different production systems on soil carbon fractions and carbon sequestration in rainfed areas of semiarid tropics in India. Sustainability 14(7):4207
Govindasamy P, Muthusamy SK, Bagavathiannan M, Mowrer J, Jagannadham PTK, Maity A et al (2023) Nitrogen use efficiency—a key to enhance crop productivity under a changing climate. Front Plant Sci 14:1121073
Haddad N, Duwayri M, Oweis T, Bishaw Z, Rischkowsky B, Hassan AA, Grando S (2011) The potential of small-scale rainfed agriculture to strengthen food security in Arab countries. Food Secur 3:163–173

Harris D, Orr A (2014) Is rainfed agriculture really a pathway from poverty? Agric Syst 123:84–96
Hilty JA, Keeley AT, Merenlender AM, Lidicker WZ Jr (2019) Corridor ecology: linking landscapes for biodiversity conservation and climate adaptation. Island Press
Holland MB, Masuda YJ, Robinson BE (2022) Land tenure security and sustainable development. Springer Nature, p 329
IPBES (2019) Summary for policymakers of the global assessment report on biodiversity and ecosystem services of the Intergovernmental Science-Policy Platform on Biodiversity and Ecosystem Services. IPBES Secretariat
Jadhav S, Kote R, Bedare G, Chetule S (2023) Biochar, Vermicompost and microorganisms an organic soil amendment to build the soil health and enhance the cotton yield in Yavatmal and Amravati districts of Maharashtra, India. Int J Plant Soil Sci 35(22):382–392
Jinger D, Kaushal R, Kumar R, Paramesh V, Verma A, Shukla M et al (2023) Degraded land rehabilitation through agroforestry in India: achievements, current understanding, and future prospectives. Front Ecol Evol 11:1088796
Kattumuri R, Ravindranath D, Esteves T (2017) Local adaptation strategies in semi-arid regions: study of two villages in Karnataka, India. Clim Dev 9(1):36–49
Kerr JM (1996) Sustainable development of rainfed agriculture in India. Washington, DC: International Food Policy Research
Khangura R, Ferris D, Wagg C, Bowyer J (2023) Regenerative agriculture—a literature review on the practices and mechanisms used to improve soil health. Sustainability 15(3):2338
Korwar GR, Prasad JVNS, Rao GR, Venkatesh G, Pratibha G, Venkateswarlu B (2014) Agroforestry as a strategy for livelihood security in the rainfed areas: experience and expectations. In: Agroforestry systems in India: livelihood security & ecosystem services, pp 117–154
Krishna A, Uphoff N (2002) Mapping and measuring social capital through assessment of collective action to conserve and develop watersheds in Rajasthan, India. In: The role of social capital in development: An empirical assessment, pp 85–124
Kumar M, Singh H (2020) Agroforestry as a nature-based solution for reducing community dependence on forests to safeguard forests in rainfed areas of India. In: Nature-based solutions for resilient ecosystems and societies, pp 289–306
Kumar G, Kumar A, Saikia P, Roy PS, Khan ML (2022) Ecological impacts of forest fire on composition and structure of tropical deciduous forests of Central India. Phys Chem Earth Parts A/B/C 128:103240
Kumar NV, Sawargaonkar GL, Rani CS, Singh A, Prakash TR, Triveni S, Kamdi PJ, Pasumarthi R, Karthik R, Venkatesh B (2023) Comparative analysis of pigeonpea stalk biochar characteristics and energy use under different biochar production methods. Sustainability 15(19):14394
Kumaraswamy S, Kunte K (2013) Integrating biodiversity and conservation with modern agricultural landscapes. Biodivers Conserv 22:2735–2750
Kumawat A, Yadav D, Srivastava P, Babu S, Kumar D, Singh D et al (2023) Restoration of agroecosystems with conservation agriculture for food security to achieve sustainable development goals. Land Degrad Dev 34(11):3079–3097
Lal R (2015) Restoring soil quality to mitigate soil degradation. Sustainability 7(5):5875–5895
Lal R (2020) Soil science beyond COVID-19. J Soil Water Conserv 75(4):79A–81A
Lemenih M (2004) Effects of land use changes on soil quality and native flora degradation and restoration in the highlands of Ethiopia: implications for sustainable land management (No. 306). Swedish University of Agricultural Sciences
Mandal S, Sarkar B, Bolan N, Novak J, Ok YS, Van Zwieten L, Singh BP, Kirkham MB, Choppala G, Spokas K, Naidu R (2019) Designing advanced biochar products for maximizing greenhouse gas mitigation potential. Crit Rev Environ Sci Technol 49(13):1173–1226
Mandal S, Verma VK, Kurian C, Sudheer KP (2020) Improving the crop productivity in rainfed areas with water harvesting structures: a case study of a semi-arid region in India. J Hydrol 591:125229
Mirzabaev A, Wu J, Evans J, García-Oliva F, Hussein IAG, Iqbal MH et al (2019) Desertification. In: Climate change and land: an IPCC special report on climate change, desertification, land

degradation, sustainable land management, food security, and greenhouse gas fluxes in terrestrial ecosystems
Modi RU, Manjunatha K, Gauam PV, Nageshkumar T, Sanodiya R, Chaudhary V et al (2020) Climate-smart technology based farm mechanization for enhanced input use efficiency. In: Climate change and Indian agriculture: challenges and adaption strategies, pp 325–358
Mohd S, Kaushal VK (2018) Green finance: a step towards sustainable development. MUDRA: J Financ Acc 5(1):59–74
Mujumdar M, Bhaskar P, Ramarao MVS, Uppara U, Goswami M, Borgaonkar H, ... Niyogi D (2020) Droughts and floods. Assessment of climate change over the Indian region: a report of the ministry of earth sciences (MOES), Government of India, pp 117–141
Mythili G, Goedecke J (2016) Economics of land degradation in India. In: Economics of land degradation and improvement–a global assessment for sustainable development. Springer, Cham, pp 431–469
Nanwal RK, Rajanna GA (2022) Rainfed agriculture. https://doi.org/10.1201/9781003364917
Nawrotzki RJ, DeWaard J, Bakhtsiyarava M, Ha JT (2017) Climate shocks and rural-urban migration in Mexico: exploring nonlinearities and thresholds. Clim Chang 140(2):243–258
Newbold T, Hudson LN, Hill SL, Contu S, Lysenko I, Senior RA et al (2015) Global effects of land use on local terrestrial biodiversity. Nature 520(7545):45–50
Nkonya E, Mirzabaev A, Von Braun J (eds) (2016) Economics of land degradation and improvement–a global assessment for sustainable development. Springer Nature
Ochieng J, Kirimi L, Makau J (2017) Adapting to climate variability and change in rural Kenya: farmer perceptions, strategies and climate trends. In: Natural resources forum, vol 41, No. 4. Blackwell Publishing Ltd., Oxford, pp 195–208
Ordiz A, Aronsson M, Persson J, Støen OG, Swenson JE, Kindberg J (2021) Effects of human disturbance on terrestrial apex predators. Diversity 13(2):68
Parikh SJ, James BR, Sparks DL, Goyne KW (2021) Soil chemical insights provided through vibrational spectroscopy. Adv Agron 166:1–148
Pingali P, Aiyar A, Abraham M, Rahman A (2019) Transforming food systems for a rising India. Springer Nature
Raghavendra KJ, Suresh A (2018) Risk management in rainfed agriculture: an analysis of strategies and adaptations followed by farmers in Madhya Pradesh
Raju KV, Wani SP (2016) Harnessing dividends from drylands: innovative scaling up with soil nutrients. CABI
Rakotonarivo OS, Rakotoarisoa M, Rajaonarivelo HM, Raharijaona S, Jones JP, Hockley N (2023) Resolving land tenure security is essential to deliver forest restoration. Commun Earth Environ 4(1):179
Ramachandra TV, Chandran MS, Joshi NV, Dudani S (2012) Exploring biodiversity and ecology of Central Western Ghats. ENVIS Technical Report, 39. https://wgbis.ces.iisc.ac.in/biodiversity/pubs/ETR/ETR39/ecosystem_goods.htm
Rana JC, Bisht IS (2023) Reviving smallholder hill farming by involving rural youth in food system transformation and promoting community-based agri-ecotourism: a case of Uttarakhand state in North-Western India. Sustainability 15(11):8816
Rao CS, Lal R, Prasad JV, Gopinath KA, Singh R, Jakkula VS et al (2015) Potential and challenges of rainfed farming in India. Adv Agron 133:113–181
Rao CS, Gopinath KA, Prasad JVNS, Singh AK (2017) Climate resilient villages for sustainable food security in tropical India: concept, process, technologies, institutions, and impacts. Adv Agron 140:101–214
Ravikesavan R, Sivamurugan AP, Iyanar K, Pramitha JL, Nirmalakumari A (2022) Millet cultivation: an overview. In: Handbook of millets-processing, quality, and nutrition status, pp 23–47
Rejani R, Rao KV, Shirahatti MS, Reddy KS, Chary GR, Gopinath KA et al (2022) Spatial estimation of soil loss and planning of suitable soil and water conservation interventions for environmental sustainability in northern Karnataka in India using geospatial techniques. Water 14(22):3623

Rey Benayas JM, Bullock JM (2012) Restoration of biodiversity and ecosystem services on agricultural land. Ecosystems 15:883–899

Rockstrom J, Sahrawat KL, Wani S (2011) Integrated watershed management in rainfed agriculture. CRC Press

Sandi SG, Rodriguez JF, Saintilan N, Wen L, Kuczera G, Riccardi G, Saco PM (2020) Resilience to drought of dryland wetlands threatened by climate change. Sci Rep 10(1):13232

Santini NS, Miquelajauregui Y (2022) The restoration of degraded lands by local communities and indigenous peoples. Front Conserv Sci 3(1):Article 873659. https://doi.org/10.3389/fcosc.2022.873659

Sharma KD (2011) Rain-fed agriculture could meet the challenges of food security in India. Curr Sci 100(11):1615–1616

Sharma KL, Mandal B, Venkateswarlu B (2012) Soil quality and productivity improvement under rainfed conditions–Indian perspectives. In: Resource management for sustainable agriculture, pp 203–230

Sharma NK, Singh RJ, Mandal D, Kumar A, Alam NM, Keesstra S (2021) Increasing farmer's income and reducing soil erosion using intercropping in rainfed maize-wheat rotation of Himalaya, India. Agric Ecosyst Environ 318:107484

Singh VP (1996) Agro-ecological analysis for sustainable development of rainfed environments in India. J Indian Soc Soil Sci 44(4):601–615

Singh R, Garg KK, Wani SP, Tewari RK, Dhyani SK (2014) Impact of water management interventions on hydrology and ecosystem services in Garhkundar-Dabar watershed of Bundelkhand region, Central India. J Hydrol 509:132–149

Singh C, Rahman A, Srinivas A, Bazaz A (2020) Risks and responses in rural India: implications for local climate change adaptation action. Clim Risk Manag 30:100226

Smith M (2000) The application of climatic data for planning and management of sustainable rainfed and irrigated crop production. Agric For Meteorol 103(1–2):99–108

Srinivasarao C, Lal R, Kundu S, Babu MP, Venkateswarlu B, Singh AK (2014) Soil carbon sequestration in rainfed production systems in the semiarid tropics of India. Sci Total Environ 487:587–603

Stern CV, Sheikh PA, Brass CT (2011) Adaptive management for ecosystem restoration: analysis and issues for Congress. Congressional Research Service, Washington, DC

Suresh A, Raju SS, Chauhan S, Chaudhary KR (2014) Rainfed agriculture in India: an analysis of performance and implications. Indian J Agric Sci 84(11):1415–1422

Uprety Y, Asselin H, Bergeron Y, Doyon F, Boucher JF (2012) Contribution of traditional knowledge to ecological restoration: practices and applications. Ecoscience 19(3):225–237

Venkateswarlu B, Prasad JVNS (2012) Carrying capacity of Indian agriculture: issues related to rainfed agriculture. Curr Sci 102:882–888

Venkateswarlu B, Shanker AK (2009) Climate change and agriculture: adaptation and mitigation strategies. Indian J Agron 54(2):226–230

Wani SP (2008) Taking soil science to farmers' doorsteps through community watershed management. J Indian Soc Soil Sci 56(4):367–377

Wani SP, Rockström J, Oweis T (eds) (2009a) Rainfed agriculture: unlocking the potential. CABI

Wani SP, Sreedevi TK, Rockström J, Ramakrishna YS (2009b) Rainfed agriculture–past trends and future prospects. In: Rainfed agriculture: unlocking the potential, vol 7, pp 1–33

Wani SP, Chander G, Sahrawat KL, Pardhasaradhi G (2015) Soil-test-based balanced nutrient management for sustainable intensification and food security: case from Indian semi-arid tropics. Commun Soil Sci Plant Anal 46(sup1):20–33

Webb NP, Marshall NA, Stringer LC, Reed MS, Chappell A, Herrick JE (2017) Land degradation and climate change: building climate resilience in agriculture. Front Ecol Environ 15(8):450–459

Existing Plantations in Northeast India: A Promising Approach to Counteract Land Degradation

4

Pentile Thong, Uttam Thangjam, and Uttam Kumar Sahoo

Abstract

In Northeast India, improper land use practices have severely degraded soil health, causing significant erosion and loss of soil organic carbon, despite forestry being the dominant land use. This chapter highlights the effectiveness of plantation initiatives, including multipurpose tree plantations, farm plantations, agroforestry, and managed secondary forests, as strategies to mitigate land degradation and promote sustainable land management. However, the expansion of monoculture plantations threatens native forests, emphasizing the need for regulatory measures and the promotion of mixed cropping systems like agroforestry, which effectively supports land rehabilitation. Although plantation-based restoration boosts carbon sequestration, balancing it with biodiversity conservation remains a challenge. The future of land restoration in the region is promising, with increased government focus on afforestation in degraded forests, agroforestry, and community engagement to ensure sustainable restoration by leveraging traditional knowledge. As the region faces challenges like deforestation and erosion, responsible plantation-based restoration projects hold potential for a sustainable future.

Keywords

Restoration · Plantation · Agroforestry · Secondary forest · Community

P. Thong
Department of Forestry, School of Earth Sciences & Natural Resource Management, Mizoram University, Aizawl, Mizoram, India

Landstack Advisory Services Pvt Ltd, Kalinga Nagar, Bhubaneshwar, India

U. Thangjam · U. K. Sahoo (✉)
Department of Forestry, School of Earth Sciences & Natural Resource Management, Mizoram University, Aizawl, Mizoram, India

V. C. Pandey (ed.), *Plantation-Based Land Restoration*,
https://doi.org/10.1007/978-981-95-2860-8_4

4.1 Introduction

Land degradation is a complex global challenge that has garnered increased attention in recent years (Prokop 2020). There are multiple definitions of land degradation (Nedd and Anandhi 2024), with the latest from the United Nations Convention to Combat Desertification (UNCCD) (2024) describing it as a "decline in biological or economic productivity and the complexity of agroecological systems resulting from land use or triggered by various human activities". Studies estimate that 19–29% of the Earth's surface is affected by land degradation, impacting approximately 3.2 billion people and causing significant economic costs (Kimura 2018; Zvoleff et al. 2020). Both natural and human-driven factors contribute to this degradation, which is further exacerbated by climate change (Zvoleff et al. 2020; Kimura 2018). Land degradation manifests through deforestation, desertification, rapid climate changes, waterlogging, salinization, erosion, and the loss of organic matter. However, land degradation can be mitigated by identifying sustainable land management (SLM) practices and adopting precise methodologies for its assessment (Mahata and Sharma 2021). Despite the critical impact of land degradation, particularly in developing countries, investments in sustainable land management remain low (Chen et al. 2022). Over the past two decades, research on land degradation has surged, with China, the United States, and the United Kingdom leading in publication volume and international collaboration (Xie et al. 2020).

Numerous international initiatives, including the United Nations Framework Convention on Climate Change (UNFCCC), the United Nations Convention to Combat Desertification (UNCCD), and Disaster Risk Reduction (DRR), have been launched with the objective of tackling climate change mitigation and combatting landscape degradation. Nonetheless, the ongoing degradation persists, and this bears significant repercussions for changes in land use. Recent studies highlight that land degradation continues to be a major concern in India. The accurate estimate potential of degraded land is significantly hampered by absence of, and often, unreliable information. Latest report by SAC-ISRO (2021) states that the current estimate of the country's degraded land is about 97.8 million hectares, equivalent to 29.7% of the country's total area. Another study by Sreenivas et al. (2021) stated that by 2015–2016, approximately 91.2 million hectares (27.77% of India's land area) were degraded. Edrisi et al. (2023a, b) further reported that by 2019, 49.24 million hectares of deforested and degraded land existed in India, marking an 11.21% increase between 2005 and 2019. Despite this, restoration efforts under current policies led to a 5.08% increase in forest cover, surpassing the rate of degradation. A net decrease of 1% in degraded land, along with a 1.83% rise in forest cover during the 2015–2019 period, reflects the positive outcomes of national and global restoration initiatives. Predictive models show that, under a "business as usual" scenario, 12.1 million hectares of degraded land could be restored by 2030, with 6.27 million hectares being converted into cropland. In a sustainable restoration scenario, an additional 18.31 million hectares could be restored by 2030. Recent research has focused on restoration methods such as agroforestry, afforestation, organic amendments, and agroecological interventions to address the issue of land degradation (Singh et al. 2024).

Land degradation in Northeast India is a significant concern, driven by population pressure and changes in land use practices. The region, once environmentally friendly, now faces serious ecological challenges due to deforestation, industrialization, and flooding (Vanlaltanpuia et al. 2023). The shift from forests to cultivation has led to increased overland flow and soil erosion, resulting in the creation of new landforms and altered sediment transport patterns (Ray et al. 2021; Lalthakimi et al. 2023). Additionally, human activities like natural resource mining, stone quarrying, oil exploration, and various industrial operations further exacerbate land degradation in this fragile ecosystem (Haokip et al. 2023). Bharadwaj et al. (2024) have called for sustainable management of land and its resources in this region through plantation of crops and fruit trees. Agroforestry systems offer promising solutions to land degradation in Northeast India by enhancing soil quality, crop yields, and ecosystem services. These systems can reduce soil erosion and runoff by up to 94% and 78%, respectively, while improving soil physical, chemical, and biological properties (Jinger et al. 2023). Long-term studies in similar soil conditions have demonstrated that agroforestry systems combining forest trees, fruit trees, and legume crops can provide multiple ecosystem services, including higher biomass production, carbon sequestration, and economic benefits (Biswas et al. 2022). These findings suggest that agroforestry is a viable strategy for rehabilitating degraded lands and addressing food security concerns in Northeast India.

In similar line with the United Nation initiative of designating the period from 2021 to 2030 as the “Decade on Ecosystem Restoration”, the Government of India has launched multiple initiatives to make substantial contributions towards both national and global responsibilities. For instance, in alignment with the Bonn Challenge commitment, India is dedicated to restoring 21 million hectares of degraded and deforested land by 2030. Furthermore, India’s Nationally Determined Contribution (NDC) is directed at enhancing forest and tree cover with the goal of sequestering an additional 2.5–3 billion tons of CO_2 equivalent by 2030. Additionally, the Green India Mission is focused on safeguarding, restoring, and augmenting India’s diminishing forest cover while addressing the challenges posed by climate change, among other objectives. This chapter aims to evaluate the effectiveness of plantation initiatives in Northeast India as a strategy for mitigating land degradation and promoting sustainable land management. It will (i) evaluate various plantation practices, such as agroforestry, farm plantation, multipurpose tree plantation, and managed secondary forest with plantation, and their role in land restoration; (ii) investigate the ecological impacts of plantations on soil organic carbon and macronutrient assessment specifically nitrogen (N), phosphorus (P), and potassium (K); (iii) discuss the potential of agroforestry for rehabilitation of degraded land; and (iv) discuss the challenges of plantation-based land restoration in Northeast India.

4.2 Types of Degraded Land

Degraded lands are areas where natural productivity has been lost due to anthropogenic activities or environmental factors. Different types of degraded lands in India include forests, mined areas, and abandoned and/or agricultural fallow (Singh and Tewari 2022). Mining wastelands, prevalent in regions such as the Deccan Plateau, Gujarat, and Rajasthan, result from extensive mining operations that deplete natural resources, leaving behind barren, often toxic landscapes unsuitable for cultivation (Balasubramanian 2015). In Northeast India, agricultural fallow lands, locally known as jhum fallows, are common (Thong et al. 2020). These lands, if not managed with appropriate soil conservation practices, lose their fertility and remain unproductive for extended periods. Furthermore, anthropogenic pressures such as population growth, urbanization, and imbalances in food demand and supply contribute to deforestation, leading to soil degradation, biodiversity loss, and a reduced capacity for the land to recover. Additionally, naturally degraded lands, such as barren regions, are widespread in India, where poor soil quality and extreme climatic conditions inhibit vegetation growth. Addressing these degraded lands is essential for environmental restoration and sustainable land management practices.

4.3 Tree Plantation in Degraded Lands

Tree planting stands as the most prevalent active strategy for restoring forests (Holl and Aide 2011). The Global Forest Resource Assessment has reported the dynamics in the annual establishment of new forest plantations worldwide. Up to the year 2000, the rate reached 4.4 million hectares (Mha) annually, of which around 3 Mha per year were deemed successful (FAO 2000). Asia and South America led in this endeavour, with prominent species being *Pinus* spp. and *Eucalyptus* spp. Subsequently, from 2000 to 2010, the rate of new plantation creation increased to 5 Mha per year (FAO 2010). However, during the period from 2010 to 2015, it decreased to 3.3 Mha per year due to reduced planting activities in East Asia, Europe, North America, South Asia, and Southeast Asia (FAO 2015). Plantation is frequently regarded as a prudent and sustainable choice for landscape restoration. When dealing with degraded sites, the practice of planting trees offers several advantages. Trees, through their shading of the soil, help reduce soil temperatures, facilitating the accumulation of organic matter and slowing down potential soil compaction. This, in turn, enhances the micro-climate of the soil, as it introduces biomass and stimulates biotic activities. As the soil condition improves, it becomes conducive to supporting a variety of plant species. In regions lacking natural sources of regeneration or when existing regeneration is incapable of colonization, plantation plays a pivotal role in promoting forest structure and facilitating natural succession (Parrotta 1993).

While many understand the direct advantages of trees in terms of providing fruits, berries, and nuts for food security, what's often less recognized are the indirect benefits that trees offer in enhancing agroecological functions. Trees play a crucial role in improving soil fertility by promoting biological nitrogen fixation in the soil, enhancing agroecosystem functions by facilitating increased water infiltration, and fostering biodiversity through the provision of shade and attraction of forest-dependent species and organisms, both above and below the ground (Leakey 2014). Trees engage in the absorption of carbon dioxide from the atmosphere during photosynthesis and store carbon within their biomass and soils. Restored tree plantations can serve as valuable carbon sinks, thus contributing to global effort aimed at mitigating greenhouse gas emissions.

The selection of the most suitable tree species is a critical step in the restoration of degraded ecosystems (Mukhopadhyay et al. 2013). Soil nutrients can serve as a functional indicator of soil development following reclamation. The slightly acidic to neutral pH conditions found beneath various plantation forests create a favourable environment for increased nutrient availability, litter decomposition, and microbial activity, particularly by bacteria and fungi that break down organic matter and release nitrogen (Chaubey et al. 2004). A study conducted on the restoration of degraded coal mines through plantation forestry (Bohre et al. 2014) indicated that over time, the bulk density of the reclaimed sites gradually decreased, while soil organic carbon, pH, electrical conductivity (EC), water retention capacity, and nutrient levels increased with the age of the plantations. Additionally, research into the impact of tree plantation on urban degraded land revealed that plantation sites had higher levels of carbon stored in the soil compared to fallow grassland. Furthermore, in the case of teak plantation, the soil carbon storage nearly reached the levels observed in natural mixed vegetation (Singh et al. 2021).

4.4 Typology of Planted Forests/Plantations

The significant role of planted forests in forest landscape restoration is unquestionable. Nevertheless, a more consistent understanding of what qualifies as a planted forest is essential. While the definitions of "forest" and "forest plantation" provided by the FAO (2000) are comprehensive enough to encompass most types of planted forests, many governments tend to interpret these definitions in a manner that associates planted forests with recently established, large-scale plantations. Consequently, the Centre for International Forestry Research (CIFOR) has devised "a typology of planted forests" based on factors such as the nature, purpose (end-use), management intensity of these planted forests, and historical context of the lands on which these planted forests are established (Table 4.1).

4.5 Criteria for Selection of Species for Degraded Lands

The process of selecting plant species for restoring degraded lands in India and other tropical regions is guided by several critical factors. Native or indigenous species are often prioritized due to their dual economic and ecological value. These species are known to enhance soil moisture retention, a vital component in restoring

Table 4.1 Typology of planted forest and its description (Ingles et al., 2002)

Typology	Description
1. Industrial plantation Timbre Biomass Food Others	Intensively managed forest stands created through planting or seeding during the afforestation and reforestation processes in order to supply material for sale locally or outside the immediate area. Individual stands or compartments typically have uniform spacing and an equitable age class and: All planted stands of imported species One or two indigenous species Usually large-scale or supporting one of the few prominent large-scale industrial businesses in the area
2. Home and farm plantation Fuelwood Timbre Fodder Orchard Forest garden Others	Managed forest created by planting, or seeding, in the course of afforestation or regeneration, with even age classes and uniform spacing, for local sale or subsistence. Often on a small size and selling, if at all, in a dispersed market
3. Agroforestry	
4. Managed secondary forest with planting	Managed forest where extra planting or seeding is used to maintain the productivity and composition of the forest

degraded ecosystems (Maiti et al. 2021). Additionally, species with deep or extensive root systems are preferred for their ability to stabilize soil, preventing further erosion and degradation (Talema et al. 2017). Beyond physical benefits, species are also selected based on their functional traits, which support key ecosystem processes such as nutrient cycling and biodiversity enhancement (Giannini et al. 2017).

Another important aspect of species selection is their role in facilitating natural succession, whereby plants help to regenerate the ecosystem over time, promoting resilience and stability. The carbon sequestration potential of species is a particularly important criterion, especially in the context of mitigating climate change, while their economic value makes them attractive for local communities who rely on forest resources for livelihoods (Sharma and Sunderraj 2005).

A holistic approach to restoration emphasizes the integration of scientific expertise with active community engagement. Successful projects take into account local preferences and needs when selecting species, ensuring that restoration efforts align with the socioeconomic realities of the region (Sacande and Berrahmouni 2016). Community participation not only increases the likelihood of project success but also fosters long-term stewardship of restored lands. Ruderal species, which were previously overlooked, have gained recognition for their potential to thrive in degraded conditions, contributing to early successional stages and offering ecosystem services such as soil improvement and habitat provision. Their adaptability to climate change further enhances their suitability for restoration projects (Ranđelović et al. 2024).

Therefore, species selection in land restoration requires a careful balance of ecological, economic, and social factors, ensuring that the chosen species can support ecosystem recovery while benefiting local communities and addressing future environmental challenges.

4.6 Existing Plantation Typologies in Northeast India

The northeastern part of India, comprising the states of Arunachal Pradesh, Assam, Manipur, Meghalaya, Mizoram, Nagaland, Sikkim, and Tripura, lies between 22°05′ and 29°30′ N latitudes and 87°55′ and 97°24′ E longitudes. Forestry is the major land use pattern and covers around 65% of its total geographical area, followed by agricultural land use (16%) (Saha et al. 2012).

Despite being a high rainfall location, this region is characterized by severe water scarcity for much of the year, heavy soil, and carbon loss. An estimation of about 601 million tonnes of soil have been transferred, resulting in environmental deterioration (Saha et al. 2012). Widespread deforestation combined with shifting cultivation practices has led to severe soil loss (200 t/ha/year), significantly degrading the soil physical health in this region. As per Wasteland Atlas of India (2019), the total area under wasteland during 2015–2016 in the region is 22.25% of the total geographical area, which is higher than the national average of 16.96%.

The soil organic carbon (SOC) pool, encompassing both its quantity and quality, serves as the fundamental component of soil (Dent et al. 2011; Manlay et al. 2007). Soil organic carbon is undeniably a pivotal indicator for assessing the health and fertility of lands worldwide. It stands as the most dependable metric for tracking soil degradation, caused by accelerated erosion (Rajan et al. 2010). The depletion of the SOC pool represents a global concern and a primary driver of soil degradation. Consequently, it is imperative to devise strategies for augmenting SOC levels and, ideally, maintaining them above the critical threshold of 10–15 grams per kilogram (1.0–1.5%). This range is crucial in mitigating the risks of soil degradation and reversing the trends of deterioration. Indigenous tree plantations have demonstrated significant potential in restoring nutrient cycling and soil organic matter in relatively short timeframes (de Moraes et al. 2008). Nitrogen, phosphorus, and potassium (NPK) are vital macronutrients for plant growth and productivity, and the levels of these nutrients are key to maintaining soil health and fertility (Sheba et al. 2019). When combined with agroecological practices such as crop rotation and intercropping, the composition of NPK can enhance the soil's physical, chemical, and biological properties, leading to increased crop yields, improved soil organic matter, higher aboveground carbon stocks, and reduced soil acidity (Ntamwira et al. 2023).

Given the pressing issues of deteriorating soil quality and significant land degradation, there is a compelling requirement for practical solutions to restore and preserve soil resources in the humid subtropical climate of Northeast India. These solutions should be capable of ensuring sustained long-term soil productivity and enhancing food security for the indigenous and local communities. Therefore, taking into account the different land conditions, elevations, and topographies of this region, the local population has embraced the following long-term strategies for the conservation of resources and the enhancement of soil health:

1. *Multipurpose Tree (MPT) Plantation*

In addition to serving as a source of fuel, fodder, timbre, and various other products, multipurpose tree species (MPTs) also contribute to the enhancement of soil health and other ecological conditions. Farmers in this region integrate a variety of tree species into different land uses. However, the selection of priority species varies from one state to another and even from one location to another within a state, influenced by the ethnic diversity and dietary preferences of the tribal communities. Multipurpose tree species offer extensive surface cover, consistent leaf litterfall, and a robust root system, resulting in 96.2% increase in soil organic carbon, 10.9% increase in porosity, 24.0% improvement in aggregate stability, and 33.2% increase in available soil moisture. Simultaneously, they reduce bulk density by 15.9% and erosion ratio by 39.5% (Saha et al. 2007). Few of multipurpose tree plantations existing in this region are shown in Table 4.2.

A study conducted by Edrisi et al. (2023a, b) highlights the substantial potential of teak tree species for rejuvenating marginal and degraded lands in the arid and semiarid regions of North India. Notably, this afforestation approach offers the additional benefit of a significant biomass supply. Furthermore, bamboo emerges as an excellent choice for eco-rejuvenating wastelands, demonstrating the capability to restore soil organic carbon (SOC) levels to nearly 1% within a year. This underscores the use of bamboo in eco-rejuvenation technology as a powerful tool for promoting sustainable rural development and nurturing native biodiversity on degraded lands in Central India (Singh et al. 2020).

Broadly, research has indicated higher SOC content under deciduous trees compared to conifers, such as pine (Oostra et al. 2006; Vesterdal et al. 2008). However, in Nigde Province, Turkey, it was observed that pine trees had significantly higher SOC values compared to bare land soils (Korkanç 2014). An assessment of SOC stocks in oil palm plantations of 6, 12, and 18 years in Andhra Pradesh, India, revealed an enhancement of 20%, 18%, and 45% in SOC stocks when compared to fallow land (Behera et al. 2020). In Yunnan Province, China, Zeng et al. (2021) successfully transformed rubber plantations into jungle rubber, resulting in a remarkable 28% increase in SOC levels. These findings underscore the potential of adopting multi-strata and multi-species rubber agroforestry, such as jungle rubber, as a promising approach for rejuvenating existing monoculture rubber plantations, with evident benefits for soil quality and ecological restoration.

Understanding nutrient dynamics is crucial for sustainable management of plantations and maintaining high productivity, particularly in addressing potential age-related declines in forest productivity (Moya et al. 2013). Various vegetation types, such as natural forests, rubber, and teak plantations, influence soil fertility differently across seasons (Ray and Thomas 2012). Bamboo species exhibit diverse patterns of nutrient accumulation and efficiency, with phosphorus frequently being the

Table 4.2 SOC stock (Mg ha^{-1}) and NPK content (ppm) of different MPT plantations

Plantation	Depth	SOC stock	N	P	K	Source
Teak	0–45	44.66				Singh et al. (2018a, b)
	0–45	44.65				Kenye et al. (2019a, b)
	0–45	65.28				Sahoo et al. (2019)
	0–30			0.012	0.026	Lalmuansangi et al. (2022)
Bamboo	0–45	29.83				Singh et al. (2018a, b)
	0–45	29.83				Kenye et al. (2019a, b)
	0–45	34.02				Sahoo et al. (2019)
		38.78				
		26.39				
	0–45	34.67				Sahoo et al. (2021a, b)
	0–45	29.83				Bordoloi et al. (2022)
	0–30			0.011	0.035	Lalmuansangi et al. (2022)
Pine	0–45	26.05				Singh et al. (2018a)
	0–45	26.05				Singh et al. (2018b)
	0–100	9.01				Kenye et al. (2019a, b)
	0–45	55				Sahoo et al. (2021a, b)
	0–30		0.17	0.014		Tripathi et al. (2004)
			0.14	0.013		
			0.17	0.014		
Oil palm	0–45	36.73				Singh et al. (2018a)
	0–40	37.48				Singh et al. (2018b)
		36.68				
		34.88				
		30.52				
		27.69				
		23.79				
		26.21				
		28.2				
		26.94				
		30.29				
		30.56				
	0–45	36.71				Kenye et al. (2019a, b)
	0–30	29.78				Sahoo et al. (2023)
		37.44				
	0–20			0.018	0.093	Behara et al. (2015)
	20–40			0.012	0.074	
Rubber	0–45	45				Sahoo et al. (2021a, b)
	0–30			0.079	0.027	Lalmuansangi et al. (2022)
	0–30		0.37	0.083	0.031	Mishra and Francaviglia (2021)

limiting nutrient (Singh and Rai 2012). In oil palm plantations, nutrient concentrations in leaves vary spatially, requiring site-specific fertilization approaches (Behera et al. 2016). Teak plantations show age-related declines in foliar nitrogen, phosphorus, and potassium, while other nutrients remain relatively stable over time (Siddiqui et al. 2009).

2. *Farm Plantation*

Farm plantations in Northeast India represent a true boon to the region, providing sustenance, income, and ecological benefits. The cultivation of a diverse range of crops, alongside the preservation of traditional and indigenous practices, contributes not only to local livelihoods but also to the environmental well-being of this culturally rich and ecologically diverse part of the country. Listed below are some of the prevailing plantations in the region (Table 4.3).

Orchard farming in India has a rich and longstanding tradition, deeply intertwined with the country's agricultural landscape. Orchards are a testament to India's diverse agroclimatic zones and have made significant contributions to both the nation's economy and its cultural heritage. A study on the carbon status of major fruit orchard soils in Maharashtra recorded the highest SOC stock in mango orchard soils followed by orange and pomegranate orchard soil (Zade et al. 2020). In the Konkan region spanning Maharashtra, Karnataka, and Goa, a transformative initiative took barren wastelands in hilly slopes and turned them into thriving mango orchards. This innovative approach not only established robust vegetation on previously exposed barren laterite rocks but also provided vital protection against soil erosion. The soil carbon stocks in these three Konkan regions were found to vary within the range of 60.5 to 64.5 Mg ha^{-1} (Ganeshamurthy et al. 2019). Furthermore, the research by Rodrigues et al. (2018) emphasizes the potential of mango orchards

Table 4.3 SOC stock (Mg ha^{-1}) and NPK content (ppm) of different farm plantations

Plantation	Depth	SOC stock	N	P	K	Source
Orange	0–45	39.49				Singh et al. (2018a, b)
	0–45	49.07				Sahoo et al. (2019)
		29.9				
		24.04				
	0–100	100.3				Sahoo et al. (2021a, b)
	0–30	42.66				Sahoo et al. (2023)
		53.06				
	0–30			0.013	0.071	Lalmuansangi et al. (2022)
	0–20		0.19	0.07		Sharma et al. (2004)
Areca nut	0–45	44.52				Singh et al. (2018a, b)
		29.07				
	0–45	59.97				Sahoo et al. (2019)
	0–30	44.71				Sahoo et al. (2023)
		41.37				
	0–60		0.18	0.017	0.086	Acharya et al. (2015)
	0–30		0.12	0.013	0.086	Paul et al. (2020)
	30–60		0.09	0.008	0.076	
Mango	0–45	16				Singh et al. (2018a, b)
	0–45	16				Sahoo et al. (2019)
Coffee	0–45	33.38				Singh et al. (2018a, b)
	0–45	33.38				Sahoo et al. (2019)

as a valuable alternative for rehabilitating degrading areas. Beyond their fruit production, these orchards significantly contribute to nutrient cycling and soil protection, highlighting their multifaceted benefits for ecological restoration and land management.

Studies conducted by Dabi et al. (2021) and Pradeep et al. (2020), have consistently shown that areca nut plantations accumulate a significantly higher soil organic carbon (SOC) stock compared to coconut, orange, and rubber plantations and other land management systems. This underscores the remarkable capacity of areca nut cultivation to enhance soil carbon levels. Additionally, coffee plantations in Karnataka exhibit higher SOC stock, surpassing even natural forests and sacred groves (Maheswarappa et al. 2021). This observation highlights the pivotal role of coffee cultivation in promoting soil health and carbon sequestration, which is integral to the sustainable management of these ecosystems.

In areca nut plantations, high content of NPK has proven to significantly boost crop yields compared to the use of organic nutrition alone. This is particularly evident in the enhanced concentrations of nitrogen (N) and potassium (K) in the leaves, which are essential for the healthy growth and productivity of the plants (Sujatha and Bhat 2016). Similarly, in orange cultivation in Kanpur, a balanced supply of NPK has been shown to improve various growth parameters, including increased chlorophyll content, which is critical for photosynthesis and overall plant vigour (Silas and Kumar 2023). Additionally, Kazi et al. (2012) found that when NPK fertilizers are supplemented with multi-micronutrients, there is a marked improvement in not only the yield but also the quality attributes of the fruit, such as size, taste, and nutritional value. This combination approach has led to a notable enhancement in both the productivity and the market value of sweet orange orchards.

According to climate change projections for the years 2021–2035 and 2036–2050, and taking into account the anticipated alterations in temperature and rainfall patterns, the most effective choices for enhancing soil organic carbon (SOC) stocks involve the restoration of secondary forests following the abandonment of shifting cultivation and orange plantations (Sahoo et al. 2023).

3. *Agroforestry*

Agroforestry practices in this region display a wide range of diversity when it comes to crop composition and crop rotation. The effectiveness of these agroforestry systems in mitigating soil erosion, preserving soil organic matter and its physical attributes, enhancing nitrogen levels through nitrogen-fixing trees, and facilitating efficient nutrient cycling is well-acknowledged (Misra 2011). These systems have evolved to become a well-established approach to integrated land management. They serve not only for renewable resource production but also hold considerable ecological significance (Saha et al. 2012). Table 4.4 provides a representation of soil organic carbon (SOC) stock in some of the existing agroforestry practices in the region.

The quantity of soil organic carbon (SOC) in agroforestry systems displays variations influenced by different regions, specific agroforestry systems, and soil depths

Table 4.4 SOC stock (Mg ha^{-1}) and NPK content (ppm) of different agroforestry practices

Agroforestry	Depth	SOC stock	N	P	K	Source
Home garden	0–100	142.6				Singh and Sahoo (2018)
		224.2				
	0–45	59.09				Singh et al. (2018a, b)
		42.63				
	0–45	50.76				Kenye et al. (2019a, b)
	0–100	110.75				Singh and Sahoo (2021)
		173.74				
	0–45	61				Sahoo et al. (2019)
		60.15				
		52.94				
		49.24				
		35.18				
		43.46				
	0–15		0.12	0.056		Rocky (2013)
	15–30		0.11	0.047		
Agroforestry	0–45	60				Sahoo et al. (2021a, b)
	0–10		0.19	0.010	0.22	Thangjam (2019)
	10–20		0.17	0.009	0.193	
Oil palm + maize	0–30	39				Ahirwal et al. (2022a)
Oil palm + turmeric		30				
Oil palm + pineapple		38				

(Negash 2013). Nair et al. (2010) documented a broad range in the potential of soil carbon stock under agroforestry, spanning from 30 to 300 Mg C ha^{-1} down to a depth of 1 m. Many studies have primarily focused on assessing changes in the topsoil layer, specifically the 0–30 cm depth, where the largest carbon pools are typically found (Makumba et al. 2007; Oelbermann and Voroney 2007). In Africa, home gardens in Ethiopia recorded higher SOC stock than woodlot, exclosure, and cultivated cropland (Maryo et al. 2023; Seid 2022; Semere 2019), while home gardens in Tanzania had the highest SOC stock followed by tea plantation and cropland.

It is noteworthy that SOC stocks in agroforestry systems are notably higher in comparison to SOC stocks in other ecosystems and soil types. This underscores the substantial carbon sequestration potential and environmental benefits associated with agroforestry practices, and therefore, the SOC stock of prevalent agroforestry systems is shown in Table 4.5.

Effective NPK management significantly enhances soil fertility and crop productivity in agroforestry systems and home gardens across India. In the Central Himalayan region, nitrogen mineralization and fine root biomass varied among agroforestry systems, with higher levels of mineral nitrogen in agri-horticulture systems (Karki et al. 2022). In Uttar Pradesh, increasing NPK levels led to improved growth and yield of linseed under a guava-based agroforestry system, yielding the best results (Srivastava et al. 2024). In South Andaman, spice trees in home gardens

Table 4.5 Variation in SOC stock of prevalent agroforestry systems in different agroclimatic region

Agroclimatic region (states)	Agroforestry systems	SOC ($Mg\ ha^{-1}$)	
		Range (mean)	SD
Northern Himalayas (Himachal Pradesh, Jammu and Kashmir, Uttarakhand)	Agri-silviculture	22.28–142.9 (58.07)	33.6
	Agri-horticulture	43.67–151.7 (64.34)	33.33
	Silvipasture	16.2–109.7 (47.63)	34.69
Indo-Gangetic region (Punjab, Haryana, Uttar Pradesh, Bihar)	Agri-silviculture	4.25–48.98 (15.25)	12.52
Eastern and Northeastern India (West Bengal, Odisha, Assam, Sikkim, Manipur, Mizoram, Nagaland, Meghalaya)	Home gardens	42.8–119.5 (52.15)	22.27
	Plantation crop-based agroforestry	30.56–176.74 (96.53)	71.57
	Boundary plantation	48.18–55.73 (51.95)	3.28
	Block plantation	–	
Western and Central India (Rajasthan, Gujarat, Maharashtra, and Madhya Pradesh)	Agri-silviculture	4.28–24.13 (12.26)	7.08
	Agri-horticulture	–	–
	Block plantation	0.1–63.80 (14.55)	19.57
Southern India (Karnataka, Andhra Pradesh, Tamil Nadu, Kerala)	Agri-silviculture	1.23–77.56 (17.08)	26.33
	Plantation crop-based agroforestry	61.26–71.39 (65.82)	3.65
	Block plantation	–	–
	Coffee plantation	78.70–170.43 (125.29)	34.66

enriched soil nutrient levels, comparable to native evergreen forests (Pandey and Singh 2009). Additionally, in Mizoram, nutrient release and leaf litter decomposition patterns vary across common home garden tree species (Upadhyaya et al. 2012).

4. *Managed Secondary Forest with Plantation*

The management of degraded ecosystems prioritizes forest plantation due to its capacity to offer a wide spectrum of ecosystem services, including soil erosion control, increased biomass, carbon pool enhancement, and biodiversity restoration (Huston and Marland 2003). As per FAO (2020), planted forests constitute approximately 7% of the world's total forest area, with Asia alone accounting for 62% of the global planted forest area. The selection of species for planted forests remains a topic of ongoing debate (Subiakto et al. 2016). However, the utilization of mixed

Table 4.6 SOC of managed fallows and planted forest

Managed fallows	Depth	SOC stock	Source
Planted forest	0–45	58.84	Gogoi et al. (2021)
Quercus forest	0–45	175	Gogoi et al. (2022)
Quercus forest	0–30	90	Ahirwal et al. (2022a)

species, indigenous varieties, and long-lived tree species has demonstrated a positive impact on biodiversity enhancement (Aubin et al. 2008; Brockerhoff et al. 2008; Hartley 2002). The jhum fallows in this region are managed and converted to plantations and/or dominant species forest, most commonly oak/*Quercus* forest (Table 4.6). In fact, oak forests represent the climax vegetation in the Eastern Himalayan region, characterized by the prevalence of one or more oak tree species. These forests thrive within an elevation range spanning from 800 m to 2750 m (Champion and Seth 1968).

Over time, introducing plantation trees on the slopes has been a prevalent approach to curbing shifting cultivation (jhum) and replacing it with economically viable alternatives. Plantations of trees like rubber, areca nut, and teak are now considered as more advantageous land use practices. Notably, these planted forests store significantly higher biomass and carbon when compared to some widely practised monoculture plantations in Northeast India, including rubber, oil palm, and areca nut (Brahma et al. 2018; Dey 2005; Singh et al. 2018a, b).

Changes in land use and land cover have a profound impact on soil organic carbon (SOC), microbial activity, and respiration (Gispert et al. 2013). A study focused on the transition from jhum cultivation to various plantations revealed the following trend in SOC stock: jhum < teak plantation < areca nut plantation < rubber plantation < secondary forest (Lungmuana et al. 2019). Hence, it becomes imperative to understand the fate of degraded jhum land and determine which land use and management practices enhance soil quality. This knowledge can serve as a foundation for devising strategies aimed at preserving and improving soil quality. In doing so, we can sustain agricultural production and livelihood security in these delicate ecosystems, securing the well-being of the local communities and the environment.

4.7 Agroforestry for Rehabilitation of Degraded Lands in India

Indian traditional agroforestry systems encompass a wide spectrum of practices, ranging from jhum to complex home gardens. They include sparse tree stands in farmlands to dense, multi-layered homesteads. Major benefit of agroforestry is its ability to mitigate climate change by sequestering carbon in both plant biomass and soil. The carbon sequestration potential varies from semiarid to temperate regions (Murthy et al. 2013). Furthermore, agroforestry systems have the capacity to generate significant quantities of leaf litter, which serves as mulch and, upon decomposition, enriches the soil by forming humus. This contributes to the enhancement of the physical, chemical, and biological properties of the soil (Jama et al. 2006).

The enhanced soil quality in agroforestry systems is driven by several factors, including biological nitrogen fixation, nutrient recycling from deeper soil layers, accumulation of soil organic matter from plants, and increased microbial and enzyme activity, thereby improving soil health and fertility (Dollinger and Jose 2018).

Agroforestry can therefore be viewed as a versatile solution to various challenges faced by the agricultural sector, one of which is its nexus with land rehabilitation as discussed below:

4.7.1 Agroforestry for Areas Impacted by Soil Erosion

In agroforestry practices, the canopy of trees plays a vital role in preventing soil erosion. It accomplishes this primarily by intercepting rainfall, which reduces the impact of raindrops and mitigates their erosive potential (Kaushal et al. 2017). The litter produced by trees contributes to the formation of water-stable aggregates, which, in turn, decreases surface runoff volume. Roots and stems of trees act as barriers, preventing sediments from moving downhill and aiding in water infiltration. Deep tree roots provide stability to slopes, averting shallow landslides. Over the long term, vegetation significantly influences water and sediment fluxes by enhancing soil water infiltration, as well as the stability and cohesiveness of soil aggregates (Zuazo and Pleguezuelo 2009).

Various agroforestry techniques are utilized to effectively manage erosion, such as windbreaks, shelterbelts, hedgerow (alley) cropping, multilayer tree gardens, home gardens, and the integration of planting crops with trees (Young 1997). According to McDonald et al. (2002), these agroforestry practices can reduce soil erosion and surface runoff when compared to control soils. The incorporation of tree leaf litter into the soil, along with its decomposition and the activity of tree roots, enhances the retention, infiltration, and storage of surface runoff (Dass et al. 2011).

4.7.2 Agroforestry for Areas Impacted by Mining, Industrial Waste, and Waterlogging

When selecting species for the rehabilitation of mined areas, it's crucial to choose those that are nitrogen-fixing, fast-growing, and well-suited to arid zone climates characterized by high temperature and possess drought-tolerant root systems. Additionally, these species should have significant socioeconomic value. Such species play a pivotal role in increasing microbiological activity in soil, resulting in nutrient mineralization, enhanced below-ground biodiversity, and improved nutrient cycling, and maintaining the stability of the soil matrix (Datar et al. 2011).

Agroforestry-focused wastewater reuse presents a promising approach to address the increasing drought in agriculture (Ranjan 2021). In the case of soils contaminated with heavy metals like Zn, Pb, and Cd, species like *Conocarpus lancifolius* are effective in rehabilitation due to their enhanced biomass production, efficient water usage, metal tolerance, and translocation capabilities (Rasheed et al. 2020). Furthermore, Hussain et al. (2021) highlighted certain species that exhibit substantial biomass growth and defence mechanisms, making them suitable for planting in soils irrigated with industrial wastewater (Hussain et al. 2021).

Fast-growing tree systems with short rotation periods can effectively manage waterlogging in canal-irrigated regions and can be used for bio-drainage (Uthappa et al. 2015). Additionally, multifunctional woodlots, agri-silviculture, agro-horticulture, and silvipasture systems have gained popularity as effective strategies for addressing saline and waterlogged conditions (Behera et al. 2015).

4.7.3 Agroforestry for Areas Impacted by Saline, Sodic, and Acidic Soils

In saline soils, the cultivation of salt-tolerant crops like pearl millet and mustard alongside *Eucalyptus tereticornis* and *Melia composita* has shown promise as intercropping (Banyal et al. 2017). In regions with an abundance of saline water, the practice of saline aquaforestry has displayed significant prospects in addressing the salinity (Banyal et al. 2018). Furthermore, there is significant potential for biofuel production, particularly through energy plantations. A study by Muchane et al. (2020) found that agroforestry systems can effectively reduce soil acidity, increasing soil pH compared to monoculture systems, though results may vary by soil type. Trees that are rich in phosphorus or base cations, such as *Gmelina arborea* with high levels of calcium and magnesium, have proven effective in ameliorating soil phosphorus and pH (Wong et al. 2004).

4.7.4 Agroforestry for Biologically Degraded Lands

Restoring soil fertility in degraded lands can be achieved through careful species selection, as certain species play a vital role in nutrient release that benefits crops, even when trees are not fully harvested. A study by Acharya and Kafle (2009) identified six multipurpose tree species with high nutrient release rates when grown on abandoned lands, underscoring the importance of thoughtful species selection for soil rejuvenation. Additionally, setting up rubber-based agroforestry systems is an effective management strategy for mitigating the negative impacts of monoculture (Liu et al. 2019).

4.8 Scenario of Plantation-Based Restoration in Northeast India

Plantation-based restoration in Northeast India holds significant potential for both carbon sequestration and biodiversity conservation. Research indicates that converting degraded forests and grasslands to rubber plantations and jhum agroforestry can substantially boost carbon stocks, with sequestration rates of 4–5 Mg ha^{-1} per year (Brahma et al. 2018). The area under rubber plantations in the region has expanded to 51,510 hectares, contributing an estimated 136 tonnes of carbon per hectare, with soil accounting for 92.7 tonnes and an additional 2.40 tonnes added through litter-fall and undergrowth vegetation. Altogether, rubber plantations in the region store about seven million tonnes of carbon. However, the unchecked expansion of rubber monocultures poses a threat to native forests, underscoring the need for regulatory measures and the promotion of mixed cropping systems (Chakraborty et al. 2018). In particular, oil palm intercropped with pineapple has been found to produce higher yields and carbon stocks compared to monoculture plantations (Ahirwal et al. 2022b). For instance, monoculture oil palm plantations in Mizoram sequestered 49.90 Mg C ha^{-1} over 11 years (Singh et al. 2018a, b). Other restoration efforts have targeted endangered species like *Aquilaria malaccensis*, *Gleditsia assamica*, and *Gymnocladus assamicus* through ex situ conservation in home gardens and plantations (Choudhury and Khan 2010). Projects using native species have seen varying success, with survival rates between 9.1% and 94.3% after 18 months, particularly among animal-dispersed, medium- to large-seeded primary forest species, suggesting their promise for future restoration (Borawake et al. 2021). While plantation-based restoration enhances carbon sequestration, balancing it with biodiversity conservation remains a key challenge in the region.

4.9 Challenges of Plantation-Based Land Restoration in Northeast India

Plantation-based land restoration in Northeast India may face several challenges, which can hinder its effectiveness and long-term success. The following are some of the issues that need further attention:

1. *Species Selection:* Choosing the right tree species adapted to the local environment is critical. Selecting the wrong species can lead to poor growth, susceptibility to pests and diseases, and reduced ecological benefits.
2. *Invasive Species:* The introduction of non-native species can become invasive and threaten local biodiversity. Careful consideration is required to ensure that plantation species do not harm native ecosystems.
3. *Land Tenure and Ownership:* Land tenure issues and unclear ownership can impede plantation establishment. Resolving these disputes is essential to avoid conflicts and ensure the sustainability of the project.
4. *Climate Change:* Changing climate patterns can impact the growth and health of plantation trees. Droughts, extreme weather events, and altered rainfall patterns can affect tree survival and growth.
5. *Illegal Logging and Poaching:* Plantations can be vulnerable to illegal logging and poaching, which can undermine conservation efforts and disrupt the economic benefits for local communities.

6. *Community Engagement:* Effective community engagement and participation are crucial for the success of plantation-based restoration. Building trust, addressing local needs, and ensuring equitable distribution of benefits can be challenging.
7. *Management and Maintenance:* Maintaining plantations is labour-intensive and requires ongoing resources. Securing funding for long-term management and addressing issues like disease control and fire prevention are essential.
8. *Market Access:* Ensuring that the timbre and non-timbre forest products from these plantations have access to markets can be challenging, especially for remote areas in Northeast India.
9. *Infrastructure and Accessibility:* Limited infrastructure and access to remote plantation sites can hinder management and harvesting activities.
10. *Government Policies and Regulations:* Inconsistent or unclear government policies and regulations can create obstacles. Clarity and supportive policies are necessary for the success of plantation-based land restoration.
11. *Monitoring and Evaluation:* Adequate monitoring and evaluation systems are crucial to assess the effectiveness of plantations. The lack of resources and expertise can impede these efforts.

Overcoming these challenges requires a holistic and collaborative approach involving local communities, governments, conservation organizations, and researchers. It necessitates a focus on sustainable management, biodiversity conservation, and the empowerment of local communities to ensure the long-term success of plantation-based land restoration in Northeast India.

4.10 Policy Initiatives and Legal Framework for Land Restoration Through Plantation in Northeast India

Several national- and state-level policies, laws, and initiatives focus on land restoration through afforestation and plantation, particularly in the ecologically sensitive northeast region, such as:

1. *National Forest Policy (1988)*—This initiative aims to increase forest cover to 33% of the nation's land area. Towards achieving this, reforestation and protection of existing forests, especially in ecologically sensitive zones, are carried out.
2. *Forest (Conservation) Act, 1980*—This regulation governs the use of forest land for non-forest purposes and requires compensatory afforestation when such land is diverted. It helps to mitigate deforestation by ensuring that any forest land lost to development projects is replaced through compensatory planting.
3. *Green India Mission (GIM)*—As part of the National Action Plan on Climate Change (NAPCC), it focuses on enhancing forest cover and improving ecosystem services, including biodiversity conservation and carbon sequestration. Degraded lands, particularly in Assam, Manipur, and Arunachal Pradesh, have been converted to plantation under this initiative.

4. *Compensatory Afforestation Fund Act (CAMPA), 2016*—This act establishes a fund for financing afforestation activities in degraded forest areas, using money collected from diverting forest land for development.
5. *National Agroforestry Policy (2014)*—Under this policy, agroforestry practices are adopted to restore soil fertility and reduce land degradation, particularly in areas practising shifting, thereby reducing pressure on forests.
6. *National Afforestation Programme (NAP)*—This scheme promotes tree plantation and forest regeneration across the northeastern states, involving local communities through Joint Forest Management Committees (JFMCs).
7. *The Scheduled Tribes and Other Traditional Forest Dwellers (Recognition of Forest Rights) Act, 2006 (FRA)*—This right allows tribal communities in states like Arunachal Pradesh, Nagaland, and Mizoram to sustainably manage and restore forests through plantations and agroforestry.
8. *National Bamboo Mission (NBM)*—This initiative encourages the planting and use of bamboo for ecological restoration, including the rehabilitation of degraded lands.
9. *National Mission for Sustainable Agriculture (NMSA)*—This initiative promotes sustainable agricultural practices that include agroforestry and the integration of tree plantations to restore degraded land.

Besides, each state has its forest policy, which aims to maintain ecological balance, preserve biodiversity, restore degraded lands, and enhance forest cover by promoting afforestation, protecting existing forests, and fostering community participation in forest management.

Mizoram, in particular, initiated the New Land Use Policy (NLUP) to formalize forest land titles and replace shifting cultivation with settled farming, primarily through palm oil plantations. This policy has led to the allocation of bamboo forests for shifting cultivation, which are then promoted for palm oil plantations by the Agriculture Department. In addition, plantations of rubber, orange, pineapple, passion fruit, mango, grape, areca nut, and banana are also carried out under this policy.

4.11 Future Outlook for Land Restoration Through Plantation

Land restoration approaches not only restore biodiversity and ecological functioning but also address rural poverty by providing goods and services to local communities (Lamb et al. 2005). Although newly planted forests may differ from the native forest composition, they can improve ecosystem services and enhance biodiversity conservation (Chazdon 2008). The recovery potential of degraded lands is evident, with studies showing increased species diversity after protection periods ranging from 5 to 20 years (Raizada et al. 2004). However, successful restoration requires adaptive management, community participation, and consideration of site-specific factors such as soil degradation and residual vegetation. Therefore, the future outlook for land restoration through plantation in Northeast India can be determined based on the following factors, namely:

1. *Increased Government Focus*—National- and state-level initiatives like CAMPA and the Green India Mission are key drivers for afforestation and land restoration, targeting degraded lands and increasing forest cover, particularly in ecologically sensitive areas. Additionally, policies under the National Action Plan on Climate Change (NAPCC) focus on forest restoration and land rehabilitation, with an emphasis on regions such as Assam, Manipur, and Arunachal Pradesh.
2. *Agroforestry and Sustainable Practices*—Agroforestry, which integrates trees into farming systems, is seen as a key strategy to restore soil fertility and reduce land degradation in areas practising shifting cultivation (jhum). This can help reduce pressure on natural forests while promoting sustainable land use.
3. *Community Engagement*—Many programmes emphasize community participation in reforestation and forest management, which can positively impact indigenous and forest-dependent communities in the northeast, where traditional knowledge is essential. Initiatives like Joint Forest Management (JFM) are key in involving local communities to ensure that restoration efforts are both effective and sustainable.
4. *Challenges of Climate Change and Development Pressure*—While the region holds vast potential for reforestation, it faces challenges such as increased deforestation due to infrastructure development and climate change impacts like erratic rainfall and rising temperatures, which could affect the success of plantation efforts.
5. *Biodiversity Restoration*—Future efforts will not only aim at increasing tree cover but also at restoring biodiversity by replanting native species, which support ecosystem resilience and services like carbon sequestration and water retention.
6. *International and Private Sector Support*—International funding and collaboration, along with the involvement of private companies in carbon credit markets and corporate social responsibility (CSR) projects, may further boost restoration efforts in the region.

4.12 Conclusion

In conclusion, plantation-based land restoration in Northeast India presents a promising and multifaceted solution to address the pressing issue of land degradation in this ecologically rich region. The most suitable model for land restoration in this region integrates agroforestry, community-based forest management, and ecological restoration using native species, ensuring both environmental sustainability and community participation. Agroforestry systems, especially in regions practising jhum (shifting cultivation), can restore soil fertility and reduce land degradation while providing sustainable livelihoods for local communities. Joint Forest Management (JFM) is critical in engaging indigenous and forest-dependent communities, leveraging their traditional knowledge to enhance restoration efforts. The strategic cultivation of native tree species not only combats soil erosion and desertification but also contributes to the conservation of biodiversity, carbon sequestration, and the overall resilience of the local ecosystems. Moreover, it fosters sustainable livelihoods for communities by generating economic opportunities through the utilization of timbre and non-timbre forest products. As Northeast India faces environmental challenges, such as deforestation and soil erosion, the promotion and responsible management of plantation-based land restoration projects hold the potential to play a pivotal role in ensuring a more sustainable and vibrant future for this unique and vital part of the country. Continued investment and collaborative efforts are essential to further enhance the success and impact of such initiatives in the region.

References

Acharya AK, Kafle N (2009) Land degradation issues in Nepal and its management through agroforestry. J Agric Environ 10:133–143

Acharya GC, Paul SC, Chakrabarty R, Ray AK (2015) Effect of organic and inorganic sources of nutrients on soil fertility status of arecanut (Areca catechu) in North-East India. Indian J Agric Sci 85(10):1335–1341

Ahirwal J, Gogoi A, Sahoo UK (2022a) Stability of soil organic carbon pools affected by land use and land cover changes in forests of eastern Himalayan region, India. Catena 215:106308

Ahirwal J, Sahoo UK, Thangjam U, Thong P (2022b) Oil palm agroforestry enhances crop yield and ecosystem carbon stock in northeast India: Implications for the United Nations sustainable development goals. Sustain Prod Consum 30:478–487

Aubin I, Messier C, Bouchard A (2008) Can plantations develop understory biological and physical attributes of naturally regenerated forests? Biol Conserv 141(10):2461–2476

Balasubramanian A (2015) Wastelands in India. In country-wide class room educational TV programme-Gyan Darshan. University of Mysore, Karnataka

Banyal R, Rajkumar, Kumar M, Yadav RK, Dagar JC (2017) Agroforestry for rehabilitation and sustenance of saline ecologies. In: Agroforestry: anecdotal to modern science, pp 413–454

Banyal R, Yadav RK, Sheoran P, Bhardwaj AK, Rajkumar PK, Tolia R (2018) Managing saline soils of Indo-Gangetic plains with Eucalyptus and Melia based agroforestry systems. Indian J Ecol 45(4):841–849

Behara SK, Rao BN, Suresk K, Manoja K (2015) Soil Nutrient Status and Leaf Nutrient Norms in Oil Palm (Elaeis guineensis Jacq.) Plantations Grown on Southern Plateau of India. Proc Natl Acad Sci India Sect B Biol Sci. https://doi.org/10.1007/s40011-015-0508-y

Behera L, Nayak MR, Patel D, Mehta A, Sinha SK, Gunaga R (2015) Agroforestry practices for physiological amelioration of salt affected soils. J Plant Stress Physiol 1:13–18

Behera SK, Suresh K, Rao BN, Mathur RK, Shukla AK, Manorama K, Ramachandrudu K, Harinarayana P, Prakash C (2016) Spatial variability of some soil properties varies in oil palm (Elaeis guineensis Jacq.) plantations of west coastal area of India. Solid Earth 7(3):979–993

Behera SK, Shukla AK, Suresh K, Manorama K, Mathur RK, Kumar A, Harinarayana P, Prakash C, Tripathi A (2020) Oil palm cultivation enhances soil pH, electrical conductivity, concentrations of exchangeable calcium, magnesium, and available sulfur and soil organic carbon content. Land Degrad Dev 31(18):2789–2803

Bharadwaj K, Boruah P, Bathari M, Maibangsa M, Terangpi H (2024) Sustainable Management of Land, Water and Crop in Hills of North East India. Int J Environ Clim Change 14(2):847–855

Biswas B, Chakraborty D, Timsina J, Bhowmick UR, Dhara PK, Lkn DKG, Sarkar A, Mondal M, Adhikary S, Kanthal S, Patra K (2022) Agroforestry offers multiple ecosystem services in degraded lateritic soils. J Clean Prod 365:132768

Bohre P, Chaubey OP, Singhal PK (2014) Biomass production and carbon sequestration by Pongamia pinnata (Linn) Pierre in tropical environment. Int J Bio-Sci Bio-Technol 6:129–140
Borawake N, Datta A, Naniwadekar R (2021) Tropical forest restoration in the Eastern Himalaya: evaluating early survival and growth of native tree species. Ecol Restor 39(3):52–63
Bordoloi R, Das B, Tripathi OP, Sahoo UK, Nath AJ, Deb S, Das DJ, Gupta A, Devi NB, Charturvedi SS, Tiwari BK (2022) Satellite based integrated approaches to modelling spatial carbon stock and carbon sequestration potential of different land uses of Northeast India. Environ Sustain Indicators 13:100166
Brahma B, Pathak K, Lal R, Kurmi B, Das M, Nath PC, Nath AJ, Das AK (2018) Ecosystem carbon sequestration through restoration of degraded lands in Northeast India. Land Degrad Dev 29(1):15–25
Brockerhoff EG, Jactel H, Parrotta JA, Quine CP, Sayer J (2008) Plantation forests and biodiversity: oxymoron or opportunity? Biodivers Conserv 17:925–951
Chakraborty K, Sudhakar S, Sarma KK, Raju PLN, Das AK (2018) Recognizing the rapid expansion of rubber plantation–a threat to native forest in parts of northeast India. Curr Sci 114:207–213
Champion HG, Seth SK (1968) A revised survey of the forest types of India. Manager of Publications
Chaubey OP, Ansari AA, Pandey A (2004) Sustainable management in degraded forests under joint forest management: an eco-silvicultural approach. In: Challenges in sustainable development, p 94
Chazdon RL (2008) Beyond deforestation: restoring forests and ecosystem services on degraded lands. Science 320(5882):1458–1460
Chen L, Msigwa G, Yang M, Osman AI, Fawzy S, Rooney DW, Yap PS (2022) Strategies to achieve a carbon neutral society: a review. Environ Chem Lett 20(4):2277–2310
Choudhury B, Khan ML (2010) Conservation and management of endangered plant species: a case study from Northeast India. Bioremed Biodivers Bioavailab 4(1):47–53
Dabi H, Bordoloi R, Das B, Paul A, Tripathi OP, Mishra BP (2021) Biomass, carbon stock and soil physicochemical properties in plantation of East Siang district, Arunachal Pradesh, India. Environ Chall 4:100191
Dass A, Sudhishri S, Lenka NK, Patnaik US (2011) Runoff capture through vegetative barriers and planting methodologies to reduce erosion, and improve soil moisture, fertility and crop productivity in southern Orissa, India. Nutr Cycl Agroecosyst 89:45–57
Datar A, Patrick A, David M (2011) Post-mined land rehabilitation in India: cataloguing plant species used in land revegetation. In: Vikukis N, Nichols O (eds) Australian mine rehabilitation workshop. JK Tech, pp 197–208
de Moraes LFD, Campello EFDC, Pereira MG, Loss A (2008) Soil characteristics in degraded area restoration at Reserva Biológica de Poço das Antas. RJ Ciência Florestal 18(2):193–206
Dent D, Boincean BP, Krupenikov IA, Krupenikov IA, Boincean BP, Dent D (2011) Humus–Guardian of fertility and global carbon sink. In: The Black Earth: ecological principles for sustainable agriculture on chernozem soils, pp 39–50
Dey SK (2005) A preliminary estimation of carbon stock sequestrated through rubber (Hevea brasiliensis) plantation in north eastern region of India. Indian Forester 131(11):1429–1436
Dollinger J, Jose S (2018) Agroforestry for soil health. Agrofor Syst 92:213–219
Edrisi SA, Bundela AK, Verma V, Dubey PK, Abhilash PC (2023a) Assessing the impact of global initiatives on current and future land restoration scenarios in India. Environ Res 216:114413
Edrisi SA, Singh A, Dubey PK, Abhilash PC (2023b) Tectona grandis L.f. mediated restoration of marginal lands in Eastern Uttar Pradesh, North India. Land Degrad Dev 34(4):929–942
FAO (2000) The Global Forest Resource Assessment (GFRA, 2000). United Nations Food and Agriculture Organization (FAO), Rome
FAO (2010) The Global Forest Resource Assessment (GFRA, 2010). United Nations Food and Agriculture Organization (FAO), Rome
FAO (2015) The Global Forest Resource Assessment (GFRA, 2015). United Nations Food and Agriculture Organization (FAO), Rome

FAO (2020) The Global Forest Resource Assessment (GFRA, 2020). United Nations Food and Agriculture Organization (FAO), Rome

Ganeshamurthy AN, Ravindra V, Rupa TR, Bhatt RM (2019) Carbon sequestration potential of mango orchards in the tropical hot and humid climate of Konkan region, India. Curr Sci 116(8):1417–1423

Giannini TC, Giulietti AM, Harley RM, Viana PL, Jaffe R, Alves R, Pinto CE, Mota NF, Caldeira CF Jr, Imperatriz-Fonseca VL, Furtini AE (2017) Selecting plant species for practical restoration of degraded lands using a multiple-trait approach. Austral Ecol 42(5):510–521

Gispert M, Emran M, Pardini G, Doni S, Ceccanti B (2013) The impact of land management and abandonment on soil enzymatic activity, glomalin content and aggregate stability. Geoderma 202:51–61

Gogoi A, Ahirwal J, Sahoo UK (2021) Plant biodiversity and carbon sequestration potential of the planted forest in Brahmaputra flood plains. J Environ Manag 280:111671

Gogoi A, Ahirwal J, Sahoo UK (2022) Evaluation of ecosystem carbon storage in major forest types of Eastern Himalaya: Implications for carbon sink management. J Environ Manag 302:113972

Haokip IC, Devi MH, Sunil BH, Dixit M, Tasung A, Gadi Y (2023) Land degradation in the North Eastern Region of India and its mitigation. Biological Forum Int J 15(12):473–479

Hartley MJ (2002) Rationale and methods for conserving biodiversity in plantation forests. For Ecol Manag 155(1–3):81–95

Holl KD, Aide TM (2011) When and where to actively restore ecosystems? For Ecol Manag 261(10):1558–1563

Hussain Z, Rasheed F, Tanvir MA, Zafar Z, Rafay M, Mohsin M, Pulkkinen P, Ruffner C (2021) Increased antioxidative enzyme activity mediates the phytoaccumulation potential of Pb in four agroforestry tree species: a case study under municipal and industrial wastewater irrigation. Int J Phytoremediation 23(7):704–714

Huston MA, Marland G (2003) Carbon management and biodiversity. J Environ Manage 67:77–86

Ingles A, Shepherd G, Applegate G, Parrotta J, Poulsen J, Evans J, Bazett M, Dudley N, Nasi R, Mansourian S, et al. (2002) Typology of planted forests. Presented, UNFF 2. 4–15 March 2002. Bogor, Indonesia: Center for International Forestry Research (CIFOR)

Jama, B., Kwesiga, F. and Niang, A., 2006. Agroforestry innovations for soil fertility management in sub-Saharan Africa: prospects and challenges ahead. World Agroforestry into the Future, 60

Jinger D, Kaushal R, Kumar R, Paramesh V, Verma A, Shukla M, Chavan SB, Kakade V, Dobhal S, Uthappa AR, Roy T (2023) Degraded land rehabilitation through agroforestry in India: Achievements, current understanding, and future prospectives. Front Ecol Evol 11:1088796

Karki H, Bargali K, Bargali SS (2022) Dynamics of fine roots and soil nitrogen in Mangifera indica-based agroforestry systems in the Central Himalayan region, India. Land Degrad Dev 33(17):3523–3538

Kaushal R, Kumar A, Alam NM, Mandal D, Jayaparkash J, Tomar JMS, Patra S, Gupta AK, Mehta H, Panwar P, Chaturvedi OP (2017) Effect of different canopy management practices on rainfall partitioning in Morus alba. Ecol Eng 102:374–380

Kazi SS, Ismail S, Joshi KG (2012) Effect of multi-micronutrient on yield and quality attributes of the sweet orange. Afr J Agric Res 7(29):4118–4123

Kenye A, Sahoo UK, Singh SL, Anudip G (2019a) Effect of four land uses on soil edaphic properties and soil organic carbon stock in Mizoram, North-East India. Indian Forester 145(12):1139–1146

Kenye A, Sahoo UK, Singh SL, Gogoi A (2019b) Soil organic carbon stock of different land uses of Mizoram, Northeast India. AIMS Geosci 1:25–40

Kimura R (2018) Global distribution of degraded land area based on dust erodibility determined from satellite data. Int J Remote Sens 39(18):5859–5871

Korkanç SY (2014) Effects of afforestation on soil organic carbon and other soil properties. Catena 123:62–69

Lalmuansangi, Zothanzama J, Lallawmkima, Lalbiakmawia B (2022) Soil quality of northeast India – a review. Int J Res Eng Sci 10(11):111–118

Lalthakimi C, Singh NS, Vanlalfakawma DC, Upadhyay KK, Tripathi SK (2023) Land use change effects on soil physical and biochemical properties during wet and dry season in forest and shifting cultivation (Jhum) Sites in Northeast India. Environ Ecol 41(4A):2584–2594

Lamb D, Erskine PD, Parrotta JA (2005) Restoration of degraded tropical forest landscapes. Science 310(5754):1628–1632

Leakey RR (2014) The role of trees in agroecology and sustainable agriculture in the tropics. Annu Rev Phytopathol 52:113–133

Liu C, Jin Y, Hu Y, Tang J, Xiong Q, Xu M, Bibi F, Beng KC (2019) Drivers of soil bacterial community structure and diversity in tropical agroforestry systems. Agric Ecosyst Environ 278:24–34

Lungmuana, Singh SB, Choudhury BU, Saha S, Hnamte V (2019) Transforming jhum to plantations: Effect on soil microbiological and biochemical properties in the foot hills of North Eastern Himalayas, India. Catena 177:84–91

Mahata S, Sharma VN (2021) The global problem of land degradation: a review. Natl Geogr J India 67:216–231

Maheswarappa V, Vasudeva R, Hegde R, Devagiri GM, Channabasappa KS, Khaple AK (2021) Soil health status in coffee-based agroforestry systems in Western Ghats of Karnataka, India. In: Forest resources resilience and conflicts. Elsevier, pp 289–297

Maiti SK, Bandyopadhyay S, Mukhopadhyay S (2021) Importance of selection of plant species for successful ecological restoration program in coal mine degraded land. In: Phytorestoration of abandoned mining and oil drilling sites. Elsevier, pp 325–357

Makumba W, Akinnifesi FK, Janssen B, Oenema O (2007) Long-term impact of a gliricidia-maize intercropping system on carbon sequestration in southern Malawi. Agric Ecosyst Environ 118:237–243

Manlay RJ, Feller C, Swift MJ (2007) Historical evolution of soil organic matter concepts and their relationships with the fertility and sustainability of cropping systems. Agric Ecosyst Environ 119(3–4):217–233

Maryo M, Wolde A, Negash M (2023) Woody species diversity and carbon stock potentials in homegarden agroforestry and other land use systems, northern Ethiopia. Heliyon 9(9):e19243

McDonald MA, Healey JR, Stevens PA (2002) The effects of secondary forest clearance and subsequent land-use on erosion losses and soil properties in the Blue Mountains of Jamaica. Agric Ecosyst Environ 92(1):1–19

Mishra G, Francaviglia R (2021) Land uses, altitude and texture effects on soil parameters. A comparative study in two districts of Nagaland, Northeast India. Agriculture 11:171

Misra PK (2011) Soil fertility management in agroforestry system. Int J Biotechnol Biochem 7(5):637–644

Moya JF, Murillo R, Portuguez E, Fallas JL, Rios V, Kottman F, Verjans JM, Mata R, Alvarado A (2013) Nutrient concentration age dynamics of teak (Tectona grandis Lf) plantations in Central America. For Syst 22(1):123–133

Muchane MN, Sileshi GW, Gripenberg S, Jonsson M, Pumariño L, Barrios E (2020) Agroforestry boosts soil health in the humid and sub-humid tropics: A meta-analysis. Agric Ecosyst Environ 295:106899

Mukhopadhyay S, Maiti SK, Masto RE (2013) Use of Reclaimed Mine Soil Index (RMSI) for screening of tree species for reclamation of coal mine degraded land. Ecol Eng 57:133–142

Murthy IK, Gupta M, Tomar S, Munsi M, Tiwari R, Hegde GT, Ravindranath NH (2013) Carbon sequestration potential of agroforestry systems in India. J Earth Sci Clim Change 4(1):1–7

Nair PR, Nair VD, Kumar BM, Showalter JM (2010) Carbon sequestration in agroforestry systems. Adv Agron 108:237–307

Nedd R, Anandhi A (2024) A synthesis on land degradation in the context of sustainable development goals. Land Degrad Dev 35(13):3937–3964

Negash M (2013) The indigenous agroforestry systems of the South-Eastern Rift Valley escarpment. Ethiopia: Their biodiversity, carbon stocks, and litterfall

Ntamwira J, Ocimati W, Blomme G, Lubobo AK, Mwarabu Lolonga Pyame D, Dhed'a Djailo B (2023) Innovative agroecological practices can restore degraded farmlands and revive crop yields. Front Sustain Food Syst 7:1017341
Oelbermann M, Voroney RP (2007) Carbon and nitrogen in a temperate agroforestry system: Using stable isotopes as a tool to understand soil dynamics. Ecol Eng 29:342–349
Oostra S, Majdi H, Olsson M (2006) Impact of tree species on soil carbon stocks and soil acidity in southern Sweden. Scand J For Res 21(5):364–371
Pandey CB, Singh L (2009) Soil fertility under homegarden trees and native moist evergreen forest in South Andaman, India. J Sustain Agric 33(3):303–318
Parrotta JA (1993) Secondary forest regeneration on degraded tropical lands: the role of plantations as "foster ecosystems". In: Restoration of tropical forest ecosystems: proceedings of the symposium held on October 7–10, 1991. Springer Netherlands, pp 63–73
Paul SC, Acharya GC, Chakraborty R, Maheswarappa HP, Hussain M, Ray AK (2020) Sustainability of soil health and system productivity through arecanut based cropping system in the NE region of India. J Plant Crop 48(2):111–119
Pradeep KS, Kumar A, Avinash RK, Karthika KS (2020) Assessment of the status of soil organic carbon stocks under natural forest and plantation ecosystems in southern Karnataka, India. J Plantation Crops 48(3):237–246
Prokop P (2020) Remote sensing of severely degraded land: detection of long-term land-use changes using high-resolution satellite images on the Meghalaya Plateau, northeast India. Remote Sens Appl Soc Environ 20:100432
Raizada A, Dhyani SK, Sharda VN (2004) Potential and prospects of vegetation recovery in degraded lands in India–a review. Indian Forester 130(40):441–455
Rajan K, Natarajan A, Kumar KA, Badrinath MS, Gowda RC (2010) Soil organic carbon–the most reliable indicator for monitoring land degradation by soil erosion. Curr Sci 99:823–827
Ranđelović D, Jakovljević K, Šinžar-Sekulić J, Kuzmič F, Šilc U (2024) Recognising the role of ruderal species in restoration of degraded lands. Sci Total Environ 938:173104
Ranjan R (2021) A potential PES mechanism for agroforestry-led industrial wastewater remediation using short-rotation trees. Water Econ Policy 7(02):2150006
Rasheed F, Zafar Z, Waseem ZA, Rafay M, Abdullah M, Salam MMA, Mohsin M, Khan WR (2020) Phytoaccumulation of Zn, Pb, and Cd in Conocarpus lancifolius irrigated with wastewater: does physiological response influence heavy metal uptake? Int J Phytoremediation 22(3):287–294
Ray JG, Thomas BT (2012) Fertility characteristics of oxic dystrustepts under natural forest, rubber, and teak plantations in different seasons, Kerala, South India. Commun Soil Sci Plant Anal 43(17):2247–2261
Ray P, Chattaraj S, Bandyopadhyay S, Jena R, Singh S, Ray S (2021) Shifting cultivation, soil degradation, and agricultural land-use planning in the northeastern hill region of India using geo-spatial techniques. Land Degrad Dev 32:3870–3892
Rocky P (2013) Socio-economic and ecological studies in home gardens of undivided Aizawl district of Mizoram, India. Doctoral dissertation, Mizoram University
Rodrigues J, Miranda I, Sousa A (2018) Microclimate in understories of a mango orchard and a degraded area in the Eastern Amazon. Ambi-Agua-An Interdicip J Appl Sci 13(3)
SAC (2021) Desertification and land degradation atlas of India (Assessment and analysis of changes over 15 years based on remote sensing). Space Applications Centre, Ahmedabad
Sacande M, Berrahmouni N (2016) Community participation and ecological criteria for selecting species and restoring natural capital with native species in the Sahel. Restor Ecol 24(4):479–488
Saha R, Tomar JMS, Ghosh PK (2007) Evaluation and selection of multipurpose tree for improving soil hydro-physical behaviour under hilly eco-system of north east India. Agrofor Syst 69(3):239–247
Saha R, Chaudhary RS, Somasundaram J (2012) Soil health management under hill agroecosystem of North East India. Appl Environ Soil Sci 2012(1):696174

Sahoo UK, Singh SL, Gogoi A, Kenye A, Sahoo SS (2019) Active and passive soil organic carbon pools as affected by different land use types in Mizoram, Northeast India. PLoS One 14(7):e0219969

Sahoo UK, Nath AJ, Lalnunpuii K (2021a) Biomass estimation models, biomass storage and ecosystem carbon stock in sweet orange orchards: implications for land use management. Acta Ecol Sin 41(1):57–63

Sahoo UK, Tripathi OP, Nath AJ, Deb S, Das DJ, Gupta A, Devi NB, Charturvedi SS, Singh SL, Kumar A, Tiwari BK (2021b) Quantifying tree diversity, carbon stocks, and sequestration potential for diverse land uses in Northeast India. Front Environ Sci 9:436

Sahoo UK, Ahirwal J, Giri K, Mishra G, Francaviglia R (2023) Modeling land use and climate change effects on soil organic carbon storage under different plantation systems in Mizoram, Northeast India. Agriculture 13(7):1332

Seid YM (2022) Biomass and soil carbon stocks of selected Agroforestry practice and cultivated land, in Tehuledere District, South Wollo Zone, Ethiopia

Semere M (2019) Biomass and soil carbon stocks assessment of agroforestry systems and adjacent cultivated land, in Cheha Wereda, Gurage Zone, Ethiopia. Int J Environ Sci Nat Resour 20(4):556043

Sharma D, Sunderraj SW (2005) Species selection for improving disturbed habitats in Western India. Curr Sci 88:462–467

Sharma P, Rai SC, Sharma R, Sharma E (2004) Effects of land-use change on soil microbial C, N and P in a Himalayan watershed. Pedobiologia 48(1):83–92

Sheba MC, Devaki R, Uma RN (2019) Case study on the soil physical parameters disparity and NPK concentrations in regions found in and around Pachapalayam, Coimbatore, Tamil Nadu. In: IOP conference series: materials science and engineering, vol 705, No. 1. IOP Publishing, p 012052

Siddiqui M, Shah A, Yaqoob S (2009) Chronosequence and crown strata effects on foliar nutrient concentrations in teak (Tectona grandis Lf) plantations. Pak J Bot 41(6):3023–3034

Silas VJ, Kumar J (2023) Effect of Different NPK Levels and Chlorophyll Content on Growth and Development of Sweet Orange (Citrus sinesis Osbeck). Int J Environ Clim Change 13(12):522–526

Singh KA, Rai AK (2012) Nutrient accumulation, distribution and use efficiency in different bamboo plant species in north-eastern hills region of India. Indian J Agric Sci 82(2):134

Singh SL, Sahoo UK (2018) Assessment of biomass, carbon stock and carbon sequestration potential of two major land uses of Mizoram, India. Int J Ecol Environ Sci 44(3):293–306

Singh SL, Sahoo UK (2021) Tree species composition, diversity and soil organic carbon stock in homegardens and shifting cultivation fallows of Mizoram, Northeast India. Vegetos 34(1):220–228

Singh K, Tewari SK (2022) Does the road to land degradation neutrality in India is paved with restoration science? Restor Ecol 30(5):e13585

Singh SL, Sahoo UK, Gogoi A, Kenye A (2018a) Effect of land use changes on carbon stock dynamics in major land use sectors of Mizoram, Northeast India. J Environ Prot 9(12):1262–1285

Singh SL, Sahoo UK, Kenye A, Gogoi A (2018b) Assessment of growth, carbon stock and sequestration potential of oil palm plantations in Mizoram, Northeast India. J Environ Prot 9(09):912

Singh L, Sridharan S, Thul ST, Kokate P, Kumar P, Kumar S, Kumar R (2020) Eco-rejuvenation of degraded land by microbe assisted bamboo plantation. Ind Crop Prod 155:112795

Singh S, Singh MK, Kumar CM, Soni P, Ghoshal N (2021) Impact of tree plantations on the dynamics of soil aggregates in urban degraded lands in the dry tropics. Land Degrad Dev 32(17):5108–5117

Singh K, Agrawala M, Dhyani S, Mishra N, Bauddh K, Byun C (2024) Land degradation and development in India: Status quo, challenges and opportunities. Land Degrad Dev 35(8):2675–2683

Sreenivas K, Sujatha G, Mitran T, Suresh KJR, Ravisankar T, Rao PVN (2021) Decadal changes in land degradation status of India. Curr Sci 121(4):25

Srivastava VK, Pandey R, Yadav CJ, Sushma M, Khatri G (2024) Effect of different fertilizer levels on growth and yield of linseed (Linum usitatissimum L.) under Guava (Psidium guajava L.) Based Agroforestry SYSTEM. Int J Environ Clim Change 14(4):203–208
Subiakto A, Rachmat HH, Sakai C (2016) Choosing native tree species for establishing man-made forest: a new perspective for sustainable forest management in changing world. Biodivers J Biol Divers 17(2):620
Sujatha S, Bhat R (2016) Impact of organic and inorganic nutrition on soil–plant nutrient balance in arecanut (Areca catechu L.) on a laterite soil. J Plant Nutr 39(5):714–726
Talema A, Poesen J, Muys B, Reubens B, Dibaba H, Diels J (2017) Multi-criteria-based plant species selection for gully and riverbank stabilization in a sub-humid tropical area. Land Degrad Dev 28(5):1675–1686
Thangjam U (2019) Provenance trial of Parkia timoriana (DC.) Merr. in North-east India. Doctoral dissertation, Mizoram University
Thong P, Sahoo UK, Thangjam U, Pebam R (2020) Pattern of forest recovery and carbon stock following shifting cultivation in Manipur, North-East India. PLoS One 15(10):e0239906
Tripathi OP, Pandey HN, Tripathi RS (2004) Distribution, community characteristics and tree population structure of subtropical pine forest of Meghalaya, northeast India. Int J Ecol Environ Sci 29(207):207–214
United Nation Convention to Combat Desertification (UNCCD) (2024) https://www.unccd.int/news-stories/press-releases/desertification-and-drought-day-2024-united-land-our-legacy-our-future
Upadhyaya K, Sahoo UK, Vanlalhriatpuia K, Roy S (2012) Decomposition dynamics and nutrient release pattern from leaf litters of five commonly occurring homegarden tree species in Mizoram, India. J Sustain For 31(8):711–726
Uthappa AR, Bana OPS, Kumar M, Kanwal M (2015) Soil physico-bio-chemical properties as influenced by varying tree densities in poplar (Populus deltoides Bartr. ex Marsh.) based agroforestry system. Indian J Agrofor 17(1):81–90
Vanlaltanpuia, Rao C, Lalrinpuia V (2023) Causes and consequences of environmental degradation in Mizoram, Northeast India. Disaster Adv 16(10):41–48
Vesterdal L, Schmidt IK, Callesen I, Nilsson LO, Gundersen P (2008) Carbon and nitrogen in forest floor and mineral soil under six common European tree species. For Ecol Manag 255(1):35–48
Wasteland Atlas of India (2019) National Remote Sensing Centre (NRSC), Hyderabad, India
Wong MT, Hairiah K, Alegre J (2004) Managing soil acidity and aluminium toxicity in tree-based agroecosystems. In: Below-ground interactions in tropical agroecosystems: Concepts and models with multiple plant components. CABI Publishing, Wallingford, pp 143–156
Xie H, Zhang Y, Wu Z, Lv T (2020) A bibliometric analysis on land degradation: current status, development, and future directions. Land 9(1):28
Young A (1997) Agroforestry for soil management, 2nd edn. CAB International
Zade SP, Bhosale SL, Gourkhede PH (2020) Carbon status in major fruit orchard soils of Parbhani District of Maharashtra, India. Int J Curr Microbiol App Sci 9(3):1969–1979
Zeng H, Wu J, Zhu X, Singh AK, Chen C, Liu W (2021) Jungle rubber facilitates the restoration of degraded soil of an existing rubber plantation. J Environ Manag 281:111959
Zuazo VHD, Pleguezuelo CRR (2009) Soil-erosion and runoff prevention by plant covers: a review. In: Sustainable agriculture, pp 785–811
Zvoleff A, Noon M, Daldegan G, Gonzalez-Roglich M (2020) Forging the path to achieving land degradation neutrality: global patterns and drivers of land degradation at global scales. In: EGU general assembly conference abstracts, p 11497

Indigenous Agroforestry and Restoration of Shifting Cultivated Lands in the Indian East Himalayan Region

5

Animekh Hazarika and Arun Jyoti Nath

Abstract

Shifting cultivation, also known as slash-and-burn agriculture or jhum, has traditionally been practised by indigenous communities in the Indian East Himalayas. While it has provided subsistence for these communities over generations, shifting cultivation has several adverse impacts on the environment and socio-economic fabric of the region. To address these adverse impacts, some ethnic communities in the Indian East Himalayan region have evolved a more sustainable and resilient agricultural system, benefiting both the environment and local communities. Indigenous agroforestry practices play a crucial role in the sustainable land management and restoration of shifting cultivated lands in the Indian East Himalayan region. This chapter explores the traditional ecological knowledge and practices employed by indigenous communities in the region to enhance soil fertility, biodiversity conservation, and resilience to climate change through agroforestry systems. Key indigenous tree species and their roles in agroforestry are discussed, along with the socio-economic implications and policy recommendations for promoting these practices.

Keywords

Biodiversity conservation · Deforestation · Ecosystem resilience · Food security · Sustainable land management · Traditional ecological knowledge

A. Hazarika · A. J. Nath (✉)
Department of Ecology and Environmental Science, Assam University, Silchar, Assam, India

V. C. Pandey (ed.), *Plantation-Based Land Restoration*,
https://doi.org/10.1007/978-981-95-2860-8_5

5.1 Introduction

The Indian East Himalayan region is renowned for its rich biodiversity and unique socio-cultural landscapes shaped by indigenous communities practising shifting cultivation (Nath et al. 2020). The shifting cultivation (slash-and-burn agriculture) has been a traditional practice among various indigenous groups in the region for centuries (Fig. 5.1). While it has sustained the livelihoods and cultural practices of indigenous communities, this traditional agricultural method also poses significant environmental and social challenges (Ramakrishnan 2016). Shifting cultivation involves clearing patches of forest by cutting and burning vegetation. This process leads to deforestation and loss of natural habitats for wildlife, disrupting ecological balance and reducing biodiversity. The burning of vegetation removes the protective cover of the forest floor, making the soil vulnerable to erosion by wind and rain (Nath et al. 2016). Fragmentation of forests due to shifting cultivation reduces connectivity between habitats, affecting wildlife movement and gene flow among species. This fragmentation can lead to localized extinctions and loss of biodiversity (Devi et al. 2021). Deforestation and altered land use patterns affect local hydrology, leading to changes in water availability, quality, and runoff patterns. This can contribute to erosion, sedimentation of water bodies, and reduced water infiltration rates (Nath et al. 2016). Deforestation through shifting cultivation releases stored carbon into the atmosphere, contributing to greenhouse gas emissions and global climate change. Loss of organic matter and nutrients exacerbates soil degradation, reducing fertility over time (Laskar et al. 2021). Shifting cultivation practices often

Fig. 5.1 Shifting cultivation in Indian East Himalayan region

yield unpredictable harvests due to soil degradation and dependence on natural cycles. This can contribute to food insecurity and poverty among communities relying on subsistence agriculture (Nath et al. 2022). Reliance on shifting cultivation as a primary agricultural practice can render communities vulnerable to economic shocks and external market forces. Limited access to markets and infrastructure further restricts income-generating opportunities. Therefore, increasing population pressure, land degradation, climate change, and other social issues have posed significant challenges to this traditional practice.

In response, indigenous communities have developed indigenous agroforestry systems that not only sustain their livelihoods but also promote ecosystem resilience and biodiversity conservation (Reang et al. 2021, 2022a). For instance, the ethnic "Hmar" community in the Indian East Himalayan region has transitioned to pineapple (*Ananas comosus*) agroforestry systems, while the "Pnar" community has moved towards cultivating economically valuable betel leaf (*Piper betle*), a more sedentary form of agriculture. Furthermore, ethnic communities integrate Himalayan alder trees (*Alnus nepalensis*) in fallow lands of shifting cultivated lands. These agroforestry systems are vital for ensuring economic well-being and providing essential ecosystem benefits while respecting cultural identities (Reang et al. 2022a). Therefore, it is essential to gain a comprehensive understanding of these systems and conduct scientific studies on the functionalities of tree components to inform policies for sustainable land management. This chapter highlights the uniqueness of these indigenous agroforestry systems, the services provided by tree components, their socio-economic implications, and the opportunities and challenges associated with the indigenous agroforestry systems to successfully promote indigenous knowledge-based land management strategies.

5.2 Indigenous Agroforestry Practices

Indigenous agroforestry systems in the Indian East Himalayan region integrate trees, crops, and livestock synergistically to maximize land productivity and ecological benefits. These systems typically involve:

5.2.1 Alder-Based Agroforestry Systems

Alder-based agroforestry systems represent an innovative and sustainable approach to land management that integrates the cultivation of alder trees with crops and livestock (Rathore et al. 2010) (Fig. 5.2). This system is particularly relevant in regions like the Indian East Himalayan region, where traditional agricultural practices such as shifting cultivation have led to soil degradation and loss of ecosystem resilience (Singh et al. 2014). Alder trees are unique in their ability to fix nitrogen through a symbiotic relationship with nitrogen-fixing bacteria (*Frankia* spp.) in their root nodules. This biological nitrogen fixation enriches the soil with nitrogen, improving soil fertility and reducing the need for synthetic fertilizers (Saikia and

Fig. 5.2 Alder-based agroforestry system of Indian East Himalayas. (Photo credit: Krishna Giri)

Jain 2007). The deep root systems of alders also contribute to soil structure improvement, erosion control, and enhanced water retention capacity. Alders are typically interplanted with crops such as maize, beans, or vegetables (Nath et al. 2020). The trees provide shade and nutrient cycling benefits, while the crops benefit from improved soil fertility and moisture retention. Diversified cropping systems in alder-based agroforestry ensure food security by reducing risks associated with mono-cropping and enhancing overall crop yields. Integrating alder trees with livestock grazing allows for efficient land use. Alders can provide fodder and shade for livestock, while the livestock contribute to nutrient cycling through manure deposition. Alder leaves and branches are often used as organic mulch or green manure, further enhancing soil fertility and organic matter content. This practice supports sustainable soil health management without relying heavily on external inputs. Additionally, alder trees provide multiple products such as timbre, firewood, and medicinal products, offering additional income opportunities for farmers.

5.2.2 *Piper*-Based Agroforestry Systems

Piper betle, commonly known as betel leaf or paan, has potential applications in agroforestry systems aimed at land restoration (Reang et al. 2024) (Fig. 5.3). This approach integrates the cultivation of betel leaf with tree species to rehabilitate degraded lands, improve soil health, and enhance ecosystem resilience. Agroforestry systems incorporating betel leaf promote biodiversity by providing habitat and food

Fig. 5.3 *Piper* agroforestry system of Indian East Himalayas

sources for beneficial insects, birds, and microorganisms. This biodiversity supports ecosystem resilience and contributes to the overall ecological balance (Hazarika et al. 2024a). Betel leaf plants have a fibrous root system that helps bind soil particles, reduce erosion, and improve soil structure. Their roots also contribute organic matter to the soil, enhancing fertility and water retention capabilities (Pradhan et al. 2013). Betel leaf cultivation offers regular harvests and market demand, providing economic opportunities for farmers engaged in land restoration activities (Reang et al. 2024). Income from betel leaf sales can support livelihoods and incentivize sustainable land management practices. Beyond economic benefits, betel leaf holds cultural significance in many societies across Asia. Betel leaf cultivation in agroforestry systems helps preserve cultural traditions and indigenous knowledge related to its use in rituals, social ceremonies, and traditional medicine practices. Betel leaf agroforestry systems contribute to climate resilience by enhancing soil carbon sequestration, enhancing water use efficiency, and lowering greenhouse gas emissions linked to synthetic inputs. Betel leaf plants contribute to water conservation by reducing runoff and enhancing infiltration, thus minimizing soil erosion and improving overall water quality.

Intercropping betel leaf with nitrogen-fixing plants like legumes or other fast-growing species can help restore soil fertility. These plants contribute nitrogen to the soil, reducing the need for synthetic fertilizers and supporting sustainable agricultural practices (Hazarika et al. 2024a). Farmers clear densely forested areas, retaining only trees and tall plants for shade and support for betel vines. To enrich soil nutrients, trimmed twigs, leaves, and branches are left to dry. Pollarding allows sufficient sunlight to reach the forest floor, aiding new vine growth. Weeding is done twice a year, with removed weeds used as mulch to retain soil moisture and enhance soil nutrients, essential for healthy betel vine growth.

5.2.3 Pineapple-Based Agroforestry Systems

Pineapple-based agroforestry systems offer a sustainable and multifunctional approach to land restoration in the Indian East Himalayan region (Fig. 5.4). Pineapple-based agroforestry presents a promising approach to land restoration, combining the cultivation of pineapple with other crops and trees to rehabilitate degraded lands, improve soil health, and enhance ecosystem resilience (Reang et al. 2021, 2022a). This region, characterized by diverse landscapes and challenging terrain, can benefit significantly from integrated agroforestry systems (Nath et al. 2021) that incorporate pineapple cultivation. Pineapple is commonly intercropped with crops like ginger, turmeric, and various vegetables. This practice optimizes land use, increases biodiversity, and provides farmers with diversified income sources throughout the year (Reang et al. 2021). Integrating pineapple cultivation with shade-providing trees such as timbre species, fruit trees (e.g. citrus, guava), or multipurpose trees (e.g. *Parkia timoriana*) enhances ecosystem services (Reang et al. 2022a). Depending on space availability on the farm, naturally growing tree species like *Albizia procera* in pineapple agroforestry systems are sometimes replaced with economically valuable species like *Aquilaria malaccensis* and *Hevea*

Fig. 5.4 Pineapple agroforestry system of Indian East Himalayas

brasiliensis, and fruit trees are introduced (Reang et al. 2022a). These changes provide substantial economic benefits, offering shade for pineapple plants, reducing soil erosion, and serving as additional income sources through timbre, fruit, or fodder production (Reang et al. 2022a).

5.3 Role of Indigenous Tree Species

Indigenous agroforestry systems provide diverse ecosystem services (Fig. 5.5). Indigenous communities in the Indian East Himalayan region have identified and cultivated a range of tree species that are integral to their agroforestry systems. These species possess multidimensional utility and functions:

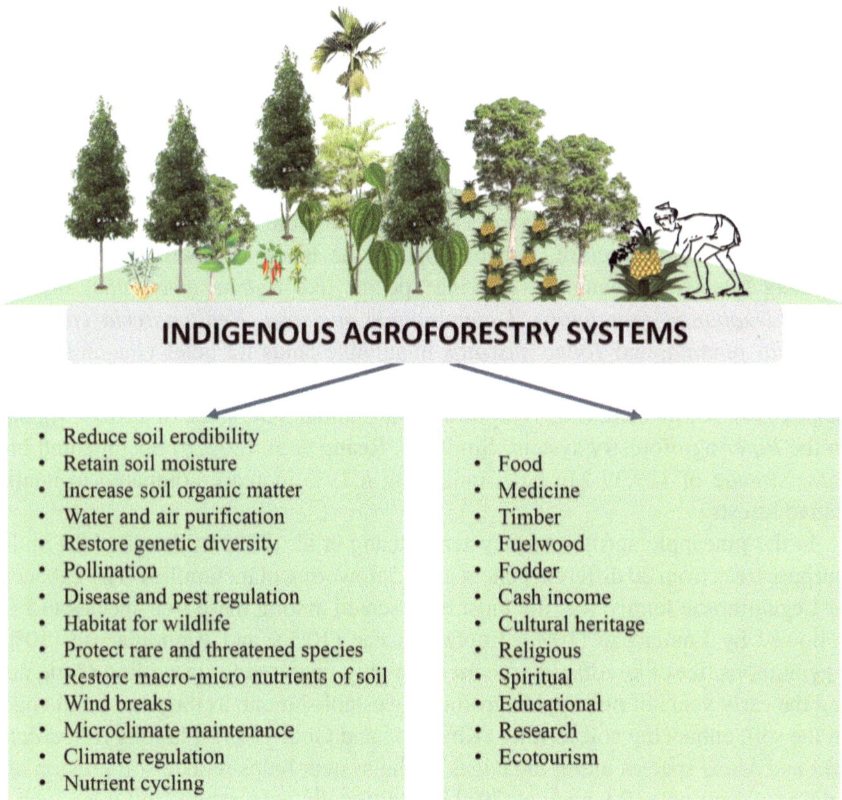

Fig. 5.5 Diverse ecosystem services provisioned by indigenous agroforestry systems

5.3.1 Soil Fertility Enhancement

Indigenous tree species in agroforestry offer a wide range of benefits, with soil fertility enhancement being a key contribution that provides substantial economic returns to farmers. The Himalayan alder, a non-leguminous tree species, can fix atmospheric nitrogen (N_2) through its association with the *Frankia* sp. and actinobacteria (Nath et al. 2020). This species contributes to a soil organic carbon (SOC) stock of 97 Mg ha^{-1} in the surface soil (0–30 cm) (Giri 2019), as well as 3.37–13.56 Mg ha^{-1} of litter and 48.3–184.8 kg ha^{-1} of nitrogen in the soil (Rathore et al. 2010). Mishra et al. (2018) observed a lower bulk density in soils under alder-based systems due to the substantial litterfall and soil organic matter (SOM) content, indicating improved soil fertility. Additionally, Himalayan alder-based systems help maintain soil physical properties and macro-micronutrient levels comparable to those of forested lands in the Indian East Himalayas, enhancing soil fertility, productivity, and soil microbial populations (Mishra et al. 2018).

In the context of *Piper* agroforestry systems, 102 tree species across 38 families of flowering plants have been identified, with the Fabaceae or Leguminosae family (10.5% of all families) being the most prevalent and contributing substantially to enhancing the soil fertility (Reang et al. 2024). Of these species, 99% serve as shade trees for cultivating betel leaf plants, and 87.3% are used as companion or support structures for betel vines. These trees create a favourable microclimate, maintaining soil moisture and structure, which supports the healthy growth of betel vines. Farmers selectively plant fast-growing species like *Aglaia spectabilis*, *Bombax ceiba*, *Duabanga grandiflora*, *Lagerstroemia speciosa*, *Neolamarckia cadamba*, *Spondias pinnata*, and *Toona australis* in suitable lands for betel vine cultivation, significantly improving soil health through leaf litter and decomposition (Reang et al. 2024). Kurmi et al. (2024) reported a substantial SOC stock of 110.74 Mg ha^{-1} in the *Piper* agroforestry system. Similarly, Reang et al. (2022b) documented biomass storage of 129.39 Mg ha^{-1}, indicating a 19% increase compared to nearby native forests.

In the pineapple agroforestry system, Reang et al. (2022a) identified 20 multipurpose trees from 20 different genera and 15 flowering plant families. The Fabaceae or Leguminosae family was the most represented among these tree species (15%), followed by Lamiaceae (10%), Euphorbiaceae (10%), and Bignoniaceae (10%). Leguminous trees like *Albizia procera* are highly favoured for providing shade during the early years of pineapple agroforestry establishment, as they can fix nitrogen in the soil, enhancing soil fertility (Khaleque and Gold 1993). Planting *Areca catechu* and *Musa* species along the edges of the system helps reduce soil erosion and retain soil nutrients (Reang et al. 2021). Additionally, integrating fruit trees such as *Mangifera indica* and cash crops like *Hevea brasiliensis* and *Aquilaria malaccensis* contributes to soil fertility through biomass decomposition and nutrient cycling. Cultivating tree bean (*Parkia timoriana*) can enhance soil fertility by fostering beneficial bacteria populations, including *Pseudomonas fluorescens*, *Agrobacterium fabrum*, *Bacillus subtilis*, etc. These bacteria may help develop native bacterial bioinoculants, which can establish and nurture new plantlets and improve degraded soil

(Singh 2019). Hazarika et al. (2024b) outlined SOC stocks ranging from 169.7 Mg C ha^{-1} to 177 Mg C ha^{-1} in pineapple agroforestry systems of different ages, with stratification ratios (SR >2) for SOC and macronutrients indicating better soil quality, signifying improved soil health comparable to nearby forested lands. The clay and modified clay ratios decreased with the increasing age of pineapple agroforestry stands (Hazarika et al. 2024b), attributed to the accumulation of organic matter in the topsoil, which protects soil structure and improves the stability of soil aggregates (Tarafdar and Ray 2005). This underscores the importance of indigenous systems for soil conservation (Hazarika et al., 2024b).

5.3.2 Biodiversity Conservation

Indigenous agroforestry supports a diverse range of crops, allowing farmers to tailor their systems to their preferences. For instance, farmers in Khonoma village, Nagaland, grow local vegetable varieties under alder trees and cultivate indigenous paddy varieties using a terrace system (Nath et al. 2020).

In *Piper* agroforestry systems, Reang et al. (2024) identified 102 tree species. According to IUCN categories, 46.8% of these species were classified as "least concern", 3.8% lacked sufficient data, and 43.8% had not been assessed. Notably, three species—*Dimocarpus longan*, *Dipterocarpus turbinatus*, and *Aquilaria malaccensis*—were classified as "near threatened", "vulnerable", and "critically endangered", respectively. The Margalef richness index (R) and Shannon-Wiener diversity index (H) were 10.38 and 3.65, respectively, in *Piper* agroforestry systems over 35 years old, surpassing those of nearby forested lands (Reang et al. 2024). Rare and threatened primate species, such as *Hoolock hoolock* and *Macaca mulatta*, visit these systems, underscoring their role in habitat restoration and species conservation (Hazarika et al. 2024a).

The pineapple agroforestry system has a Shannon-Wiener diversity index (H) ranging from 1.04 to 1.86, which is lower than that of natural forested lands in the Indian East Himalayas. This is due to the planting of pineapples in rows with 0.7–0.8 m spacing, leaving less space for integrating more trees (Reang et al. 2021). However, this system plays a crucial role in conserving the economically important tree bean (*Parkia timoriana*) and critically endangered *Aquilaria malaccensis*, which is no longer found in the wild (Reang et al. 2022a). Additionally, the pineapple agroforestry system serves as a foraging ground for rare and threatened species such as *Hystrix indica*, *Macaca mulatta*, and *Sus scrofa cristatus*.

5.3.3 Climate Resilience

Alder trees, as nitrogen-fixing species in agroforestry systems, are fast-growing and thrive in moist conditions. They play a crucial role in mitigating soil erosion and maintaining soil moisture through their extensive root systems, which aid in soil stabilization and water retention. The canopy of alder trees moderates microclimates by reducing temperature extremes and shielding understory crops from harsh weather conditions (Rana et al. 2018).

In *Piper* agroforestry systems, the density and growth of natural trees enhance soil fertility through biomass decomposition and provide adequate sunlight for the betel vine growing beneath them (Hazarika et al. 2024a). These trees protect soil from erosion on hill slopes with their deep root systems and retain soil nutrients that might otherwise be washed away by heavy rainfall. Additionally, tree species act as barriers to high winds.

Similarly, in pineapple agroforestry systems, trees provide a suitable climate for understory pineapple suckers. Nitrogen-fixing tree species in these systems help improve soil fertility (Reang et al. 2022a). Integrating tree species along the edges helps prevent soil erosion and retain soil nutrients, protecting them from being washed away by intensified rainfall. Additionally, the presence of tree species acts as a protective barrier for crops, safeguarding them from severe pest and disease outbreaks, thus securing farmers' productivity.

5.4 Socio-economic Implications

The adoption of indigenous agroforestry practices has profound socio-economic implications for local communities in the Indian East Himalayan region:

5.4.1 Food Security

Indigenous agroforestry practices enhance food security by diversifying and stabilizing food production. The integration of tree species with perennial and annual crops meets farmers' needs and provides economic returns. Improved soil fertility, facilitated by alder trees, significantly boosts the productivity of local vegetables and paddy grown under the alder canopy, ensuring a steady food supply and income generation (Nath et al. 2020).

In *Piper* agroforestry systems, tree species create a favourable microclimate for betel vines, helping to mitigate climate variability. The healthy growth and continuous harvesting of betel leaves provide farmers with a stable income source. Additionally, wild fruits and medicinal plants like *Baccaurea ramiflora*, *Citrus hystrix*, *Garcinia xanthochymus*, and *Garuga pinnata* accomplish the nutritional needs of farmers and can be sold in markets to supplement household income (Reang et al. 2024).

In pineapple agroforestry systems, the robust growth of pineapples ensures substantial income per cropping cycle, providing farmers with reliable economic returns. Fruit trees such as *Mangifera indica*, *Litchi chinensis*, and *Oroxylum indicum* contribute to dietary requirements, while *Parkia timoriana*, known for its nutritional value and income potential, benefits farmers and ensures food security. Additionally, selling various products like *Areca catechu*, *Aquilaria malaccensis*, and *Hevea brasiliensis* from the system helps farmers secure their food supply.

Moreover, agroforestry systems fulfil household fuelwood needs, showcasing their multifunctional nature.

5.4.2 Livelihoods

Farmers practising alder agroforestry systems typically grow a variety of vegetables, including cauliflower, cabbage, chilli, cucumber, French bean, ginger, maize, potato, pea, sesame, soybean, squash, tomato, millets, and paddy in some regions (Nath et al. 2020). The products from these systems help farmers secure their food supply and generate income by selling them in local markets. Additionally, the branches of alder trees are used as fuelwood, reducing household energy expenses (Rana et al., 2018). The cultivation of ghost pepper (*Capsicum chinense*) is particularly noteworthy, with a market price of US$2.40–3.60 per kg. Farmers also grow sesame (*Sesamum indicum*) in their shifting cultivation practices, selling it for approximately US$1.80–2.40 per kg.

The average annual production of *Piper* leaves is approximately 0.25 million leaves per hectare. Farmers usually sell their product in bundles known as “kuri” (6720 leaves) either to intermediaries or directly in local markets. The price per “kuri” ranges from US$19.2 to US$89.8 and can reach up to US$192.4 during less productive periods. Additionally, selling wild fruits and medicinal plants from the agroforestry system can supplement household income (Hazarika et al. 2024a).

On average, older pineapple farms (>15 years) produce 5200–5400 plants within a 0.1 hectare area. The harvested fruits are sold at local market prices ranging from US$0.12 to US$0.48 each. Pineapple farmers also sell seed pods of *Parkia timoriana*, *Areca catechu*, and other fruit products grown on the same farm. A mature *Parkia timoriana* tree can produce approximately 10,000–15,000 pods, yielding an annual income of US$100–200 per tree (Rocky et al. 2004).

Furthermore, *Aquilaria malaccensis*, found in *Piper* and pineapple agroforestry systems, is highly valued for its oleoresin, which is used in the fragrance and perfume industries, incense making, ornamental displays, and ethnomedicine (Ador et al. 2021; Hazarika et al., 2023). Depending on the quality and quantity produced, individual families can earn between US$600 and US$18,000 from *A. malaccensis*, with an average annual income of US$2288 (Saikia and Khan 2014).

5.4.3 Cultural Continuity

The continuity of traditional farm management techniques is vital for sustaining both the agroforestry system and farmers’ livelihoods. These farming techniques are deeply embedded in the cultural and traditional practices of local communities in the Indian East Himalayan region, are passed down through generations, and are integral to maintaining agrobiodiversity and cultural identity. Knowledge of soil productivity and farm management practices such as zero tillage and mulching, often considered climate-smart, is essential for climate-resilient agriculture

(Hazarika et al. 2024c). Adjustments in planting and selective harvesting of agroforestry products reflect farmers' adaptability and flexibility in response to climatic changes. Additionally, various seed storing and germination techniques highlight the unique ecological knowledge of farmers, which has been passed down through generations.

5.5 Challenges and Opportunities

5.5.1 Challenges in Promoting Indigenous Agroforestry Systems

- *Land tenure issues*

Indigenous communities in the Indian East Himalayan region often lease land from the forest department or revenue for farming, adhering to unique customary laws and regulations that vary across locations and farming communities (Reang et al. 2024). Land rights are typically not legally documented and are passed down orally from one generation to the next. Securing land tenure is crucial for maintaining indigenous agroforestry lands against changes in land use and preserving their cultural identity. Secure land tenure is also essential for accessing government agricultural schemes. Additionally, land tenure among indigenous communities faces challenges due to population growth.

- *Knowledge erosion*

The continuity of indigenous agroforestry systems among the youth faces significant challenges due to agricultural practices, market structures, income generation, and climate change. Transferring agro-ecological knowledge is difficult, especially with the rise of economically driven agriculture like monoculture plantations of *Areca catechu* and rubber. Market-oriented approaches in indigenous farming often lead to the use of pesticides and fertilizers to meet market demands and maximize economic benefits, which diminishes interest in traditional farming knowledge and practices.

- *Climate change impact*

Indigenous agroforestry systems are susceptible to climate change, experiencing erratic temperature and precipitation patterns, as well as frequent extreme events, which can negatively affect farm productivity and affect farmers' livelihoods. For instance, changes in precipitation and temperature regimes can impact the size, colour, and flavour of pineapples (Williams et al. 2018). Increased weed infestation also poses a critical challenge to farmers in indigenous agroforestry systems. Additionally, prominent diseases such as "root rot" and "leaf spot" are serious concerns in the *Piper* agroforestry system, leading to declines in crop yield and causing economic instability among households (Hazarika et al. 2024a).

- *Market structure*

The selling of products from indigenous agroforestry systems operates through a complex network of intermediaries. The sales process adheres to traditional practices that lack transparency and can involve coercive tactics. Crucially, these intermediaries neither provide receipts nor maintain records, which prevents farmers from securing better trade outcomes for their crops. Farmers can sell their products directly in local markets or to intermediary traders. Local wholesalers buy farm products from these farmers and then resell them to retailers. These wholesalers transport the products to district centres, where they sell them to agents. These agents further distribute the betel leaves through district-level wholesalers and retailers. Occasionally, farm products are transported to neighbouring districts, major cities within the state, or nearby states.

- *Policy support*

The National Agroforestry Policy of 2014 proposes various incentives and support structures, such as input subsidies and an interest moratorium during the gestation period, to promote agroforestry. However, in the Indian East Himalayan region, indigenous agroforestry systems lack an incentive-based mechanism. In India, the minimum support price (MSP) for major crops ensures farmers receive a minimum remuneration, allowing them to recover cultivation costs when market prices decline, thus providing a price floor during market failures. Unfortunately, such provisions are currently unavailable for indigenous agroforestry products. Moreover, indigenous agroforestry systems have not benefited from mechanisms like REDD+ (Reducing Emissions from Deforestation and Forest Degradation), which aims to conserve and enhance forest carbon stocks, or other payment for ecosystem services schemes. Additionally, tree tenure remains ineffective in many parts of Indian East Himalayan region. Securing tree tenure involves the legal capacity to register, harvest, transport, and market trees and their products. The effectiveness of tree-based systems is often limited by farmers' lack of security over their land and tree tenure (Kang and Akinnifesi 2000).

5.5.2 Opportunities for Promoting Indigenous Agroforestry Systems

- *Natural farming*

Considering the growing importance of natural farming, combining natural farming with indigenous agroforestry systems can offer numerous benefits. The synergy between natural farming principles and traditional agroforestry practices can enhance sustainability, resilience, and productivity. This integration promotes ecological balance, reduces dependency on external inputs, and supports biodiversity. Additionally, it strengthens food security, improves soil health, and mitigates climate change impacts by increasing carbon sequestration and reducing greenhouse gas emissions. Therefore, natural farming and indigenous agroforestry systems, when integrated, can create a sustainable agricultural model that respects and leverages natural processes and traditional knowledge, ultimately contributing to environmental health, economic viability, and cultural preservation.

- *Green credit fund*

Integrating green credit funds with indigenous agroforestry systems can offer substantial benefits for both the environment and local communities. Green credit funds are financial mechanisms designed to support environmentally sustainable projects. When applied to agroforestry systems, they can help promote sustainable land use, enhance biodiversity, and provide economic incentives for farmers. Therefore, incorporating indigenous agroforestry systems under green credit fund will provide an excellent opportunity to the agroforestry managers to diversify their income sources.

- *Capacity building*

Capacity building among farmers and youth is crucial to promoting indigenous agricultural systems and ensuring their future sustainability. This can be achieved by disseminating knowledge about soil health and soil cards, the importance of land and tree tenure, sustainable land preparation and management techniques, zero tillage, and climate-smart agriculture. Additionally, it is important to educate the youth on indigenous farming techniques and effectively promote these methods. Institutions like the Indian Council of Agricultural Research (ICAR), regional agricultural centres, and the National Bank for Agriculture and Rural Development (NABARD) could support the establishment of small industries based on produce from indigenous agroforestry systems, thereby boosting the rural economy and sustaining these practices.

- *Market access*

Successfully selling products from indigenous agroforestry systems requires a fundamental understanding of market dynamics and connectivity. These systems and their market chains often face challenges such as limited market facilities and accessibility, leading to higher labour and transportation costs that affect economic well-being. To help farmers gain market recognition, it is essential to promote produce from indigenous systems through agencies or by obtaining geographical indications. Strong market networks enhance climate adaptability (Hazarika et al. 2024c) and secure better trade-offs for crops. Adequate market structures and international export networks are essential for this process.

5.6 Conclusions

Indigenous agroforestry practices in the Indian East Himalayan region offer a sustainable pathway for restoring shifting cultivated lands. By enhancing soil fertility, promoting biodiversity conservation, and providing socio-economic benefits to local communities, these practices contribute to resilience against climate change and support Sustainable Development Goals. However, numerous challenges must be addressed for the successful promotion of indigenous agroforestry systems in the region, including labour-intensive practices, economic viability, knowledge and skill requirements, and institutional and policy barriers. Additionally, concerted efforts are needed to overcome these challenges and leverage opportunities for scaling up these practices through supportive policies, capacity building, and improved market access.

5.7 Recommendations

Based on the findings presented, the following recommendations are proposed:

- *Policy support*: Develop and implement policies that recognize and support indigenous agroforestry practices.
- *Research and development*: Invest in research to document and promote traditional ecological knowledge.
- *Community engagement*: Foster partnerships between government agencies, NGOs, and local communities to promote sustainable land management practices.
- *Education and awareness*: Raise awareness among stakeholders about the benefits of indigenous agroforestry for ecosystem resilience and livelihoods.

References

Ador MAH, Farabi F, Ahmed R, Khatun R, Haque MMU (2021) Agar (*Aquilaria agallocha* Roxb.) based small-scale enterprises in Bangladesh: management, production, marketing and role in socio-economic development. Trees For People 6:100141. https://doi.org/10.1016/j.tfp.2021.100141

Devi AR, Sudhakar Reddy C, Shimrah T (2021) Assessment of forest fragmentation in a traditional shifting agricultural landscape in Senapati District of Manipur, Northeast India. Environ Dev Sustain 23(7):10344–10356. https://doi.org/10.1007/s10668-020-01059-4

Giri K (2019) The first report of indigenous free-living diazotroph *Kosakonia sacchari* isolated from Himalayan Alder-based shifting cultivation system in Nagaland, India. J Soil Sci Plant Nutr 19(3):574–579

Hazarika A, Deka JR, Nath PC, Sileshi GW, Nath AJ, Giri K, Das AK (2023) Modelling habitat suitability of the critically endangered Agarwood (*Aquilaria malaccensis*) in the Indian East Himalayan region. Biodivers Conserv 32(14):4787–4803. https://doi.org/10.1007/s10531-023-02727-3

Hazarika A, Nath AJ, Pandey R, Pebam R, Devi NB, Das AK (2024a) Climate change vulnerability of tribe managing Piper agroforestry systems in the Indian sub-Himalayan region. Agric Syst 216:103914. https://doi.org/10.1016/j.agsy.2024.103914

Hazarika A, Kurmi B, Francaviglia R, Sileshi GW, Paramesh V, Das AK, Nath AJ (2024b) The transition from shifting cultivation to indigenous agroforestry as nature-based solution for land restoration in the Indian Eastern Himalayas. Ecol Indic 162:112031. https://doi.org/10.1016/j.ecolind.2024.112031

Hazarika A, Nath AJ, Reang D, Pandey R, Sileshi GW, Das AK (2024c) Climate change vulnerability and adaptation among farmers practicing shifting agriculture in the Indian Himalayas. Environ Sustain Indicators 23:100430. https://doi.org/10.1016/j.indic.2024.100430

Kang BT, Akinnifesi FK (2000) Agroforestry as alternative land-use production systems for the tropics. In: Natural resources forum, vol 24, No. 2. Blackwell Publishing Ltd., Oxford, pp 137–151. https://doi.org/10.1111/j.1477-8947.2000.tb00938.x

Khaleque K, Gold MA (1993) Pineapple agroforestry: an indigenous system among the Garo community of Bangladesh. Soc Nat Resour 6(1):71–78. https://doi.org/10.1080/08941929309380808

Kurmi B, Nath AJ, Sileshi GW, Pandey R, Das AK (2024) Impact of progressive and retrogressive land use changes on ecosystem multifunctionality: implications for land restoration in the Indian Eastern Himalayan region. Sci Total Environ 912:169197. https://doi.org/10.1016/j.scitotenv.2023.169197

Laskar SY, Sileshi GW, Pathak K, Debnath N, Nath AJ, Laskar KY, Singnar P, Das AK (2021) Variations in soil organic carbon content with chronosequence, soil depth and aggregate size under shifting cultivation. Sci Total Environ 762:143114. https://doi.org/10.1016/j.scitotenv.2020.143114

Mishra G, Giri K, Pandey S (2018) Role of *Alnus nepalensis* in restoring soil fertility: a case study in Mokokchung, Nagaland. Natl Acad Sci Lett 41:265–268. https://doi.org/10.1007/s40009-018-0668-4

Nath AJ, Brahma B, Lal R, Das AK (2016) Soil and jhum cultivation. In: Encyclopedia of soil science, 3rd edn. CRC Press, Boca Raton. https://doi.org/10.1081/E-ESS3-120053748

Nath AJ, Sahoo UK, Giri K, Sileshi GW, Das AK (2020) Incentivizing hill farmers for promoting agroforestry as an alternative to shifting cultivation in Northeast India. In: Dagar JC, Gupta SR, Teketay D (eds) Agroforestry for degraded landscapes. Springer, Singapore. https://doi.org/10.1007/978-981-15-4136-0_14

Nath AJ, Kumar R, Devi NB, Rocky P, Giri K, Sahoo UK et al (2021) Agroforestry land suitability analysis in the Eastern Indian Himalayan region. Environ Chall 4:100199. https://doi.org/10.1016/j.envc.2021.100199

Nath AJ, Reang D, Sileshi GW (2022) The shifting cultivation juggernaut: an attribution problem. Global Chall 6(8):2200051. https://doi.org/10.1002/gch2.202200051

Pradhan D, Suri KA, Pradhan DK, Biswasroy P (2013) Golden heart of the nature: *Piper betle* L. J Pharmacogn Phytochem 1(6):147–167

Ramakrishnan PS (2016) Jhum: shifting agriculture. In: Encyclopedia of soil science, 3rd edn. CRC Press, Boca Raton. https://doi.org/10.1081/E-ESS3-120016599

Rana SK, Rana HK, Shrestha KK, Sujakhu S, Ranjitkar S (2018) Determining bioclimatic space of Himalayan alder for agroforestry systems in Nepal. Plant Divers 40(1):1–18. https://doi.org/10.1016/j.pld.2017.11.002

Rathore SS, Karunakaran K, Prakash B (2010) Alder based farming system a traditional farming practices in Nagaland for amelioration of jhum land. Indian J Tradit Knowl 4:677–680

Reang D, Hazarika A, Sileshi GW, Pandey R, Das AK, Nath AJ (2021) Assessing tree diversity and carbon storage during land use transitioning from shifting cultivation to indigenous agroforestry systems: implications for REDD+ initiatives. J Environ Manag 298:113470. https://doi.org/10.1016/j.jenvman.2021.113470

Reang D, Nath AJ, Sileshi GW, Hazarika A, Das AK (2022a) Post-fire restoration of land under shifting cultivation: a case study of pineapple agroforestry in the Sub-Himalayan region. J Environ Manag 305:114372. https://doi.org/10.1016/j.jenvman.2021.114372

Reang D, Sahoo UK, Giri K, Hazarika A, Nath AJ (2022b) Ethnic homestead forests of North-East India revealed as diverse land-use systems. Agrofor Syst 96(3):465–478. https://doi.org/10.1007/s10457-021-00700-z

Reang D, Hazarika A, Sileshi GW, Nath AJ, Paramesh V, Singha WR, Das AK (2024) Piper agroforestry in the Indian Himalayas: indigenous peoples' practices, policies and incentives. CABI Agric Biosci 5(1):9. https://doi.org/10.1186/s43170-024-00214-5

Rocky P, Sahoo UK, Thapa HS (2004) Livelihood generation through tree bean (*Parkia roxburghii* g. Don) in Imphal. J Non-Timber For Prod 11(2):135–139

Saikia SP, Jain V (2007) Biological nitrogen fixation with non-legumes: an achievable target or a dogma? Curr Sci 92:317–322

Saikia P, Khan ML (2014) Homegardens of upper Assam, Northeast India: a typical example of on farm conservation of Agarwood (*Aquilaria malaccensis* Lam.). Int J Biodivers Sci Ecosyst Serv Manag 10(4):262–269. https://doi.org/10.1080/21513732.2014.973449

Singh MK (2019) Potential of underutilized legume tree *Parkia timoriana* (DC.) Merr. In Eco-restoration of Jhum fallows of Manipur. J Pharmacogn Phytochem 8(2):1685–1687

Singh AK, Arunachalam A, Ngachan SV, Mohapatra KP, Dagar JC (2014) From shifting cultivation to integrating farming: experience of agroforestry development in the northeastern Himalayan region. In: Dagar J, Singh A, Arunachalam A (eds) Agroforestry systems in India: livelihood security & ecosystem services, Advances in agroforestry, vol 10. Springer, New Delhi. https://doi.org/10.1007/978-81-322-1662-9_3

Tarafdar PK, Ray R (2005) Effect of trees on improvement of physical environment and fertility in soils of West Bengal. Bull Natl Inst Ecol 16:129–136

Williams PA, Larbi RT, Yeboah I, Frempong GK (2018) Smallholder farmers experiences of climate variability and change on pineapple production in Ghana: examining adaptation strategies for improved production. J Agric Ext Rural Dev 10(2):35–43. https://doi.org/10.5897/JAERD2017.0919

Remediation of Saline Soil Through Halophytes

6

M. L. Dotaniya, C. K. Dotaniya, Mahesh Kumar,
D. R. Chaudhary, Kuldeep Kumar, R. K. Doutaniya,
H. M. Meena, R. C. Sanwal, Rajhans Verma,
Laxman Kumawat, Harpreet Singh, and S. Rajendiran

Abstract

For sustainable agricultural production and soil health, plant nutrients are vital for crop growth, and their application needs to be carefully balanced. The availability of nutrients in the soil is shaped by both management practices and the soil's physico-chemical properties. High concentrations of salt ions, such as calcium (Ca), sodium (Na), and chloride (Cl), along with their associated carbonate

M. L. Dotaniya (✉)
ICAR-Directorate of Rapeseed-Mustard Research, Bharatpur, India

C. K. Dotaniya · R. C. Sanwal
Department of Soil Science & Agricultural Chemistry, SKRAU, Bikaner, India

M. Kumar · H. M. Meena
ICAR-Central Arid Zone Research Institute, Jodhpur, India

D. R. Chaudhary
Central Salt and Marine Chemicals Research Institute, Bhavnagar, India

K. Kumar
ICAR- Indian Institute of Soil and Water Conservation, RC, Kota, Rajasthan, India

R. K. Doutaniya
Department of Agronomy, SKN College of Agriculture, Jobner, Rajasthan, India

R. Verma
Department of Soil Science & Agricultural Chemistry, SKN College of Agriculture, Jobner, Rajasthan, India

L. Kumawat
Junagadh Agricultural University, Junagadh, India

H. Singh
Regional Research Station, PAU, Gurdaspur, Punjab, India

S. Rajendiran
ICAR – Indian Institute of Horticultural Research, Bengaluru, India

V. C. Pandey (ed.), *Plantation-Based Land Restoration*,
https://doi.org/10.1007/978-981-95-2860-8_6

and bicarbonate salts, can limit the availability of essential plant nutrients. The issue of excessive salt in soil and water is greatly aggravated in arid and semi-arid regions and has impacted on 6.74 million hectares (Mha) of Indian land. Sodic areas predominate on around 3.77 Mha of the salt-affected Indian soils, causing an annual loss of farm production of 11.2 million tonnes (mt) estimated at US$ 2.3 billion. Salt ions hinder mineralization in saline/alkaline soils and reduce microbial diversity and population. According to studies, crop plants have a restricted ability to absorb nutrients under saline-sodic soils. A sustainable way to treat salt-affected soils is to use halophytes, which are plants that are naturally suited to saline conditions. They can absorb and sequester excess salts because of their innate salt tolerance, which lowers soil salinity and increases land production. In addition to reducing soil salinity, halophytes for saline soil restoration are an economical and ecologically sound method that also has other advantages. Development of genetic salt tolerant crop varieties is one of the potential ways for escaping the salt toxicity issues. However, some of the fern and shrubs are capable to survive in the saline soil that can be adopted for sustaining the crop production in the salt affected regions.

Keywords

Crop production · Halophytes · Plant nutrients · Soil health

6.1 Introduction

Indian population is increasing at the pace of 2.4% annually and needs 350 million tonne (mt) of food grain production by 2050 (Kumar and Sharma 2020). It is a challenging task for the planner and researcher to produce from limited natural resources. However, the food bowl of India is also showing the plant nutrient deficiency and observed the decline rate of partial factor productivity. It is a very challenging task for breeder to develop new high-yielding varieties and soil scientist for efficient management of plant nutrients and natural resources. It is well known figure that India has only 2.4% world geographical to feed almost 17.8% human and 20% animal mass on the earth (Down to Earth 2024). India's agricultural sector has substantially advanced since the country's independence. The production of food grains increased significantly, rising from 50 mt in 1950 to 275 mt in 2017, an increase of over 5.5 times. India is now self-sufficient thanks to this boom, and it is even a net exporter of food grains. In 2018–2019, the nation produced 285.17 million tonnes of food grains, an unprecedented amount of wheat (116.5 mt) and rice (103.6 mt), according to Tiwari (2020). Now the total food grain production recorded was 316 Mt by the year 2021–2022. Although the extension of cultivated areas contributed to the rise of food grain production between 1949 and 1965 (Vaidyanathan 1986), the key factor was adoption of high yielding varieties and supply of fertilizers inputs in the mid-1960s. A key factor was the mid-1960s adoption of an extensive variety of high-yielding inputs. According to Dantwala (1986), this approach encompasses

Table 6.1 Saline and sodic soil area (Mha) of India (Kumar and Sharma 2020)

State	Saline	Sodic	Total
Andhra Pradesh	0.78	1.97	2.75
Andaman and Nicobar	0.08	0.00	0.08
Bihar	0.47	1.06	1.53
Gujarat	1.68	0.54	2.22
Haryana	0.49	1.83	2.32
Karnataka	0.02	1.48	1.50
Kerala	0.20	0.00	0.20
Madhya Pradesh	0.00	1.40	1.40
Maharashtra	1.84	4.23	6.07
Orissa	1.47	0.00	1.47
Punjab	0.00	0.15	0.15
Rajasthan	0.20	0.18	0.38
Tamil Nadu	0.01	0.35	0.36
Uttar Pradesh	0.22	1.35	1.57
West Bengal	0.44	0.00	0.44
Total	2.96	3.77	6.73

the use of high-yielding varieties, assured irrigation, plant protection strategies, farm mechanization and financial assistance. It was also recorded that India is having 6.74 million hectare land that is suffering due to salinity and alkalinity (Arora et al. 2013; Dotaniya et al. 2018a). Whereas, improper input usage and management practice leads 10% additional area under the salinity every year and expected approximately half of the arable land of India under salinity stress by the 2050 (Kumar and Sharma 2020). It was estimated that 44% areas in salinity fall under 12 states and in Andaman and Nicobar, whereas, sodic soils having extension in 11 states by covering 44% area (Table 6.1). Most of the area under salinity is limiting the crop yield potential. However, implementing of advanced research outputs and technologies can improve the soil health and reduce the salt toxicity (Dotaniya et al. 2022a, 2023b; Meena et al. 2023b).

Based on the soil physico-chemical properties, saline soils are classified into:

6.1.1 Saline Soil

These soils contain elevated levels of salt ions such as chlorides, sulfates, carbonates, bicarbonates, Na, Ca, and Mg, impacting water uptake and nutrient absorption kinetics. The process of salt ion deposition in soil is referred to as salinization. The degree of salinization is contingent upon the type of salt, their relative proportions in the soil, and their solubility. The solubility of calcium (Ca) and magnesium (Mg) carbonates is low (Ca, 0.013 g/L; Mg, 2.5 g/L), but it increases with rising partial pressure of CO_2 and decreasing pH in the soil (Zhao et al. 2019). Conversely, the solubility of sodium chloride (NaCl, 264 g/L), sodium sulfate (Na_2SO_4, 161 g/L), and sodium carbonate (Na_2CO_3, 178 g/L) is high at 20 °C, contributing significantly to soil salinity (Bajwa 2002). Soil classification based on analytical parameters includes pH less than 8.5 (1:2 soil/water suspension) and saturation electrical conductivity (EC_e) exceeding 4 dS/m. Soils are classified based on salinity levels as non-saline (0–2 dS/m), slightly saline (2–4 dS/m), moderately saline (4–8 dS/m), strongly saline (8–16 dS/m), and extremely saline (>16 dS/m). These soils are primarily composed of chloride and sulfate ions of calcium, magnesium, and sodium. White encrustations on the soil surface are a common indicator of salinity. While these soils may appear physically well-structured, they hinder water and nutrient absorption by plant roots due to the elevated salt concentrations. Soil salinity negatively impacts plant growth, mineralization of organic matter, and nutrient ratios in the soil solution (Dotaniya et al. 2025a) and leads to a decline in soil microbial populations and diversity (Dotaniya et. al. 2025b, c).

6.1.2 Alkali Soils

Alkaline soils, or alkali soils, are characterized by a high pH (above 8.5), poor soil structure, and limited ability to absorb water. In some calcareous soils, sodic soil layer can be observed at 0.5 to 1 meter depth. The predominant presence of sodium carbonates ($CO_3^{2-} + HCO_3^{-}$) is a key factor contributing to the unfavorable physicochemical characteristics of alkali soils. These soils possess significant sodium saturation and alkalinity, adversely impacting plant growth and crop productivity. In India, alkali soils are predominantly found in the arid and semi-arid regions (Punjab, Haryana, Uttar Pradesh, Bihar, and western Rajasthan). The formation of alkali soils occurs through an alkalinization process. These soils are classified as alkaline when they have a pH above 8.5, electrical conductivity (EC) below 4 dS/m, residual sodium carbonate (RSC) over 2.5 meq/L, and an exchangeable sodium percentage (ESP) greater than 15. Additionally, the sodium adsorption ratio (SAR) is a crucial parameter and should ideally be less than 10.

The sodium adsorption ratio (SAR) is a measure of the proportion of sodium (Na) relative to calcium (Ca) and magnesium (Mg) in soil. It is used to assess the potential for sodium-induced soil dispersion:

$$\begin{aligned} \mathrm{SAR} &= \left[\mathrm{Na}+\right] / \surd\left[\mathrm{Ca}^{2+}/2 + \mathrm{Mg}^{2+}/2\right] \\ &= \left\{\mathrm{Na}^{+}/23\right\} / \surd\left\{\mathrm{Ca}^{2+}/40 + \mathrm{Mg}^{2+}/24\right\} \end{aligned}$$

In this formula, [ ] represents concentration in milliequivalents per liter (meq/L), and { } represents concentration in milligrams per liter (mg/L). The SAR is crucial parameter and should ideally be less than 10 for good quality water. Over time, as the soil is exposed to water with a specific SAR value, the exchangeable sodium percentage (ESP) tends to align with the SAR value.

The residual sodium carbonate (RSC, meq/L): The formula for calculating the residual sodium carbonate as per Rawat et al. (2018) is as follows:

$$RSC = \left[HCO_3^- + CO_2^{-3}\right] - \left[Ca^{2+} + Mg^{2+}\right]$$
$$= \left\{HCO_3^- / 61 + CO_3^{-2} / 30\right\} - \left\{Ca^{2+} / 20 + Mg^{2+} / 12\right\}$$

which must not be much higher than 1 and preferably less than 0.5.

With the above fact, salt inhibits plant growth through two proposed theories: either by causing ion toxicity or by disrupting osmotic processes (Hao et al. 2021). Disruptions to osmotic processes frequently result in abnormalities in cell hydration and water transport, which affect how cells function. For instance, high salt concentrations in the soil can interfere with plant cell osmotic activities when there is a salinity stress. Because of the osmotic imbalance caused by the excess salts, water moves out of plant cells, resulting in dehydration and adversely affecting a number of physiological processes. Salinity hampers various physiological functions in plants, including the ability to absorb water (Azeem et al. 2023; Dotaniya et al. 2024c), reduced nutrient absorption (Liu et al. 2023), diminished photosynthetic rate (Piao et al. 2023), and decreased yield productivity (Ghonaim et al. 2021). The accumulation of sodium ions (Na^+) due to salinity induces osmotic stress and ionic toxicity, which triggers the formation of reactive oxygen species (ROS) such as hydroxyl radicals, singlet oxygen, hydrogen peroxide, and superoxide (Mangal et al. 2023). These ROS lead to oxidative stress, resulting in damage to mitochondria, chloroplasts, cell functions, and biomolecules like lipids, proteins, and nucleic acids (Gill and Tuteja 2010). In plant cells, this cascade results in lipid peroxidation, protein denaturation, and DNA mutations (Ghamarnia et al. 2023; Mohamed et al. 2021; Ghonaim et al. 2023). To combat ROS, plants produce antioxidants (Sofy et al. 2021). These antioxidants include molecules like hydrogen peroxide (H_2O_2), singlet oxygen, superoxide anion (O_2^-), and hydroxyl radicals (OH^-). When the production of ROS exceeds the plant's ability to neutralize them, oxidative stress occurs, leading to the damage of DNA, lipids, and proteins. Additionally, soluble sugars, proline, and phenolic compounds are believed to play a crucial role in enhancing a plant's tolerance to salt stress, as these substances accumulate in response to abiotic stress (Naeem et al. 2020; Naseriyeh et al. 2023). Osmolytes serve to maintain the functional equilibrium of cells, safeguarding plant cells by functioning as antioxidants, enhancing cellular redox potential, stabilizing membranes and macromolecules, and supplying energy during stress recovery (Naseriyeh et al. 2023). Additionally, plants experiencing salt stress generate proteins, potentially utilized for energy storage or the regulation of osmotic potential (Ghonaim et al. 2021).

6.2 Why Need Saline Soil Remediation

India is an extremly populated country with vast variability in climate, relief, terrain, rainfall, soil and their fertility status etc. These diversities mostly affect the crop productivity and type of food availability. Protecting agricultural productivity, ensuring food security, advancing environmental sustainability, and sustaining the livelihoods of agriculturally dependent populations are the main reasons for the necessity of saline soil rehabilitation. To mitigate the effects of soil salinity and restore soil health, a range of remediation strategies and sustainable soil management practices are employed (Saha et al. 2023). Significant portion of saline soil in India needs holistic management with respect to crop and climate. In this regard, scientific technological advancement enhanced the crop productivity and profitability over a period. The negative impacts of elevated salt concentrations on soil quality and agricultural productivity give rise to the necessity of saline soil remediation. Some of the major factors that are needed to remediate the saline soils are as follows:

6.2.1 Effect on Water and Nutrient Uptake

Salt in soils can hinder plant growth by disrupting water uptake, nutrient absorption, and overall plant development. An osmotic imbalance caused by high salt concentrations makes it challenging for plants to get water from the soil (Zhao et al. 2010). Elevated salt concentrations, especially those containing Na, might hinder plants' ability to absorb vital nutrients. Plant health may be further impacted by nutritional deficits resulting from this upset in the balance of nutrients. In order to achieve long-term agricultural sustainability and guarantee that the land remains productive for future generations, it is imperative that salinized soil be addressed.

6.2.2 Effect on Crop Yield

Given the growing global population, ensuring food security is crucial. To meet the rising demand for food, agricultural productivity must be sustained or increased, which can be partially achieved through the remediation of saline soils. Because of altered precipitation patterns and increased temperatures brought on by climate change, problems with soil salinity may worsen. Agricultural systems can enhance their resilience to climate change by addressing and improving saline soils. Excessive salt accumulation in the soil can lead to reduced crop yields and quality (Zhang et al. 2010; Dotaniya et al. 2024b). Stunted growth, decreased output, and, in extreme situations, crop loss are all possible in crop plants. High salinity adversely affects reproductive development by hindering microsporogenesis and stamen filament elongation. It also triggers programmed cell death in specific tissues, causes ovule abortion, and accelerates senescence in fertilized embryos (Ashraf 2004). Farmers' and agriculturally dependent populations' livelihoods are adversely impacted by salinized soil. However, salinity-related crop losses can lead to food insecurity, financial hardship, and a greater reliance on outside food sources.

6.2.3 Effect on Soil Health

Saline soils have the potential to change the soil's physical characteristics, leading to compaction and the development of impermeable layers. It reduces the soil microbial count, activity as a diversity (Dotaniya et al. 2018b). This hinders drainage and water infiltration, which is bad for root growth. Salinity problems can affect nearby ecosystems in addition to agricultural regions. In addition to damaging aquatic life and upsetting ecosystems, saline runoff into water bodies can also spread soil salinity. Inadequate irrigation techniques and other forms of water management are frequently associated with saline soil. Remedial actions can encourage sustainable irrigation methods and increased water efficiency.

6.3 Halophytes: Concept and Definition

Halophytes comprised two words—"halo," which means salt, and "phyte," which represents plant species. In general, halophytes are a group of plants grow in or tolerate to salted soil without affecting the plant physiology and yield. However, some of the literature indicated that halophytes are well grown in saline environment with or without abiotic stress. The most popular definition is described by Aronson (1989) as "these group of species naturally grow only in saline conditions." A major problem in agriculture is salinity stress, which has an impact on crop productivity and soil quality. Later on, Dalton et al. (1997) again redefines the halophytes based on the salinity stress index (SSI). Farmers use the SSI as a vital tool to manage this stress. This indicator helps farmers to make decisions about sustainable agricultural methods by giving them information about the degree of salinity stress on crops. One major environmental stressor affecting plant growth and development is salt stress. It increases intracellular osmotic pressure and can result in harmful Na accumulation (Zhao et al. 2021). Based on the salt concentration, these plants could tolerate to the extent of 200-500 mM salt solution. It is clear that these plant could tolerant at lower concentration of salts but at higher levels of salts, will damage their metabolic processes. Bucur et al. (1957) further categorized the halophytes into two different groups based on their adaption to an environment condition and growth and development:

6.3.1 Euhalophytes

For their growth, saline environment is exclusively mandatory. Growth and development of these plants are preferably requires sufficient amount of salt in ecosystems. It is having radicular system.

6.3.2 Neohalophytes

These plants are having potential to adapt to salinity or halophyte ecosystems. However, they may survive in saline and non-saline media. Whole plant or a specific part of the plant may have characteristics to adopt to the saline environment.

According to Nikalje et al. (2019), halophytes, or salt-tolerant plants, possess distinctive characteristics that enable them to thrive in harsh environments. These plants thrive in saline environments because they can either absorb, store, or expel excess salt (Saddhe and Kumar 2019). According to their salt needs for their growth, halophytes are divided into two categories—facultative and obligate. While obligate halophytes require a specific amount of salt for optimal growth, facultative halophytes can grow normally without it (Nikalje et al. 2019). However, Farzi et al. (2017) conducted a phytodesalination experiment using three halophytic species from the Chenopodiaceae family—*Salicornia europaea*, *Salsola crassa*, and *Bienertia cycloptera*. These salt-tolerant plants were cultivated in a constructed wetland, where an elemental analysis of the soil focused on Ca, Mg, Na, and chlorine (Cl), along with the assessment of soil electrical conductivity before and after the desalination process, indicated the successful growth of these plants across various salinity levels (Dotaniya et al. 2019). This suggests their potential as effective plants for desalination purposes. Desalination is the process of purifying brackish or saltwater to make it fit for industrial, agricultural, or human consumption by eliminating salt and other contaminants. Desalination is mostly used to create freshwater from saline but copious water sources, thereby resolving the problem of water scarcity in areas with restricted access to conventional freshwater sources. In agriculture, the utilization of saline water poses a significant challenge as it can lead to soil damage, adversely affecting crop quality and yield (Dotaniya et al. 2023a). The imperative arises to identify and cultivate resilient crops capable of withstanding the stress imposed by salinity (Hussain et al. 2022). Environmental conditions, marked by increasing water deficit and temperature stress during grain filling, significantly affect grain yield and quality, particularly influencing protein content (De Santis et al. 2021). Addressing these challenges is crucial for sustaining agricultural productivity and ensuring food security in the face of changing environmental conditions. In a related study, Suaire et al. (2016) assessed the salt tolerance species (*Atriplex halimus* and *Atriplex hortensis*) by evaluating parameters such as germination rate, survival rate, growth rate, and NaCl accumulation. The results led the authors to recommend *Atriplex* species as promising candidates for phytodesalination, especially in mitigating road runoff pollution caused by de-icing salts (Suaire et al. 2016). Many researchers identified the halophytes and their role in desalinization process (Table 6.2).

Table 6.2 List of halophyte plants

S. no.	Plant species	Identified by
1	*Atriplex nummularia*	de Souza et al. (2014)
2	*Sulla carnosa*	Jlassi et al. (2013)
3	*Tecticornia indica, Suaeda fruticosa, Sesuvium portulacastrum*	Zorrig et al. (2012)
4	*Artemisia argyi, Limonium bicolor, Melilotus suaveolens, Salsola collina*	Sakai et al. (2012)
5	*Atriplex prostrata, Distichlis spicata, Juncus torreyi Coville, Phragmites australis, Spartina alterniflora Loisel, Schoenoplectus tabernaemontani, Typha angustifolia, Typha latifolia*	Rozema et al. (2014)
6	*Halogeton glomeratus*	Li et al. (2019)
7	*Echinochloa stagnina*	Ado et al. (2016)
8	*Salsola soda* and *Portulaca oleracea*	Karakas et al. (2017)
9	*Arthrocnemum macrostachyum*	Barcia-Piedras et al. (2019)
10	*Sesuvium verrucosum, Bacopa monnieri*	Lastiri-Hernández et al. (2019)
11	*Puccinellia nuttalliana, Typha latifolia*	Xu et al. (2019)
12	*Atriplex nummularia*	Rocha de Moura et al. (2019)
13	*Cakile maritima*	Arbelet-Bonnin et al. (2019)
14	*Sesuvium portulacastrum, Tecticornia indica, Suaeda fruticosa*	Rabhi et al. (2010)
15	*Salicornia europaea, Salsola crassa, Bienertia cycloptera*	Farzi et al. (2017)
16	*Atriplex halimus, Atriplex hortensis*	Suaire et al. (2016)
17	*Salsola soda, Portulaca oleracea*	Karakas et al. (2016)
18	*Sesuvium portulacastrum*	Muchate et al. (2016)
19	*Schoenoplectus tabernaemontani, Andropogon gerardii, Typha angustifolia, Elymus canadensis, Panicum virgatum, Spartina pectinata, Distichlis spicata*	Pouladi et al. (2016)
20	*Juncus torreyi Coville, Schoenoplectus tabernaemontani, Typha latifolia*	Rozema et al. (2016)
21	*Phragmites australis, Puccinellia nuttalliana* (salt accumulators), *Spartina pectinata*	McSorley et al. (2016)
22	*Chenopodium quinoa*	Jaikishun et al. (2019)
23	*Salicornia europaea*	Muscolo et al. (2014)
24	*Suaeda salsa*	Song and Wang (2015)

Modified from Saddhe et al. (2020)

6.4 Role of Halophytes in Agricultural Production System

In agricultural production systems, halophytes—plants that can withstand high salinity levels in their growing environment may play a significant roles especially in the regions where soil salinity is a problem. Halophytes have the following important functions and advantages in agriculture:

6.4.1 Soil Salinity Management

Soil salinity management is a great challenge, in which reduction of soluble salt below the harmful level or mobilization of excess salt to avoid the salt toxicity for a period of plant growth, to sustain the plant productivity. Many mechanisms are involved, some of which are listed below:

6.4.1.1 Phytoremediation Mechanism

It is a technique that uses plants to absorb and remove salts from the soil, helping to restore soil health and improve its quality. Halophytes are useful in this regard. In the long term, this can aid in lowering soil salinity levels.

6.4.1.2 Salt Exclusion

Certain halophytes possess systems that prevent salt from entering their shoots, thereby averting the buildup of salts in delicate plant tissues. This characteristic may help to keep the soil's salt balance more in the advantageous range.

6.4.2 Biodiversity and Ecosystem Restoration

Halophytes, especially in saline or coastal locations, can be useful in recovering degraded ecosystems. By establishing habitats for other creatures acclimated to these circumstances, they promote biodiversity. Often, halophytes have characteristics that increase their resistance to environmental stressors like drought and extreme heat. Including halophytes in agricultural systems can help them become more resilient to climate change. Salt-tolerant plants known as halophytes flourish in environments with extremely saline soil or water ($\geq$200 *m*M NaCl equivalent) (Mann et al. 2023). Halophytes provide a variety of methods for improving biodiversity, repairing salty and damaged habitats, and producing long-term environmental advantages. Because of their special adaptations, they can not only survive in highly salinized environments but also aid in the restoration of damaged ecosystems.

6.4.3 Alternative Crops for Saline Environments

A few halophytes are being investigated as possible crops for saline farming. Even though halophytes have a lot of potential applications in agriculture, it's vital to remember that not all of them are appropriate for large-scale production. More study is required to determine which halophytes are suited for particular settings and uses in order to generate crops with desirable qualities. A thorough grasp of halophytes' biology, agronomy, and possible benefits to sustainable agriculture is necessary for integrating them into agricultural systems. In regions where traditional crops might struggle, these plants can be grown and offer an alternate source of revenue and food production. This is especially true in areas where the soil is highly salted. In areas prone to salinity, several halophytes can serve as cattle fodder crops (Dotaniya et al. 2022b). These plants support sustainable livestock production in difficult settings by giving animals a wholesome feeding source.

6.4.4 Bioenergy Production

Certain halophytes have the capability to be utilized as biomass in the production of bioenergy. They can be grown on marginal soils that are unsuitable for traditional crops because of their capacity to thrive in saline settings. High-salinity marginal lands that would be unprofitable for traditional agriculture can now be used productively thanks to halophytes. This can increase the amount of land that can be farmed and support sustainable land use. The biomass production capacities of few species of halophytes are listed in Table 6.3. A variety of factors, including cultivation techniques, environmental circumstances, and genetic variability within species, might affect the actual production of biomass.

6.4.5 Soil Organic Carbon Improvement

Halophyte biomass can be used for a variety of things, such as soil enhancement, food, fodder, and bioenergy production. Through their capacity to accumulate and decompose biomass, halophytes can add to the amount of organic matter; however, other crop residues are also beneficial for detoxification (Dotaniya et al. 2024a). Thus, the fertility and structure of the soil may be enhanced (Meena et al. 2023a). This comprises the entire bulk of the halophyte's leaves, stalks, flowers, and seeds. Different halophyte species have different growth forms and patterns, which affect how much biomass they produce above ground. However, certain halophytes, especially those with deep root systems, can greatly increase below-ground biomass and aid in enhancing soil structure. Halophytes grow and break down into organic matter, which they then add to the soil. The biomass that the halophytes create and the rate at which it decomposes afterward determine how much organic matter needs to

Table 6.3 Biomass production potential of halophytes

Halophyte species	Biomass Aboveground (kg/m^2)	Biomass Below-ground (kg/m^2)	Utilization	References
Salicornia bigelovii	5.2	3.1	Biofuel production	Smith et al. (2020)
Atriplex halimus	4.7	2.5	Forage and soil improvement	Jones et al. (2020)
Suaeda maritima	3.5	2.0	Erosion control	Garcia et al. (2020)
Distichlis spicata	6.0	4.2	Livestock forage	Brown and White (2020)

be given to the soil. Halophytes add organic matter to the soil, which improves its structure, ability to hold onto water, and availability of nutrients. It supports general soil health and microbiological activity, especially on marginal or saline fields. The composting of halophyte biomass enriches the soil with organic matter. The quality of the plant material, microbial activity, and environmental factors can all affect how quickly plants decompose. Mangroves are a special kind of carbon storage. An estimated 4.03 pentagram (pg) of carbon is stored globally in mangrove biomass, with 70% of that storage occurring on coastal margins up to 10° latitude (Mitra et al. 2012). Additionally, they found that the rate of aboveground biomass varied significantly across the central and western Indian Sundarbans, which were 0.44–1.48 t/ha/month, respectively.

6.5 Mechanism of Halophyte Adoption in Saline Environment

6.5.1 Salt Exclusion and Ion Regulation

Ion transporters, including the salt overly sensitive (SOS) route, are used by halophytes to control the concentration of ions inside their cells. This process aids in removing too many Na ions, a typical salt stress.

6.5.2 Osmotic Adjustment

To maintain osmotic balance, halophytes accumulate compatible solutes such as proline and glycine betaine. These solutes aid in preserving turgor pressure and preventing water loss from cells.

6.5.3 Water Use Efficiency

Certain halophytes display a decreased transpiration rate in an effort to preserve water. This can be accomplished by controlling stomatal conductance and making structural changes to the leaves, such as making them smaller or more succulent (Ma et al. 2012).

6.5.4 Root System Adaptations

Specialized salt glands on the leaves of some halophytes, such as mangroves, actively exude excess salt. This lessens the buildup of salt in important tissues. To reach water sources in deeper soil layers and avoid the high salt concentrations close to the surface, many halophytes grow deep root systems.

6.5.5 Morphological Changes

Specific leaf adaptations, like smaller or altered leaf forms, are seen in certain halophytes. By minimizing the surface area exposed to direct sunlight, these modifications lessen the amount of water lost through transpiration. To stay hydrated and store water in the face of elevated salt concentrations, halophytes may form succulent tissues (Flowers and Colmer 2008).

6.5.6 Antioxidant Defense System

Halophytes frequently possess a strong antioxidant defense mechanism to stave off oxidative stress brought on by salt. Cells are shielded from harm by non-enzymatic antioxidants like glutathione and enzymes like superoxide dismutase.

6.5.7 Genetic Adaptations

Salt tolerance is conferred by unique genes found in halophytes. The development of crops with enhanced salt tolerance may rely on our ability to understand and manipulate these genes. Halophyte adaptability is influenced by epigenetic alterations that alter gene expression. Under salt stress, histone changes and DNA methylation play a role in regulating genes.

However, the above adaptive characteristics of halophytes are mediating the adaption in saline environment. Saddhe et al. (2020) mentioned the three major mechanisms in halophyte plant for phytodesalinization process:

(1) Salt exclusion: To prevent Na^+ uptake, several plants contain ultrafiltration mechanisms.
(2) Na^+ accumulation: Via the action of the vacuolar Na^+/H^+ antiporter, the majority of halophytes are able to collect Na^+ and Cl^- from the soil and accumulate in the vacuole. To drive Na^+ accumulation against the concentration gradient, the vacuolar V-PPase and V-ATPase maintain the proton gradient.
(3) Salt excretion: It expels extra salt from the cell as salt crystals; certain plants contain salt glands (Fig. 6.1).

6.6 Salinity Tolerance Indices

Salinity tolerance indices are essential instruments in the agricultural field that help farmers evaluate and improve their crops' capacity to flourish in salinized conditions. These indicators offer insightful information about the adaptive potential of crops, facilitating well-informed decision-making and the application of practical mitigation techniques for the effects of salinity stress. For farmers dealing with the difficulties of salinity stress in agriculture, the salinity stress index is a vital resource.

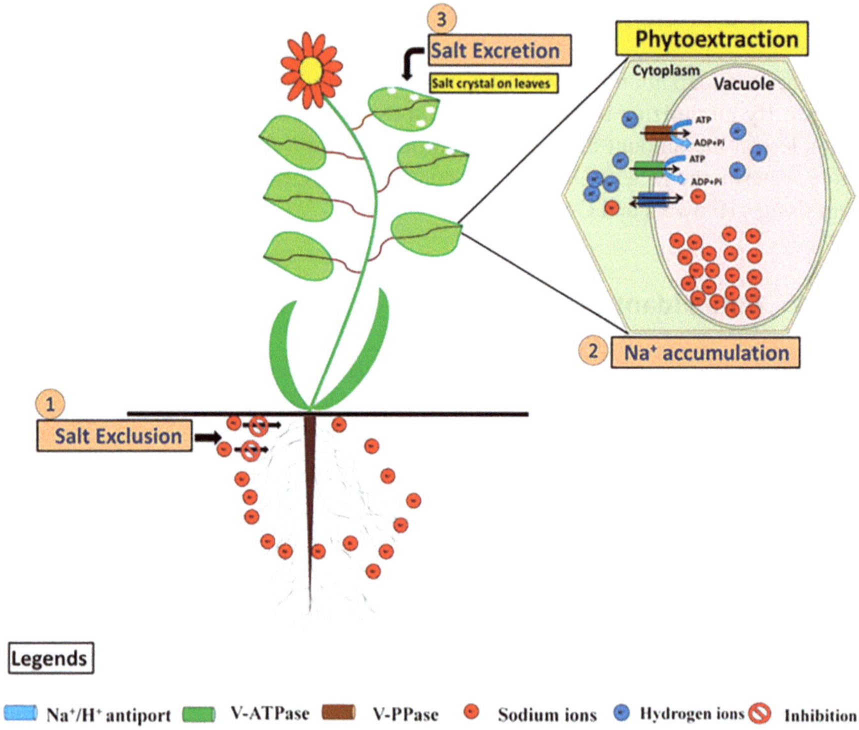

Fig. 6.1 Mechanism of halophyte adoption in saline environment (Saddhe et al. 2020)

Through proactive management of soil salinity, educated crop selection, and implementation of mitigation techniques, its application enables farmers to contribute to resilient and sustainable agricultural practices.

Different equations were used to calculate the salinity tolerance indices for assessing the salt potential in soil and their effect on crop growth:

Stress susceptibility index (SSI)	$\frac{1-\left(\frac{Y_s}{Y_p}\right)}{1-\left(\frac{\bar{Y}_s}{\bar{Y}_p}\right)}$ (Fischer and Maurer 1978)
Stress tolerance index (STI)	$\frac{Y_p \cdot Y_s}{(\bar{Y}_p)^2}$ (Fernandez 1992)
Mean productivity (MP)	$\frac{Y_p + Y_s}{2}$ (Hossain et al. 1990)

Tolerance TOL	$Y_P - Y_s$ (Hossain et al. 1990)
Yield Index (YI)	$\frac{Y_s}{\bar{Y}_s}$ (Gavuzzi et al. 1997)
Yield Stability Index (YSI)	$\frac{Y_s}{Y_p}$ (Bouslama and Schapaugh 2021)

where the Y_S and Y_P represent the average yields of genotype under both stress and non-stress conditions, respectively. $\bar{Y}_s$ and $\bar{Y}_p$ denote the average yields of all genotypes under stress and non-stress conditions, respectively.

6.6.1 Applications of Salinity Tolerance Indices

(i) Salinity tolerance indices are numerical measures that quantify the degree of a crop's tolerance to saline conditions. These indices help farmers identify and select crops or varieties that can withstand elevated soil salinity, ensuring sustainable agricultural practices.
(ii) Ion exclusion efficiency, which gauges a plant's capacity to absorb necessary ions while rejecting detrimental ones, is a crucial salinity tolerance parameter. Higher ion exclusion efficiency crops are more suited for saline soil growth.
(iii) Another important metric for assessing a plant's capacity to control water absorption and sustain turgid pressure in high-salinity environments is osmotic adjustment. Crops that possess efficient osmotic adjustment mechanisms have greater resistance to salinity-induced water stress.
(iv) The varying ion buildup in plant tissues is also taken into account by salinity tolerance indices. Certain crops effectively limit the uptake of ions in delicate areas, like shoots, while storing ions in less susceptible areas, like roots, thereby enhancing their overall ability to withstand salinity.
(v) This score focuses on how well a crop can thrive in the presence of salinity stress. Higher growth reduction tolerant crops can withstand higher salt levels and still provide bountiful yields.
(vi) Certain crops could react to salinity stress in different ways. Farmers are able to make precise judgments based on the unique characteristics of the crops they cultivate thanks to tailored salt tolerance indices for certain crops or kinds.

These indices are an essential resource for farmers looking to maximize crop yields in salinized conditions. Farmers in areas impacted by salinity stress should choose crops wisely, implement appropriate management techniques, and strive toward sustainable agriculture by comprehending and utilizing these indices.

6.7 Factor Affecting the Phytoremediation Process Under Saline Soils

The technique of using plants to eliminate, break down, or contain pollutants in soil and water is known as phytoremediation. Although phytoremediation is a promising method for environmental remediation, there are a number of variables that can affect how effective it is in salty soils (Dotaniya et al. 2022c). Optimizing phytoremediation processes in saline soils is essential to increasing the approach's effectiveness in addressing environmental contamination. This requires a thorough understanding of and careful management of these elements (Saha et al. 2017). Some of the major factors that affect the desalinization process by halophytes are listed here:

Salt tolerance of plants: Plants' efficacy in phytoremediation is strongly influenced by their capacity to flourish in salinity. Saline soils are more suitable for plants that can withstand salt.

1. Halophyte plant species: The ability of various plant species to store, withstand, and move pollutants around varies. It is essential to choose the right plant species for the particular pollutants present in saline soils.
2. Salinity levels: Plant development and metabolism are directly impacted by the soil's salt level. Because high salinity alters plant physiological processes, it can impact how effective phytoremediation procedures are.
3. Contaminant type and concentration: The effectiveness of phytoremediation and the choice of suitable plant species depend on the type and concentration of pollutants in the soil (Dotaniya et al. 2018c).
4. Soil properties: The availability and mobility of contaminants are affected by soil pH, texture, and nutrient content. Therefore, a comprehensive understanding of soil properties is essential for successful phytoremediation.
5. Climate conditions: Temperature, precipitation, sunlight, and other environmental factors can all have an impact on microbial activity and plant growth, which can then have an impact on the phytoremediation process.
6. Microbial interactions: Enhancing phytoremediation depends critically on the existence of particular microbial communities in the rhizosphere of plants. Contaminants can be broken down or transformed with the help of microbes.
7. Plant growth stage: Phytoremediation's efficacy varies according to the stage of plant growth. It's possible that some plants are more adept at storing pollutants at particular points in their life cycles.
8. Nutrient availability: Plant growth and metabolism depend on adequate food availability. Plant performance and general health might be impacted by nutrient imbalances or deficits during phytoremediation.
9. Root development: The ability of plants to reach and accumulate pollutants is influenced by the breadth and depth of their root growth. Robust root systems improve phytoremediation's effectiveness.

6.8 Conclusions

Due to their distinctive adaptations that enable them to flourish in salt settings, halophytes are important players in a variety of biological and environmental processes, particularly in areas with high salinity. Most of the developed and developing countries are having salinity problems that limit the crop yield potential. Application of halophytes mediated the salt concentration in soil and improved the soil health parameters like *Arthrocnemum macrostachyum*, *Salicornia* spp., and *Spartina* spp. Halophytes are naturally adapted to survive and reproduce in saline environments, demonstrating advanced mechanisms for salt tolerance. These include managing ion toxicity, water stress, and oxidative stress, osmotic adjustment, and maintaining ion homeostasis. Their specialized systems for coping with salt underscore their remarkable resilience in such conditions. These mechanisms encompass processes for the accumulation, exclusion, and addition of salt, allowing halophytes to maintain a balanced and functional internal environment in the face of challenging salinity levels. Different mechanisms and methods are available in remediation of saline soils, which are collectively used to effectively manage the salt toxicity areas.

References

Ado MN, Guero Y, Michot D, Soubeiga B, Kiesse TS, Walter C (2016) Phytodesalinization of irrigated saline Vertisols in the Niger Valley by *Echinochloa stagnina*. Agric Water Manag 177:229–240

Arbelet-Bonnin D, Ben-Hamed-Louati I, Laurenti P, Abdelly C, Ben-Hamed K, Bouteau F (2019) Cakile maritima, a promising model for halophyte studies and a putative cash crop for saline agriculture. Adv Agron 155:45–78

Aronson JA (1989) HALOPH a data base of salt tolerant plants of the world. Ofûce of Arid Land Studies, University of Arizona: Tucson

Arora S, Bhuva C, Solanki RB, Rao GC (2013) Halophytes for bio-saline agro-forestry and phytoremediation of coastal saline lands. J Soil Water Conserv 12(3):252–259

Ashraf M (2004) Some important physiological selection criteria for salt tolerance in plants. Flora 199:361–376

Azeem M, Pirjan K, Qasim M, Mahmood A, Javed T, Muhammad H, Yang S, Dong R, Ali B, Rahimi M (2023) Salinity stress improves antioxidant potential by modulating physio-biochemical responses in *Moringa oleifera* Lam. Sci Rep 13:2895

Bajwa MS (2002) Soil salinity and Alkalinity. In: Sekhon et al (eds) Fundamental of soil science. Indian Society of Soil Science, New Delhi, pp 291–308

Barcia-Piedras JM, Pérez-Romero JA, Mateos-Naranjo E, Camacho M, Redondo-Gómez S (2019) Effect of prior salt experience on desalination capacity of the halophyte *Arthrocnemum macrostachyum*. Desalination 463:50–54

Bouslama M, Schapaugh WT (2021) Stress tolerance in soybean. Part 1: evaluation of three screening techniques for heat and drought tolerance. Crop Sci 24:933–937

Brown TH, White PF (2020) Distichlis spicata: A halophytic forage grass for saline environments – Yield and nutritional evaluation. Range Ecol Manage 73(4):512–520. https://doi.org/10.1016/j.rama.2020.03.005

Bucur N, Dobrescu C, Gg T, Gh L, Teşu C, Dumbravă I, Afusoaie D (1957) Contribuţii la studiul halofiliei plantelor din păşuni şi fâneţe de sărătură din Depresiunea Jijia-Bahlui (partea a I-a). Stud şi Cerc (Biol şi Şt Agric). Acad RPR filiala Iaşi 8(2):277–317

Dalton FN, Maggio A, Piccinni G (1997) Effect of root temperature on plant response functions for tomato: comparison of static and dynamic salinity stress indices. Plant Soil 192:307–319

Dantwala ML (1986) Strategy of agricultural development since independence. In: Dantwala et al (eds) Indian agricultural development since Independence. Oxford and LBH Publishing Co. Pvt. Ltd, New Delhi, pp l–15

De Santis MA, Soccio M, Laus MN, Flagella Z (2021) Influence of drought and salt stress on durum wheat grain quality and composition: a review. Plants 10(12):2599

De Souza ER, dos Santos Freire MBG, de Melo DVM, de Assunção Montenegro AA (2014) Management of Atriplex nummularia Lindl. in a salt affected soil in a semi arid region of Brazil. Int J Phytoremed 16(1):73–85

Dotaniya ML, Meena VD, Rajendiran S, Coumar MV, Sahu A, Saha JK, Kundu S, Das H, Patra AK (2018a) Impact of long-term application of Patranala sewage on carbon sequestration and heavy metal accumulation in soils. J Indian Soc Soil Sci 66(3):310–317. https://doi.org/10.5958/0974-0228.2018.00038.5

Dotaniya ML, Aparna K, Dotaniya CK, Singh M, Regar KL (2018b) Role of soil enzymes in sustainable crop production. In: Khudus et al (eds) Enzymes in food biotechnology. Springer international, pp 569–589

Dotaniya ML, Meena VD, Saha JK, Rajendiran S, Patra AK, Dotaniya CK, Meena HM, Kumar K, Meena BP (2018c) Environmental impact measurements: tool and techniques. In: Martínez L, Kharissova O, Kharisov B (eds) Handbook of ecomaterials. Springer, Cham. https://doi.org/10.1007/978-3-319-48281-1_60-1

Dotaniya ML, Saha JK, Rajendiran S, Coumar MV, Meena VD, Das H, Kumar A, Patra AK (2019) Reducing chromium uptake through application of calcium and sodium in spinach. Environ Monit Assess 191:754. https://doi.org/10.1007/s10661-019-7948-4

Dotaniya ML, Meena VD, Dotaniya CK, Meena MD, Doutaniya RK, Verma R, Sanwal RC, Parewa HP, Jatav HS, Meena R, Sarkar A, Saha JK (2022a) Engineered biochar as adsorbent for removal of heavy metals from soil medium. In: Ramola S, Mohan D, Masek O, Méndez A, Tsubota T (eds) Engineered biochar. Springer, Singapore. https://doi.org/10.1007/978-981-19-2488-0_9

Dotaniya ML, Meena VD, Saha JK, Dotaniya CK, Mahmoud AED, Meena BL, Meena MD, Sanwal RC, Meena RS, Doutaniya RK, Solanki P, Lata M, Rai PK (2022b) Use of poor quality water for sustainable crop production in changing scenario of climate change. Environ Dev Sustain. https://doi.org/10.1007/s10668-022-02365-9

Dotaniya ML, Rajendiran S, Kirti S, Saha JK, Dotaniya CK, Patra AK (2022c) Immobilization of chromium bioavailability through application of organic waste to Indian mustard (*Brassica juncea*) under chromium contaminated Indian soils. Environ Monit Assess. https://doi.org/10.1007/s10661-022-10625-4

Dotaniya ML, Dotaniya CK, Yadav DK, Mahmoud AED, Doutaniya RK, Kumar D, Kumar A, Meena HM, Verma R, Kuldeep K, Meena A (2023a) Chromium dynamics in the soil-plant continuum. In: Kumar et al. (eds), Chromium in plants and environment. Springer Nature. https://doi.org/10.1007/978-3-031-44029-8

Dotaniya ML, Meena MD, Choudhary RL, Meena MK, Singh HV, Dotaniya CK, Meena LK, Dotaniya RK, Meena KN, Jat RS, Rai PK (2023b) Management of plant nutrient dynamics under alkaline soils through graded application of pressmud and gypsum. PLoS One 18(8):e0288784. https://doi.org/10.1371/journal.pone.0288784

Dotaniya ML, Nagar MC, Amita S, Dotaniya CK, Doutaniya RK (2024a) Pressmud subdue phytoremediation indices in lead-contaminated soils: a human health perspective. Bull Environ Contam Toxicol 112:16. https://doi.org/10.1007/s00128-023-03848-9

Dotaniya ML, Rajendiran S, Elanchezhian R, Dotaniya CK, Saha JK (2024b) Effect of chromium and cations concentration on stress enzymes in spinach crop grown under vertisol of central India. J Agric Food Res. https://doi.org/10.1016/j.jafr.2024.100969

Dotaniya ML, Meena MD, Choudhary RL, Meena MK, Meena VD, Singh HV, Lakaria BL, Jat RS, Rai PK, Kumar K, Doutaniya RK, Singh H (2024c) Dynamics of major plant nutrients and enzymatic activities in soil influenced by application of biochar and organic waste. PLoS One. https://doi.org/10.1371/journal.pone.0307487

Down to Earth (2024) The great divide. https://www.downtoearth.org.in/environment/the-great-divide-18236. Assessed 9 Nov 2024

Dotaniya ML, Meena MD, Choudhary RL, Dotaniya CK, Meena VD, Meena MK, Singh HV, Jat RS (2025a) Management of plant nutrients for enhancing oil content in oilseed crops - A review. J Oilseed Brassica 16(1):1–10

Dotaniya ML, Meena MD, Choudhary RL, Meena MK, Meena VD, Singh HV, Jat RS, Bhardwaj NR, Rai PK (2025b) Effect of organic sources on microbial population and enzymatic activities under salt degraded soil condition. Bull Environ Conta Toxic 1(4):11–17

Dotaniya ML, Meena MD, Choudhary RL, Meena MK, Meena VD, Singh HV, Jat RS, Doutaniya RK, Kumar K, Singh H, Rai PK (2025c) Determining application rates of FYM and pressmud to improve soil health properties in salt degraded soils. PLOS ONE 20(1):e0317463. https://doi.org/10.1371/journal.pone.0317463

Farzi A, Borghei SM, Vossoughi M (2017) The use of halophytic plants for salt phytoremediation in constructed wetlands. Int J Phytoremediation 19(7):643–650

Fernandez GCJ (1992) Effective selection criteria for assessing plant stress tolerance. Proc Symp Taiwan 25:257–270

Fischer RA, Maurer R (1978) Drought resistance in spring wheat cultivars. I. Grain yield responses. Aust J Agric Res 29:897–912

Flowers TJ, Colmer TD (2008) Salinity tolerance in halophytes. New Phytol 179:945–963

Garcia ML et al (2020) Reclamation of saline soil by planting annual euhalophyte Suaeda salsa with drip irrigation: A three-year field experiment in arid northwestern China. Ecol Engin 162:106178. https://doi.org/10.1016/j.ecoleng.2020.106178

Gavuzzi P, Rizza F, Palumbo M, Campaline RG, Ricciardi GL, Borghi B (1997) Evaluation of feld and laboratory predictors of drought and heat tolerance in winter cereals. Can J Plant Sci 77:523–531

Ghamarnia H, Basiri M, Ghobadi M, Palash M (2023) Performance of peppermint (*Mentha piperita* L.) in diferent water defcit and salinity management. Agrotech Ind Crops 3(2):84–95

Ghonaim MM, Mohamed HI, Omran AAA (2021) Evaluation of wheat (*Triticum aestivum* L.) salt stress tolerance using physiological parameters and retrotransposon-based markers. Genet Res Crop Evol 68:227–242

Ghonaim MM, Attya AM, Aly HG, Mohamed HI, Omran AAA (2023) Agro-morphological, biochemical, and molecular markers of barley genotypes grown under salinity stress conditions. BMC Plant Biol 23:526

Gill SS, Tuteja N (2010) Reactive oxygen species and antioxidant machinery in abiotic stress tolerance in crop plants. Plant Physiol Biochem 48:909–930

Hao S, Wang Y, Yan Y, Liu Y, Wang J, Chen S (2021) A review on plant responses to salt stress and their mechanisms of salt resistance. Horticulturae 7(6):132

Hossain ABS, Sears AG, Cox TS, Paulsen GM (1990) Desiccation tolerance and its relationship to assimilate partitioning in winter wheat. Crop Sci 30:622–627

Hussain MI, Al-Dakheel AJ, Chaudhry UK, Khan MI, Al Haithloul HA, Alghanem SM, Alaklabi A (2022) Morpho-physiological response of barley to assess genotypic differences of salinity tolerance under hyper arid climate. Agric Water Manag 272:107832

Jaikishun S, Li W, Yang Z, Song S (2019) Quinoa: in perspective of global challenges. Agronomy 9(4):176

Jones RT et al (2020) Atriplex halimus L.: Its biology and uses. J Arid Environ 182:104261. https://doi.org/10.1016/j.jaridenv.2020.104261

Jlassi A, Zorrig W, Khouni AE, Lakhdar A, Smaoui A, Abdelly C, Rabhi M (2013) Phytodesalination of a moderately-salt-affected soil by Sulla carnosa. Int J Phytoremediation 15(4):398–404

Karakas S, Cullu MA, Kaya C, Dikilitas M (2016) Halophytic companion plants improve growth and physiological parameters of tomato plants grown under salinity. Pak J Bot 48(1):21–28

Karakas S, Cullu MA, Dikilitas M (2017) Comparison of two halophyte species (*Salsola soda* and *Portulaca oleracea*) for salt removal potential under different soil salinity conditions. Turk J Agric For 41(3):183–190

Kumar P, Sharma PK (2020) Soil salinity and food security in India. Front Sustain Food Syst 4:533781. https://doi.org/10.3389/fsufs.2020.533781

Lastiri-Hernández MA, Alvarez-Bernal D, Bermúdez-Torres K, Cárdenas GC, Ceja-Torres LF (2019) Phytodesalination of a moderately saline soil combined with two inorganic amendments. Bragantia 78(4):579–586

Li B, Wang J, Yao L, Meng Y, Ma X, Si E, Ren P, Yang K, Shang X, Wang H (2019) Halophyte Halogeton glomeratus, a promising candidate for phytoremediation of heavy metalcontaminated saline soils. Plant Soil 442(1–2):323–331

Liu H, Todd JL, Luo H (2023) Turf grass salinity stress and tolerance–a review. Plants 12(4):925

Ma Q, Yue LJ, Zhang JL, Wu GQ, Bao AK, Wang SM (2012) Sodium chloride improves photosynthesis and water status in the succulent xerophyte *Zygophyllum xanthoxylum*. Tree Physiol 32:4–13

Mangal V, Lal MK, Tiwari RK, Altaf MA, Sood S, Kumar D, Bharadwaj V, Singh B, Singh RK, Aftab T (2023) Molecular insights into the role of reactive oxygen, nitrogen and sulphur species in conferring salinity stress tolerance in plants. J Plant Growth Regul 42(2):554–574

Mann A, Lata C, Kumar N, Kumar A, Kumar A, Sheoran P (2023) Halophytes as new model plant species for salt tolerance strategies. Front Plant Sci 14:1137211

McSorley KA, Rutter A, Cumming R, Zeeb BA (2016) Chloride accumulation vs chloride excretion: Phytoextraction potential of three halophytic grass species growing in a salinized landfill. Sci Total Environ 572:1132–1137

Meena MD, Dotaniya ML, Meena BL, Rai PK et al (2023a) Municipal solid waste: opportunities, challenges and management policies in India: a review. Waste Manag Bull. https://doi.org/10.1016/j.wmb.2023.04.001

Meena MK, Yadav BL, Jat RS, Meena MD, Choudhary RL, Dotaniya ML, Singh HV, Meena VD, Rai PK (2023b) Soil fertility, microbial dynamics and mustard yield potential mediated by addition of organic amendments under sodic irrigation water. Arch Agron Soil Sci. https://doi.org/10.1080/03650340.2023.21924892

Mitra A, Sengupta K, Banerjee K (2012) Spatial and temporal trends in biomass and carbonsequestration potential of Sonneratia apetala buch.-Hamin indian sundarbansroc. Natl Acad Sci India Sect B Biol Sci 82(2):317–323

Mohamed AH, Omar AA, Attya AM, Elashtokhy MMA, Zayed EM, Rizk RM (2021) Morphological and molecular characterization of some Egyptian sixrowed Barley (*Hordeum vulgare* L.). Plants 10(11):2527. https://doi.org/10.3390/plants10112527

Muchate NS, Nikalje GC, Rajurkar NS, Suprasanna P, Nikam TD (2016) Physiological responses of the halophyte Sesuvium portulacastrum to salt stress and their relevance for saline soil bioreclamation. Flora 224:96–105

Muscolo A, Panuccio MR, Piernik A (2014) Ecology, distribution and ecophysiology of *Salicornia europaea* l. In: Sabkha ecosystems. Springer, Dordrecht, pp 233–240

Naeem M, Basit A, Ahmad I, Mohamed HI, Wasila H (2020) Efect of salicylic acid and salinity stress on the performance of tomato. Gesunde Pfanzen 72:393–402

Naseriyeh T, Kahrizi D, Alvandi H, Aghaz F, Nowroozi G, Shamsi A, Hosseini O, Arkan E (2023) Glycyrrhizic acid delivery system Chitosan-coated liposome as an adhesive anti-infammation. Cell Mol Biol 69(4):1–6. https://doi.org/10.14715/cmb/2023.69.4.1

Nikalje GC, Bhaskar SD, Yadav K, Penna S (2019) Halophytes: prospective plants for future. In: Ecophysiology, abiotic stress responses and utilization of halophytes. Springer, Singapore, pp 221–234

Piao J, Che W, Li X, Li X, Zhang C, Wang Q, Jin F, Hua S (2023) Application of peanut shell biochar increases rice yield in saline-alkali paddy fields by regulating leaf ion concentrations and photosynthesis rate. Plant Soil 483(1):589–606

Pouladi FS, Anderson BC, Wootton B, Rozema L (2016) Evaluation of phytodesalination potential of vegetated bioreactors treating greenhouse effluent. Water 8(6):233

Rabhi M, Karray-Bouraoui N, Medini R, Attia H, Abdelly C, Smaoui A (2010) Seasonal variations in phytodesalination capacity of two perennial halophytes in their natural biotope. J Biol Res 14:181

Rawat KS, Singh SK, Gautam SK (2018) Assessment of groundwater quality for irrigation use: a peninsular case study. Appl Water Sci 8:233

Rocha de Moura ES, Rebouças Cosme C, de Sousa LT, da Silva DN, dos Santos FC, de Sousa Neto ON, de Sousa Junior FS, Costa Rebouças T (2019) Phytoextraction of salts by Atriplex Nummularia Lindl. irrigated with reject brine under varying water availability. Int. J. Phytoremediation 21(9):892–898

Rosielle AA, Hamblin J (1981) Theoretical aspects of selection for yield in stress and non-stress environment. Crop Sci 21:943–946

Rozema ER, Gordon RJ, Zheng Y (2014) Plant species for the removal of Na^+ and Cl^- from greenhouse nutrient solution. HortScience 49(8):1071–1075

Rozema ER, Gordon RJ, Zheng Y (2016) Harvesting plants in constructed wetlands to increase biomass production and Na^+ and Cl^- removal from recycled greenhouse nutrient solution. Water Air Soil Pollut 227(5):136

Saddhe AA, Kumar K (2019) Molecular cloning, expression analysis, and heterologous characterization of a novel sodium/hydrogen exchanger from a mangrove plant, Rhizophora apiculata. Plant Gene 19:100192

Saddhe AA, Manuka R, Nikalje GC, Penna S (2020) Halophytes as a potential resource for phytodesalination. In: Grigore MN (ed) Handbook of halophytes, pp 2–21. https://doi.org/10.1007/978-3-030-17854-3_92-1

Saha JK, Rajendiran S, Coumar MV, Dotaniya ML, Kundu S, Patra AK (2017) Soil pollution – an emerging threat to agriculture. Springer, Singapore, p 380. https://doi.org/10.1007/978-981-10-4274-4

Saha JK, Dotaniya ML, Coumar MV (2023) Risk assessment and remedition of contamination soil. In: Jha P et al (eds) Recent advances of soil research: IISS contribution. Indian Society of Soil Science, pp 64–79

Sakai Y, Ma Y, Xu C, Wu H, Zhu W, Yang J (2012) Phytodesalination of a salt-affected soil with four halophytes in China. J Arid Land Stud 22(1):17–20

Smith JA et al (2020) Increasing the value of Salicornia bigelovii green biomass grown in a desert environment through biorefining. Indu Crops Products 158:113022. https://doi.org/10.1016/j.indcrop.2020.113022

Sofy M, Mohamed H, Dawood M, Abu-Elsaoud A, Soliman M (2021) Integrated usage of Trichoderma harzianum and biochar to ameliorate salt stress on spinach plants. Arch Agron Soil Sci 68(14):2005–2026. https://doi.org/10.1080/03650340.2021.1949709

Song J, Wang B (2015) Using euhalophytes to understand salt tolerance and to develop saline agriculture: Suaeda salsa as a promising model. Ann Bot 115(3):541–553

Suaire R, Durickovic I, Framont-Terrasse L, Leblain JY, De Rouck AC, Simonnot MO (2016) Phytoextraction of Na^+ and Cl^- by *Atriplex halimus* L. and *Atriplex hortensis* L.: a promising solution for remediation of road runoff contaminated with de-icing salts. Ecol Eng 94:182–189

Tiwari R (2020) India clocks record foodgrain production in 2018–19. The Economic Times. Accessed 12 Feb 2020

Vaidyanathan A (1986) India's agricultural development in a regional perspective, R.C. Dutt Lectures on Political Economy, Centre for Studies in Social Sciences, Calcutta

Xu Q, Renault S, Yuan Q (2019) Phytodesalination of landfill leachate using Puccinellia nuttalliana and Typha latifolia. Int J Phytoremediation 21(9):831–839

Zhang SR, Song J, Wang H, Feng G (2010) Effect of salinity on seed germination, ion content and photosynthesis of cotyledons in halophytes or xerophyte growing in Central Asia. J Plant Ecol 3:259–267

Zhao KF, Song J, Fan H, Zhou S, Zhao M (2010) Growth response to ionic and osmotic stress of NaCl in salt-tolerant and salt-sensitive maize. J Integr Plant Biol 52:468–475

Zhao X, Conrads H, Zhao C, Ingwersen J, Stahr K, Wei X (2019) The effect of different temperature and pH levels on uptake of CO_2 in Solonchaks. Geoderma 348:60–67

Zhao S, Zhang Q, Liu M, Zhou H, Ma C, Wang P (2021) Regulation of plant responses to salt stress. Int J Mol Sci 22:4609. https://doi.org/10.3390/ijms22094609

Zorrig W, Rabhi M, Ferchichi S, Smaoui A, Abdelly C (2012) Phytodesalination: a solution for salt-affected soils in arid and semi-arid regions. J Arid Land Stud 22(1):299–302

Agroforestry for Landscape Restoration and Sustainability

7

Abhishek Raj, Manoj Kumar Jhariya, and Vimal Chandra Pandey

Abstract

Agroforestry (AF) is a brilliant land use that covers 1.6 billion ha (78% in the tropics and 22% in the temperate regions), which enhances biodiversity, land productiveness, and livelihood and maintains eco-restoration with global sustainability. Landscape restoration through AF is a genuine strategy that helps improve land quality and maximizes biodiversity along with the delivery of precious ecosystem services. Therefore, AFs (agroforestry systems) intensify environmental services, including soil, food, and climate security, to ensure net zero emissions globally. The AFs reconcile with landscape management and improve sustainability in the long term. Practicing AFs from the perspective of landscape management promotes sustainability in global production and ecosystem health. However, land degradation and desertification affect AF production, people's livelihood, and environmental health and ecosystem services. In this context, land degradation neutrality (LDN)-based policy must be framed, which can be possible through adopting AFs that makes farmer's benefits as livelihood generation at a global scale. A constructive roadmap is needed to encourage farmers to

A. Raj
Pt. Deendayal Upadhyay College of Horticulture and Forestry, Dr. Rajendra Prasad Central Agriculture University, Samastipur, Bihar, India

M. K. Jhariya (✉)
Department of Farm Forestry, Sant Gahira Guru Vishwavidyalaya, Ambikapur, Chhattisgarh, India

V. C. Pandey
CSIR – National Botanical Research Institute, Lucknow, Uttar Pradesh, India

Society for Green Development, Lucknow, Uttar Pradesh, India

V. C. Pandey (ed.), *Plantation-Based Land Restoration*,
https://doi.org/10.1007/978-981-95-2860-8_7

adopt AFs in any degraded wasteland to restore soil fertility and health of plant-soil system with a greater scale of sustainability.

Keywords

Agroforestry system · Land degradation · Landscape restoration · Ecosystem services · Sustainability

7.1 Introduction

Agroforestry (AF) integrates a woody perennial, crops, and livestock, which perform social, economic, and ecological services. It is a landscape unit that depends on sustainable land use management by combining uses of plant forestry and agricultural crops (Van Noordwijk 2019). This practice transforms agricultural land into a more productive agroecological land use system, which helps in achieving the Sustainable Development Goals (SDGs) (Waldron et al. 2017; Mosquera-Losada et al. 2018). Numerous studies are available on long-established AF practices, which are greatly connected with land use science and related policies that ensure social and ecological benefits in the long term (Van Noordwijk 2019). However, many past and recent studies are available on the multifarious benefits provided by AF at different landscape levels. These benefits and services are based on social, economic, cultural, and environmental scales (Rosenstock et al. 2019; Torralba et al. 2016; Jhariya et al. 2022a). Thus, AF ensures permanent vegetation cover by "perennialization" of agriculture, which protects the ecosystem and its sustainability. Paddy with MPTs (multipurpose trees) in Chhattisgarh state is a well-known example of an agroforestry system (AFs) (Viswanath et al. 2000; Jhariya et al. 2015; Devi et al. 2023). Similarly, West Africa-based parkland system and Germany-based Streuobst (traditional orchard meadows) are recognized AFs in the world (Fifanou et al. 2011; Plieninger et al. 2015). "Cinderella systems" is also a famous AFs and sustainable land use system, which is rediscovered recently (Nair et al. 2017).

Salinization, waterlogging, soil erosion, desertification, and acidification are a variety of land degradation due to unsustainable land use practices including deforestation, intensive farming systems, mining, and other anthropogenic system. This degraded land needs to be restored by practicing AFs, which becomes an effective land use system that contributes to food, nutritional, and climate security along with greater landscape restoration (Gupta et al. 2020; Raj et al. 2024). AF is vital for rehabilitating degraded ecosystems and enhancing soil health (Banerjee et al. 2023). This method has been effectively implemented in various countries, improving soil fertility over time. In the western Colombian Amazon, mature cacao AFs with pasture have significantly improved depleted soils (Suárez et al. 2021). Thus, AF is a powerful approach for restoring soil fertility and rehabilitating land. In this context, the present chapter discusses about global overview of land degradation, its impacts, and restoration through sustainable approaches as AFs. AFs and their different service functions are also addressed in this chapter. A sustainable landscape management (SLM)-based approach under AF technology is also briefly discussed. Various constraints, challenges, and research-based policy along with a roadmap for landscape restoration are included in this chapter.

7.2 Land Degradation in India and the World

Land degradation deprives the quality of land due to poor soil health and fertility, which affects biodiversity, water quality, food production, and the C sequestration potential of AFs (IUCN 2015; Laban et al. 2018). The land degradation percentage in the world is depicted in Fig. 7.1 (FAO 2010). Similarly, a global estimate (million ha) of land degradation is depicted in Fig. 7.2 (Gibbs and Salmon 2015; Dagar and Gupta 2016). Land degradation leads to losses of ES whose cost ranges from 6.30 to 10.60 trillion US dollars per year affecting world GDP of 10–17% (ELD 2015). The figure of land degradation is not only confined to India but also expanded throughout the world.

The extent and severity of land degradation varies due to climatic, topographic, and land use practices throughout the world (IUCN 2015; Raj et al. 2023). As per FAO and ITPS (2015), approx. 25% of global land areas were reported as degraded. Arable land, grazing land, and forests are subjected to 65%, 30%, and 20% of degradation in the region of Africa. Water erosion is another type of land degradation that prevailed in East and South Asia. Wind erosion disturbs soils by removing the top fertile organic layer and essential nutrients in the dry regions. Salt-affected soils are widely distributed in arid and semi-arid regions (Toderich et al. 2013). Land degradation affects approx. 29.3% of India's geographical areas due to

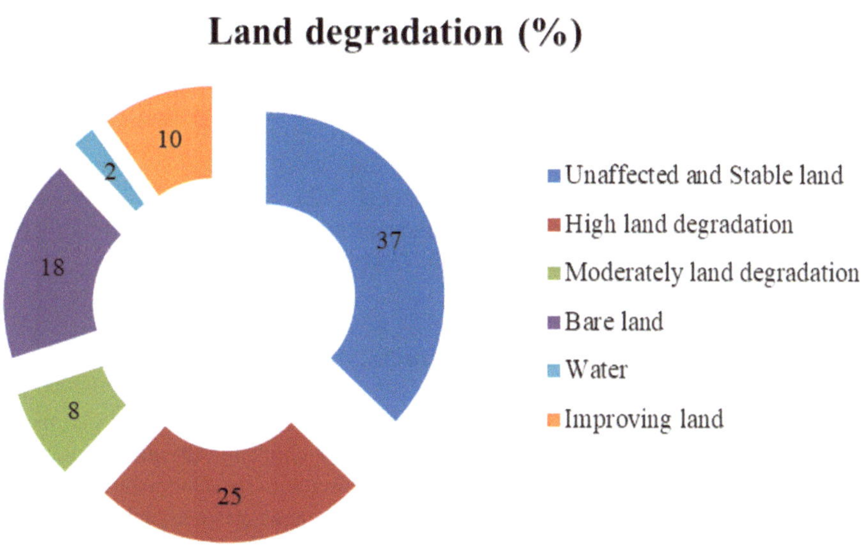

Fig. 7.1 Land degradation (%) in the world (FAO 2010)

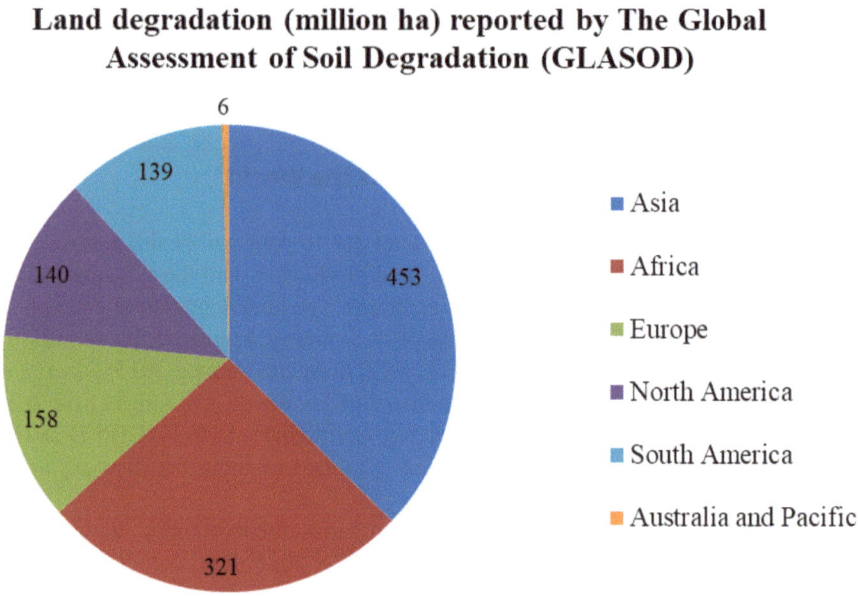

Fig. 7.2 Global estimates (million ha) of land degradation. (Gibbs and Salmon 2015; Dagar and Gupta 2016)

unsustainable land use practices, acidification, desertification, soil erosion, and mining (ISRO 2018). Landslides and water erosion disturb human life and public properties, which prevailed in the Caribbean and Latin America (FAO and ITPS 2015). Sodicity and salinity are also key threats to soil in arid and semi-arid regions, which affects one billion (approx. 7%) of global land areas (Wicke et al. 2011). India consists of approx. 18% of the global population over an area, which is just 2.42% of world spread (Bhattacharyya et al. 2015). Desertification, wind erosion, water erosion, salinity, sodicity, and other anthropogenic activities also disturb land quality enormously (AbdelRahman 2023). Rainfed farmland and forest land almost cover more than half of degraded land in India, which affects biodiversity, food security, and ecosystem services (Sreenivas et al. 2021). Energy-intensive cropping system and resource-intensive farming practices are popular in irrigated farms but observed as unsustainable in rainfed farms (Jagdish 2023).

7.3 Land Degradation, Its Impacts, and Restoration

Land degradation is a significant global environmental issue discussed nationally and internationally. Land degradation is a complex challenge driven by interconnected factors that affect ecosystem health and productivity. Understanding these factors is essential for effective mitigation. Soil characteristics such as texture, structure, and nutrient content significantly influence land degradation. Poor-quality soil can hinder plant growth and agricultural yield, while insufficient water availability can lead to drought and soil desiccation. Livestock management practices, particularly overgrazing, contribute to soil compaction, reducing water absorption and increasing erosion as vegetation is lost, leaving soil vulnerable. Human activities play a central role in land degradation. Urbanization often removes natural vegetation and disrupts ecosystems, while intensive agricultural practices can deplete soil nutrients and disturb its microbial balance. Vegetation cover is crucial in preventing land degradation, as healthy plants stabilize soil, reduce erosion, and support nutrient cycling and moisture retention. Conversely, inadequate vegetation increases erosion and degradation risks. Additionally, extreme climate conditions further exacerbate these challenges. Many factors affect various drivers for land degradation in the world. Technological, climatic, economic, cultural, policy, and institutional factors are affecting the land degradation process. Intensive agricultural practices, climate change and global warming, infrastructure development and extension, increased aridity, and deforestation activity are major drivers associated with current trends of various forms of land degradation in the world. In this context, a driver for land degradation and its associated factors is depicted in Fig. 7.3 (Gichuki et al. 2019; IUCN et al. 2015). A report indicates that the non-sustainable exploitation of 25% of Earth's land surface impacts the livelihoods of approximately 3.20 billion individuals (GEF 2022). Soil degradation reflects a reduction in the land's inherent capacity to produce naturally. This issue is predominantly caused by the overuse of fertilizers, pesticides, and practices that are not grounded in scientific agricultural principles (Alewell et al. 2020; Raj et al. 2025). Severe weather including drought, flood, and uncertain rainfall along with human activities accelerates land degradation and related consequences, which adversely affect soil quality, food, and livelihood production (El-Rawy et al. 2020; AbdelRahman et al. 2019, 2022a, b). It not only destroys human beings but also affects ecosystem goods and services. Agricultural intensification, overgrazing, deforestation, urbanization, and climate changes induce soil salinity, sodicity, desertification, and erosion promoting land degradation and related consequences (Janečková Molnárová et al. 2023). Land degradation represents damage and deprives of biological and economic efficiency of productive land in irrigated, rainfed, pasture, and forest land. It will not only affect land productiveness and soil fertility but also cause poor environmental services (UNCCD 2012). The status of land degradation, its causes, and its severity are mapped by organizations through several attempts, which are further necessary to take global action and initiatives (ADB 2008; Hossini et al. 2022). Various forms of land degradation exist and affect AF health and productivity. Desertification makes arid deserts that deprive soil fertility, AF production, human health, and ecosystem services. Desertification expanded enormously to several extents, which further affects water availability, global food production, and sustainability. Land degradation and desertification affect AF production, people's livelihood, and environmental health and ecosystem services. World food production could decline by 12% in the upcoming 25 years due to land degradation, which caused rising food prices by 30%. The annual cost of land degradation is also estimated at 3.4 trillion Euros, which caused a huge loss of economy along with ecosystem health (ELD Initiative 2013a, b). Land degradation threatens soil quality and productiveness, and

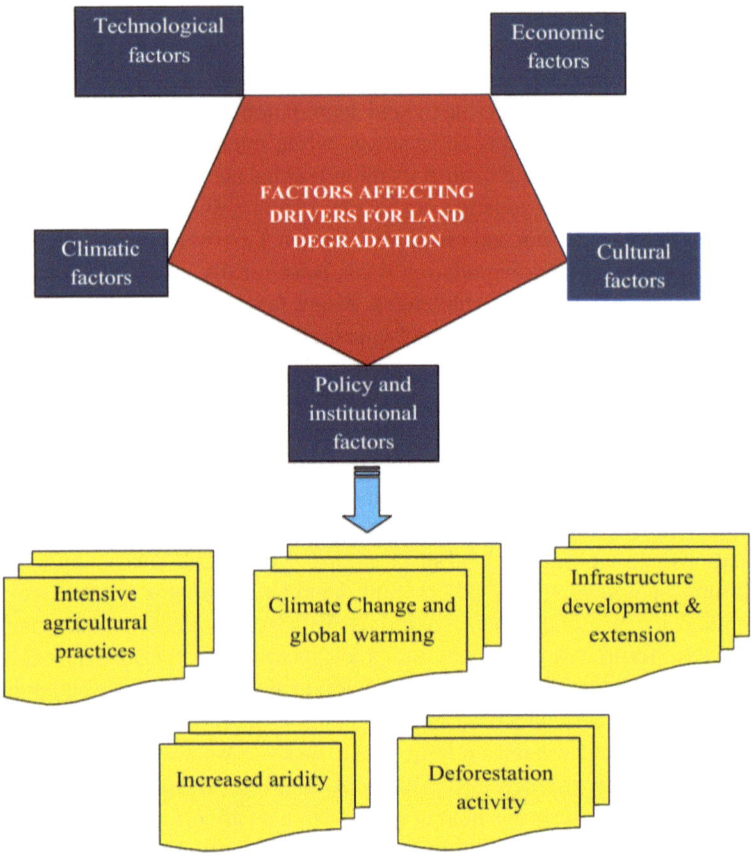

Fig. 7.3 Drivers for land degradation and its associated factors. (Gichuki et al. 2019; IUCN et al. 2015)

its restoration becomes an urgent need today. However, practicing organic farming might be useful for preventing soil/land deterioration to some extent (AbdelRahman 2023).

7.4 Land Restoration Through Sustainable Approaches

The land degradation issue is directly linked with sustainability in terms of soil sustainability, which represents soil resilience to prevent many degradation processes and maintain fertility and productiveness that provides social and ecosystem services for a long time (AbdelRahman et al. (2022a, b, c). Sustainable land management is a peculiar concept that maintains integrity between socioeconomic and ecological principles during land management for AF and other farming systems to achieve intergenerational equity (UNEP 1991; Aboelsoud et al. 2022). Proper land use management and planning are urgently needed in the current scenario of land degradation. Improper farming practices, climatic variability, and different soil properties caused an unstable ecosystem by continuous stages of degradation. Land degradation affects both AF and the surrounding environment. Therefore, sustainable approaches are needed to rectify the ongoing issue of land degradation and related deleterious impacts on our ecosystem. Sustainable AFs and other farming technologies perform better functions toward land reclamation and ecorestoration practices. Land degradation can be controlled by practicing drip irrigation in AF, less synthetic inputs of agrochemicals, and combined usage of inorganic and organic fertilizers. Eco-designing of AFs maximizes biodiversity and ensures higher soil fertility, organic carbon (C) inputs, efficient nutrient cycling, and greater food production with ecosystem services. Active engagement of local communities and stakeholders is a good intervention to combat desertification. Sustainable usage of land and water can fix plant exploitation, overgrazing, trampling of soils, and erosion that are the major causes of desertification. Sustainable farming including AFs can increase biodiversity and vegetation cover, which protects land from soil erosion and minimizes land degradation. Soil protection ensures land fertility and ecorestoration, which maintain ecological stability and environmental sustainability (AbdelRahman 2023).

7.5 Agroforestry Systems and Their Service Functions

AFs can be classified by their spatial arrangements and components, with key terms including coincident, concomitant, interpolation, intermittent, and relay cropping. Coincident systems have all components—trees, crops, and livestock—sharing the same space simultaneously, like a coffee farm with overhead shading trees. Concomitant systems involve different components coexisting for a time, typical in practices like "taungya," where crops are interplanted with young trees. Interpolation describes arrangements where different components occupy the same area but at different growth stages, such as a home garden with varying maturity levels. Intermittent systems feature one constant component, usually perennials like trees, alongside the sporadic cultivation of annual crops, such as livestock grazing under fruit trees seasonally. Lastly, relay cropping involves sowing and harvesting one crop before planting another in the same area, ensuring continuous crop cover. AF represents the domestication of both woody perennial trees and herbaceous crops. This is a sustainable land use farming system practiced by ancient humans in tropical regions. However, it is problem-solving science recognized during the last three decades due to multiple benefits and service functions. It can sequester C and produce a diverse range of environmental, social, and economic benefits. Twenty percent of the global population (approx. 1.20 billion people) depends on AFs and related products and services in urban and rural areas of developing countries (Gupta et al. 2020). Biodiversity conservation, bio-drainage, erosion control, C sequestration, soil improvement, land restoration, and climate change mitigation are key important service functions provided by AFs. Restoration of degraded land through AFs can offer many potential benefits for local communities as food production and livelihood security (Hillbrand et al. 2017). Reconciling AF production with greater biodiversity maximizes service functions at the landscape level (Rey Benayas and Bullock 2012). Farm size, types and nature of crop species, livestock including grass species, fruit and forest trees, tree-crop interactions and compatibility, and tree components are key scientific management practices involved in AFs in degraded land, which intensify ecosystem services. Provisioning services include forest wood, timber, fuelwood, fodder, medicine, cash crops, and bioenergy. Soil fertility maintenance, C sequestration, pollination services, climate regulation, biodiversity conservation, and disease and pest management are important regulating services. Therefore, AF management in degraded landscapes provides ecosystem services that restore wasteland and maintain global sustainability. In this context, AF management in degraded landscapes and ecosystem services is depicted in Fig. 7.4 (Jose 2012; Gupta et al. 2020). Moreover, AF mitigates C footprints and climate change through the C sequestration process and stores C in a plant-soil system that helps in landscape restoration (van Noordwijk 2018; Jhariya et al. 2024). AF and diverse tree-crop plantations capture more C in tropical areas than

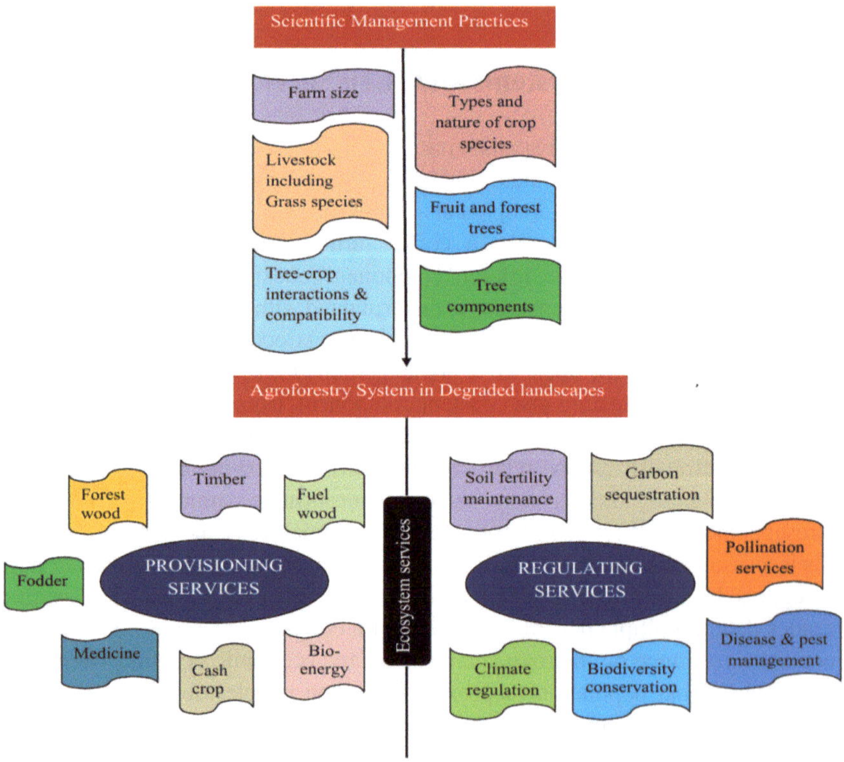

Fig. 7.4 Agroforestry management in degraded landscapes and ecosystem services. (Jose 2012; Gupta et al. 2020)

conventional farming or monoculture (Sharma et al. 2022). Chavan et al. (2023) evaluated the C storage potential of AFs in India with poplar and eucalyptus. They found poplar captured 212.75 Mg C/ha and eucalyptus captured 237.27 Mg C/ha, with annual sequestration rates of 10.31 Mg C/ha/year for poplar and 12.79 Mg C/ha/year for eucalyptus. Using both species in agriculture increased C storage by 2.8 and 2.3 times, respectively, compared to single-crop systems. AFs enhance environmental health by improving soil quality, increasing fertility, managing water, purifying the air, reducing pollution, combating climate change, and lowering C footprints. Thus, they provide various ecological benefits and play a crucial multifunctional role in ecosystem management.

7.6 Agroforestry Systems for Productivity Restoration to Degraded Land

The figure for the rising population ranges from current as 7.7 billion to 8.6, 9.7, and 10.9 billion in the upcoming years 2030, 2050, and 2100, respectively (UN 2017, 2019). This increasing population infuriates the need for sustainable and more productive uses of our land resources. Deforestation, intensive agriculture practices, high synthetic inputs, and overgrazing deprive land quality and lower productivity. These practices deprive soil fertility and direct productivity to unproductive land. It caused poor biodiversity and ecosystem services. Adopting AF may reclaim waste and degraded land along with greater delivery of sustainable ecosystem services. Traditional and modern AFs combine trees, crops, and animals/pastures that are integrated in such a way as to provide many ecosystem services along with higher productivity restoration to degraded land. Therefore, AFs can provide a sustainable alternative to low-diversity cropping systems. It presented better opportunities for degraded land restoration for agricultural uses, delivering ecosystem services and enhancing the multifunctionality of landscapes (Hillbrand et al. 2017; Park et al. 2018).

7.7 Agroforestry Systems for Soil Restoration

AF improves soil fertility by maintaining physical properties and SOC pools that further enhance biological nitrogen fixation. A close type of nutrient cycling observed under AFs helps in fewer losses of essential nutrients that ameliorate microclimate favorable for crop growth and increased biological production (Gupta et al. 2020). Better AF management can ensure C restoration in both plants and soils and enhance productivity. C_{seq} into the plant and soil can also be through sustainable AFs, which promote decarbonization and mitigate C footprints and climate change issues. Integrating fast-growing legume-based MPTs (multipurpose trees) in any component of AFs enhances the C_{seq} and stores within the tree-soil system as biomass (Gupta et al. 2013). C farming practices in AFs reduce GHG emissions, which also enhance biomass and replenish soil fertility and SOC pool (Sharma et al. 2021).

Alley cropping, home gardens, woodlots, protein banks, shelterbelt, windbreaks, taungya system, multifunctional improved fallow, and shifting cultivation perform major functions including greater potentials of C_{seq}. A higher soil C pool was reported under deeper soil layers of AFs as compared to other treeless land use practices. Similarly, drumstick-based cropping system becomes remunerative as tobacco along with greater provision of environmental services including higher SOC pools with efficient nutrient management (Singh et al. 2016). AFs potentially sequester more C than other agricultural systems due to higher diversification and variability (Sileshi et al. 2014). The rate of C_{seq} in AFs is reported maximum under fertile humid soil rather than degraded or water-limiting soils. Incorporating the quality and quantity of biomass C into the soil enhances SOC pools and builds soil organic matter in AFs (Nyamadzawo et al. 2012). In fact, the SOC value is reported higher in AFs apart from forests and followed by tree plantations and arable crops (Nair et al. 2009). C_{seq} in soils increases under adopting nitrogen-fixing tree-based high root biomass to aboveground biomass ratio effectively (Negash and Kanninen 2015). However, scientific management in AFs is needed to increase C_{seq} rates, which further improves C status in plant-soil systems. Tillage practices disturb soil aggregates, which reduce C storage and stabilizations, whereas soil C increased by adopting conservation agriculture practices. Reduced tillage with additional C inputs through integrating nitrogen-fixing cover crops becomes an innovative strategy to enhance C storage in the AFs (Garcia-Franco et al. 2015). Integrating pulse crops within any AFs builds nitrogen status by replacing inorganic nitrogen fertilizer, which further augments soil C status and mitigates footprints (Liu et al. 2016). Therefore, a sustainable land use system includes AFs, and conservation agriculture improves input use efficiency and resource conservation, which mitigate C footprints by enhancing C sequestration and decarbonization (Young 1989; Dagar and Tewari 2017).

7.8 Agroforestry Systems for Biodiversity Conservation

Agrobiodiversity refers to the variety of plants, animals, and microorganisms vital for sustaining agricultural ecosystems. This diversity is essential for ensuring food security, improving nutritional health, and promoting sustainable livelihoods. Furthermore, it plays a significant role in both adapting to and mitigating the impacts of climate change. The relationship between agro-climatic zones, agroecological zones, and agrobiodiversity is crucial. Agro-climatic zones are defined by specific soil types, rainfall patterns, temperature ranges, and water availability, which determine the types of vegetation that can flourish in a location. In contrast, agroecological zones are created by overlaying agro-climatic zones on different landforms, influencing the climate and length of the growing season. Customized AFs can be designed for specific agroecological zones. For example, arid AF utilizes drought-resistant trees to provide shade and wind protection, while date palm AF integrates the cultivation of date palms with other crops. Sustainable land intensification using AF boosts soil health, productivity, biodiversity, and ecosystem services. Plant-soil diversification and its diversity enhance productivity, profitability, and many direct and indirect ecosystem services, which make AFs sustainable. Combining MPTs or leguminous trees with plants and livestock makes ecologically sound and more profitable AFs. Diversified plants deliver diversified products or biomass, which is based on potentially more storage and sequestration of C into plant-soil system. AF diversity in different continents played a significant role in both plant-soil diversifications. Diversified AFs perform different functions under varying agroecological conditions. Models tailored to specific locations reduce competition among tree crops, improving both biodiversity and productivity (Bardule et al. 2013). Therefore, focusing on tree-crop interactions and species characteristics is vital for maximizing yields. This approach fosters increased biodiversity, which supports vital ecosystem services like food security and environmental stability (Guo 2000). Trees on farms and property lines improve adaptability and boost biodiversity and ecosystem services. AF enhances biodiversity by merging different plant and animal species within a framework of trees, crops, and livestock. However, biodiversity loss threatens the ecosystem services offered by AF, such as food security and environmental quality, with around 37,400 species currently at risk of extinction (IUCN Red List 2020). Trees create favorable conditions that enhance biodiversity, positively influencing plants, soil, water, and microclimates in sustainable AFs. Both tropical and temperate regions have seen increased biodiversity in AF (Dollinger and Jose 2018). Research by Sistla et al. (2016) indicates that AFs generally support more biodiversity than forests or tree-only systems, with higher plant and animal variety compared to forests and monocultures (Bucheli and Bokelmann 2017). Furthermore, AF in Canada exhibits greater microbial diversity than traditional cropping systems (Banarjee et al. 2015).

7.9 Sustainable Landscape Management Under Agroforestry Systems

Sustainable landscape management (SLM) fosters the stewardship of agriculture and other natural resources, resulting in enhanced biodiversity and the provision of countless ecosystem services that contribute to human well-being (Prasad et al. 2021a, b). A synergistic relationship is observed among various land use practices, which is examined from ecological, economic, cultural, and social perspectives within a viable landscape framework. Consequently, SLM represents an integrated approach that promotes improved ecosystem health and environmental sustainability (Denier et al. 2015). AF combines forestry with agricultural technologies, thereby facilitating more effective landscape management and practices (Van Noordwijk 2019). This serves as a prime example of agroecological land use systems that aids in transitioning agricultural technology closer to the Sustainable Development Goals (SDGs) while addressing C footprints and climate change issues (Mosquera-Losada et al. 2018). Thus, there exists a conceptual link between AF and SLM, which advances the UN Sustainable Development Goals across the world's production landscapes. These two essential strategies are intrinsically connected, with several key steps that intertwine AF and SLM. For instance, the implementation of AFs in regions reliant on single-crop agriculture ensures the provision of enhanced ecosystem services (Laroche et al. 2020). Likewise, the redesign of coffee-based AFs also promotes greater landscape sustainability in central Nicaragua (Andreotti et al. 2020). Therefore, a landscape approach in AF optimizes biodiversity, food production, and socioeconomic development while also enhancing social and cultural values (Arts et al. 2017). Thus, AF is deemed an advanced technology and a crucial tool for ecological health and sustainability through the integration of landscape management strategies (Angelstam et al. 2019). Moreover, an impact of different SLM practices on SOC pools for reinforcing LDN is depicted in Table 7.1 (Sanz et al. 2017; Jhariya et al. 2022b). UN SDGs are being implemented through various SLM strategies, particularly by practicing AFs in the tropical regions. Studying AF as sustainability science, promoting AFs for landscape-scale benefits, promotion of AF benefits and values, encouraging landscape governance, and supporting AF innovation analysis and design are key steps toward AF harnessing for SLM (Plieninger et al. 2020).

7.10 Agroforestry Systems for Sustainability Outcome Through Landscape Approach

Landscape approaches strengthening the goal for biodiversity conservation, food production, human well-being, livelihood promotion, and related social and cultural values (Arts et al. 2017). In such a quest, landscape assumed a better spatial unit for multifunctionality delivery and governance, which remarked potential landscape management toward higher level of sustainability (Angelstam et al. 2019). Therefore, AFs reconcile with landscape management and improve sustainability for long term. Practicing AFs in the perspective of landscape management promotes sustainability in global production and ecosystem health. Landscape approach and its management through AF can restore land quality and its fertility, which is the basis of plant-soil health and its sustainable production. AF supports ecological and social diversity in agricultural landscape of the Mediterranean region, which is based on empirical evidence (Flinzberger et al. 2020). Adopting AFs in agricultural landscape can provide more ecosystem services with greater level of sustainability. This is due to increasing biodiversity, enhanced land/soil fertility, greater C_{seq}, and climate security that improve health, income, and sustainable food production. Further, AF-based intercropping system becomes good strategies and socially appreciated due to developing agricultural landscape and its sustainable production. AF-based intercropping system delivers multifarious ecosystem services in agricultural and other degraded landscape, which is based on types of models and its adoptability, tree-crop interactions, and scientific management (Laroche et al. 2020).

Table 7.1 Impacts of different Sustainable Landscape Management (SLM) practices on SOC pools for reinforcing LDN (Sanz et al. 2017; Jhariya et al. 2022b)

Sustainable Landscape Management (SLM) practices	Impact on SOC ranges from high-scale (1) to low-scale impact (3)	Example associated with Sustainable Landscape Management (SLM) practices	Potential impacts on LDN (land degradation neutrality)
Agriculture-based Sustainable Landscape Management (SLM) practices			
Vegetation management-based SLM practices	2.40	Involve conservation agriculture system under SLM practices, which reduce soil disturbance through zero-tillage practices. Mulching practices are also observed, which ensures soil covers. Plant rotation system along with intercropping practices maximizes plant diversification	Soil improvement, fertility maintenance, better C management, erosion checking and efficient water regulation
ISFM (integrated soil fertility management)-based SLM practices	2.30		
No soil disturbance-based SLM practices	2.30		
IPM (integrated pest management)-based SLM practices	2.20		
SEC (soil erosion control)-based SLM practices	2.0		
Water management and its regulation-based SLM practices	1.60		
Pasture/grazing-based Sustainable Landscape Management (SLM) practices			
ISFM (integrated soil fertility management)-based SLM practices	2.50	AFs involve better soil-pasture and sustainable nutrient management	Better C management through efficient sequestration potential enhances SOC pools, which regulate nutrient cycling and restore degraded grazing land
Vegetation management-based SLM practices	2.30		
Grazing pressure management	2.20		
Animal waste management	2.0		
Forest-based Sustainable Landscape Management (SLM) practices			

(continued)

Table 7.1 (continued)

Sustainable Landscape Management (SLM) practices	Impact on SOC ranges from high-scale (1) to low-scale impact (3)	Example associated with Sustainable Landscape Management (SLM) practices	Potential impacts on LDN (land degradation neutrality)
Forest restoration-based SLM practices	3.0	Scientific promotion of assisted regeneration and protected forest areas establishment in varying forest vegetation localities	Biodiversity conservation and better C management through efficient sequestration potential enhance SOC pools and erosion checking and efficient nutrient cycling
Reforestation and afforestation-based SLM practices	2.80		
Minimizing deforestation under SLM practices	2.50		
Disease and pest control-based SLM practices	2.0		
SEC (soil erosion control)-based SLM practices	1.80		
Sustainable forest management (SFM) practices	1.70		
Drainage management-based SLM practices	1.0		
Mixed system-based SLM practices			
AF-based SLM practices	3.0	Fruit tree and plantation crop-based multitier AF, which also involves MPTs (multipurpose trees) and home garden system	Biodiversity conservation and better C management through efficient sequestration potential enhance SOC pools, erosion checking and efficient nutrient cycling, micro-climate amelioration, and windbreak system
Vegetation management-based SLM practices	2.30		
Grazing pressure management	2.20		

7.11 Constraints and Challenges

Land degradation and its conversion from agriculture and forestry to wasteland are ongoing challenges faced by growing world. Unsustainable land use practices deprive quality of land and plant-soil health, which becomes a major constraint behind achievement of global sustainability. Landscape degradation and its development through adopting AF is a major concern today. Significant soil erosion in Western Kenya, primarily from deforestation (0.84 t/ha) for agriculture and grass/

shrub areas (0.52 t/ha), has been observed (Kogo et al. 2020). This issue is worsened by the shift from agriculture to urban development (Lu et al. 2020). While urbanization and industrialization drive economic growth, they also lead to rural populations leaving agricultural and AF jobs for non-agricultural employment, resulting in reduced AF production and landscape degradation. Understanding the impacts of land use changes on land resource productivity and sustainability is essential, as it influences the stress on these resources (Kogo et al. 2020). Weaker policy, less scientific knowledge, people unawareness, and legal constraints are existing behind success of AF adoption for landscape management. Weak market linkages, improper investments, and poor extension strategies are major challenges for AF landscape development (Gupta et al. 2020). Farmers faced many challenges due to lack of knowledge on soil fertility and poor information for AF adoption in degraded land in many dry areas. Farmers are much concern about long-term soil recovery, replenishing methods of soil fertility, and land fertility restoration-based AF technology, which helps in augmenting food production with greater global sustainability. Thus, AF must be oriented as "sustainable production intensification," which reduces many land quality-related constraints and global challenges (AbdelRahman 2023).

7.12 Research Policy and Roadmap for Landscape Restoration

Land degradation presents significant global environmental challenges, frequently discussed on various platforms. AF is crucial for restoring degraded ecosystems and enhancing soil health. However, there is an urgent need for comprehensive policies to identify effective AF practices that maintain soil fertility, support diverse cropping systems resilient to climate change, and aid in landscape restoration. Customizing AF strategies for degraded and marginal lands could serve as a sustainable solution promoting global sustainability (Olsson et al. 2019). Nonetheless, the identification and evaluation of constraints on sustainability and the pursuit of solutions are crucial for enhancing AF productivity. Ecologically focused AFs should be integrated into contemporary and future research methodologies that sustain productivity and profitability through sustainable practices alongside landscape restoration. AFs are regarded as application-centered research that varies according to soil types, topography, and existing climatic conditions. The limited availability of resources and their improper use present significant challenges for ecologists and environmentalists. Future research directions in AF necessitate a thorough investigation into the conservation and management of natural resources, which is essential for effective AF practices on increasingly degraded lands. This approach not only aids in the restoration of degraded landscapes but also offers numerous invaluable ecosystem services that contribute to ensuring soil health, food supply, and climate stability. Moreover, it is important to develop an actionable roadmap for advancing research and development concerning the adoption of AF practices aimed at utilizing degraded lands across various spatial scales, from farm plots to entire ecosystems and landscapes, ultimately benefiting landowners and farmers. Such initiatives will also contribute to biodiversity conservation, enhance soil fertility, and improve C sequestration processes, thus addressing C emissions and climate change challenges (Gupta et al. 2020).

7.13 Conclusions

Land degradation significantly threatens the global environment, affecting livelihoods and ecosystem services. Unsustainable land use diminishes land quality and disrupts plant-soil interactions, hindering sustainability efforts. To combat this issue, rehabilitating degraded landscapes through AF is essential, as it boosts food security, enhances nutrition, and strengthens climate resilience while improving land quality and biodiversity. AF stabilizes soil, reduces erosion, improves water quality, captures C, and increases agricultural productivity, benefiting local communities. Diverse plant species enhance water infiltration, minimize runoff, and foster better microclimates and biodiversity compared to monocultures. Additionally, AF decreases nutrient runoff and soil erosion, boosting crop yields and resilience essential for food security. SLM advocates for effective agricultural practices and resource management that enhance biodiversity and ecosystem services. Therefore, implementing robust policies to promote AF on degraded lands is crucial for restoring soil fertility, improving C sequestration, and addressing climate change impacts.

References

AbdelRahman MA (2023) An overview of land degradation, desertification and sustainable land management using GIS and remote sensing applications. Rend Lincei Sci Fis Nat 34:1–42

AbdelRahman MAE, Metwaly MM, Shalaby A (2019) Quantitative assessment of soil saline degradation using remote sensing indices in Siwa Oasis. Remote Sens Appl Soc Environ 13:53–60. https://doi.org/10.1016/j.rsase.2018.10.004

AbdelRahman MAE, Aff AA, D'Antonio P, Gabr SS, Scopa A (2022a) Detecting and mapping salt-affected soil with arid integrated indices in feature space using multi-temporal Landsat imagery. Remote Sens 14:2599

AbdelRahman MAE, Metwaly MM, Aff AA, D'Antonio P, Scopa A (2022b) Assessment of soil fertility status under soil degradation rate using geomatics in West Nile Delta. Land 11:1256. https://doi.org/10.3390/land11081256

AbdelRahman MAE, Engel B, Eid MSM, Aboelsoud HM (2022c) A new index to assess soil sustainability based on temporal changes of soil measurements using geomatics– an example from El-Sharkia Egypt. All Earth. https://doi.org/10.1080/27669645.2022.2103953

Aboelsoud HM, AbdelRahman MAE, Kheir AMS, Eid MSM, Ammar KA, Khalifa TH, Scopa A (2022) Quantitative estimation of saline-soil amelioration using remote-sensing indices in arid land for better management. Land 11:1041

ADB (2008) Land Degradation in Central Asia, final report, Central Asian Countries Initiative for Land Management, Multicounty Partnership Framework Support Project, ADB TA 6356-REG

Alewell C, Ringeval B, Ballabio C, Robinson DA, Panagos P, Borrelli P (2020) Global phosphorus shortage will be aggravated by soil erosion. Nat Commun 11(1):1–12

Andreotti F, Speelman EN, Van den Meersche K, Allinne C (2020) Combining participatory games and backcasting to support collective scenario evaluation: an action research approach

for sustainable agroforestry landscape management. Sustain Sci. https://doi.org/10.1007/s11625-020-00829-3
Angelstam P, Munoz-Rojas J, Pinto-Correia T (2019) Landscape concepts and approaches foster learning about ecosystem services. Landsc Ecol 34(7):1445–1460
Arts B, Buizer M, Horlings L, Ingram V, van Oosten C, Opdam P (2017) Landscape approaches: a state-of-the-art review. Annu Rev Environ Resour 42(1):439–463
Banarjee S, Baah-Acheamfour M, Carlyle CM, Bissett A, Richardon AE, Siddique T, Bork EW, Chang SX (2015) Determinants of bacterial communities in Canadian agroforestry systems. Environ Microbiol 18:1805–1816
Banerjee A, Jhariya MK, Bargali SS, Palit D (2023) Ecorestoration for sustainability. Wiley-Scrivener, pp 1–441. https://onlinelibrary.wiley.com/doi/book/10.1002/9781119879954. ISBN: 9781119879718
Bardule A, Rancłne S, Gnjtmane I, Berzins P, Stesele V, Lazdina D, Bardulis A (2013) The effect of fertiliser type on hybrid aspen increment and seed yield of perennial grass cultivated in the agroforestry system. Agron Res 11:13–25
Bhattacharyya R, Ghosh BN, Mishra PK, Mandal B, Rao CS, Sarkar D, Franzluebbers AJ (2015) Soil degradation in India: challenges and potential solutions. Sustainability 7(4):3528–3570. https://doi.org/10.3390/su7043528
Bucheli VJP, Bokelmann W (2017) Agroforestry systems for biodiversity and ecosystem services: the case of the Sibundoy Valley in the Colombian province of Putumayo. Int J Biodivers Sci Ecosyst Serv Manag 13:380–397
Chavan SB, Dhillon RS, Sirohi C, Uthappa AR, Jinger D, Jatav HS, Chichaghare AR, Kakade V, Paramesh V, Kumari S, Yadav DK, Minkina T, Rajput VD (2023) Carbon sequestration potential of commercial agroforestry systems in indo-Gangetic Plains of India: poplar and eucalyptus-based agroforestry systems. Forests 14(3):559. https://doi.org/10.3390/f14030559
Dagar JC, Gupta SR (2016) Agroforestry: potentials for rehabilitation of degraded lands, constraints and the way forward. In: Dagar JC, Tewari JC (eds) Agroforestry research developments. Nova, New York, pp 47–97
Dagar JC, Tewari VP (2017) Agroforestry: anecdotal to modern science. Springer Nature Singapore Pte Ltd, p 879
Denier L, Scherr S, Shames S, Chatterton P, Hovani L, Stam N (2015) The little sustainable landscapes book: achieving sustainable development through integrated landscape management. Global Canopy Programme, Oxford
Devi A, Jhariya MK, Raj A, Singh KP, Singh B (2023) *Acacia nilotica*: a promising species for soil sustainability. In: Land and environmental management through forestry, pp 339–353. https://doi.org/10.1002/9781119910527.ch14
Dollinger J, Jose S (2018) Agroforestry for soil health. Agrofor Syst 92:213–219. https://doi.org/10.1007/s10457-018-0223-9
ELD Initiative (2013a) Opportunity lost: Mitigating risk and making the most of your land assets. An assessment of the exposure of business to land degradation risk and the opportunities inherent in sustainable land management. Available from: www.eld-initiative.org
ELD Initiative (2013b) The rewards of investing in sustainable land management. Interim report for the economics of land degradation initiative: a global strategy for sustainable land management. Available from: www.eld-initiative.org/
ELD Initiative (2015) The value of land: prosperous lands and positive rewards through sustainable land management. Available from www.eld-initiative.org
El-Rawy M, Mohamed AE, Ismail E (2020) Integrated use of pollution indices and geomatics to assess soil contamination and identify soil pollution source in El-Minia Governorate, Upper Egypt. J Eng Sci Technol 15(4):2223–2238
FAO (2010) The state of the world's land and water resources for food and agriculture. Food and Agriculture Organization of the United Nations, Rome
FAO and ITPS (2015) Status of the world's soil resources (SWSR) – main report. Food and Agriculture Organization of the United Nations and Intergovernmental Technical Panel on Soils, Rome

Fifanou VG, Ousmane C, Gauthier B, Brice S (2011) Traditional agroforestry systems and biodiversity conservation in Benin (West Africa). Agrofor Syst 82(1):1–13

Flinzberger L, Zinngrebe Y, Plieninger T (2020) Labelling in Mediterranean agroforestry landscapes: a Delphi study on relevant sustainability indicators. Sustain Sci. https://doi.org/10.1007/s11625-020-00800-2

Garcia-Franco N, Albaladejo J, Almagro M, Martínez-Mena M (2015) Beneficial effects of reduced tillage and green manure on soil aggregation and stabilization of organic carbon in a Mediterranean agroecosystem. Soil Tillage Res 153:66–75. https://doi.org/10.1016/j.still.2015.05.010

Gibbs HK, Salmon JM (2015) Mapping the world's degraded lands. Appl Geogr 57:12–21

Gichuki L, Brouwer R, Davies J, Vidal A, Kuzee M, Magero C, Walter S, Lara, P, Oragbade C, Gilbey B (2019) Reviving land and restoring landscapes: policy convergence between forest landscape restoration and land degradation neutrality. IUCN, Gland, viii + 34p

Global Environmental Facility (GEF) (2022) Land degradation, main issue. https://www.thegef.org/what-we-do/topics/land-degradation#:~:text=Globally%2C%20about%2025%20percent%20of,land%20area%20has%20been%20degraded.&text=If%20this%20trend%20continues%2C%2095,farmers%2C%20and%20the%20very%20poor. Accessed 30 Jan 2022

Guo Q (2000) Climate change and biodiversity conservation in Great Plains agroecosystems. Glob Environ Chang 10:289–298

Gupta A, Dhyani SK, Ramnewaj HAK, Prasad R, Alam B, Rizvi RH, Gupta G, Pandey KK, Jain A (2013) Modeling analysis of potential carbon sequestration under existing agroforestry systems in three districts of Indo-gangetic plains in India. Agrofor Syst 87(3):1129–1146

Gupta SR, Dagar JC, Teketay D (2020) Agroforestry for rehabilitation of degraded landscapes: achieving livelihood and environmental security. In: Agroforestry for degraded landscapes: recent advances and emerging challenges, vol 1, pp 23–68

Hillbrand A, Borelli S, Conigliaro M, Olivier A (2017) Agroforestry for landscape restoration: exploring the potential of agroforestry to enhance the sustainability and resilience of degraded landscapes. Food and Agriculture Organization of the United Nations, Rome. https://doi.org/10.1016/j.cosust.2013.10.013

Hossini H, Karimi H, Mustafa YT, Al-Quraishi AMF (2022) Role of effective factors on soil erosion and land degradation: a review. In: Al-Quraishi AMF, Mustafa YT, Negm AM (eds) Environmental degradation in Asia, Earth and environmental sciences library. Springer, Cham. https://doi.org/10.1007/978-3-031-12112-8_11

ISRO (2018) Desertification and land degradation atlas of selected districts of India (Based on IRS LISS III data of 2011–13 and 2003–05), vol 2. Space Applications Centre, ISRO, Ahmedabad, 145 p. www.sac.gov.in

IUCN (2015) Land degradation neutrality: implications and opportunities for conservation. www.iucn.org/ecosystem-management/global-drylands-initiative

IUCN (2020) The IUCN red list of threatened species. Version 2020–3. https://www.iucnredlist.org Downloaded on May 08, 2021

IUCN, Davies J, Gudka M, Laban P, Metternicht G, Alexander S, Hannam I (2015) Land degradation neutrality: implications and opportunities for conservation. IUCN, Gland

Jagdish P (2023) Environmental implications of soil degradation in India-a review. Agric Rev 25(1):57–63

Janečková Molnárová K, Sklenička P, Bohnet IC, Lowther-Harris F, van den Brink A, Movahhed Moghaddam S, Fanta V, Zástěra V, Azadi H (2023) Impacts of land consolidation on land degradation: a systematic review. J Environ Manag 329:117026. https://doi.org/10.1016/j.jenvman.2022.117026. (PMID: 36608617)

Jhariya MK, Bargali SS, Raj A (2015) Possibilities and perspectives of agroforestry in Chhattisgarh. In: Zlatic M (ed) Precious forests-precious Earth. InTech, Croatia, pp 237–257. https://doi.org/10.5772/60841. ISBN: 978-953-51-2175-6, 286 pages

Jhariya MK, Meena RS, Banerjee A, Meena SN (2022a) Natural resources conservation and advances for sustainability. Elsevier. https://doi.org/10.1016/C2019-0-03763-6. ISBN: 9780128229767

Jhariya MK, Raj A, Banerjee A, Meena RS, Bargali SS, Kumar S, Nema S, Poonam, Oraon PR (2022b) Plan and policies for soil organic carbon management under agroforestry system. In: Plans and policies for soil organic carbon management in agriculture. Springer Nature Singapore, Singapore, pp 191–219

Jhariya MK, Meena RS, Banerjee A, Kumar S, Raj A (2024) Agroforestry for carbon and ecosystem management. Elsevier. ISBN: 9780323953931

Jose S (2012) Agroforestry for conserving and enhancing biodiversity. Agrofor Syst 85:1–8. https://doi.org/10.1007/s10457-012-9517-5

Kogo BK, Kumar L, Koech R (2020) Impact of land use/cover changes on soil erosion in western Kenya. Sustainability 12(22):9740. https://doi.org/10.3390/su12229740

Laban P, Metternicht G, Davies J (2018) Soil biodiversity and soil organic carbon: keeping drylands alive. IUCN, Gland, viii + 24p. http://sci-hub.tw/10.2305/IUCN.CH.2018.03.en

Laroche G, Domon G, Olivier A (2020) Exploring the social coherence of envisioned landscapes featuring agroforestry intercropping systems (AIS) using rural resident visual assessments and perceptions. Sustain Sci. https://doi.org/10.1007/s11625-020-00837-3

Liu C, Cutforth H, Chai Q (2016) Farming tactics to reduce the carbon footprint of crop cultivation in semiarid areas. A review. Agron Sustain Dev 36:69. https://doi.org/10.1007/s13593-016-0404-8

Lu L, Huang A, Xu Y, Marcos-Martinez R, Duan Y, Ji Z (2020) The influences of livelihood and land use on the variation of Forest transition in a typical mountainous area of China. Sustainability 12(22):9359

Mosquera-Losada MR, Santiago-Freijanes JJ, Rois-Díaz M, Moreno G, den Herder M, Aldrey-Vázquez JA, Ferreiro-Domínguez N, Pantera A, Pisanelli A, Rigueiro-Rodríguez A (2018) Agroforestry in Europe: a land management policy tool to combat climate change. Land Use Policy 78:603–613

Nair PK, Mohan Kumar B, Nair VD (2009) Agroforestry as a strategy for carbon sequestration. J Plant Nutr Soil Sci 172:10–23

Nair PKR, Viswanath S, Lubina PA (2017) Cinderella agroforestry systems. Agrofor Syst 91:901–917

Negash M, Kanninen M (2015) Modeling biomass and soil carbon sequestration of indigenous agroforestry systems using CO2FIX approach. Agric Ecosyst Environ 203:147–155. https://doi.org/10.1016/j.agee.2015.02.004

Nyamadzawo G, Nyamugafata P, Wuta M, Nyamangara J (2012) Maize yields under coppicing and non coppicing fallows in a fallow-maize rotation system in Central Zimbabwe. Agrofor Syst 84:273–286

Olsson L, Barbosa H, Bhadwal S, Cowie A, Delusca K, Flores-Renteria D, Hermans K, Jobbagy E, Kurz W, Li D, Sonwa DJ, Stringer L (2019) Land Degradation. In: Climate Change and Land: an IPCC special report on climate change, desertification, land degradation, sustainable land management, food security, and greenhouse gas fluxes in terrestrial ecosystems. https://doi.org/10.1017/9781009157988.006

Park H, Turner N, Higgs E (2018) Exploring the potential of food forestry to assist in ecological restoration in North America and beyond. Restor Ecol 26:284–293

Plieninger T, Levers C, Mantel M, Costa A, Schaich H, Kuemmerle T (2015) Patterns and drivers of scattered tree loss in agricultural landscapes: orchard meadows in Germany (1968–2009). PLoS One 10(5):e0126178

Plieninger T, Muñoz-Rojas J, Buck LE, Scherr SJ (2020) Agroforestry for sustainable landscape management. Sustain Sci 15:1255–1266

Prasad R, Jhariya MK, Banerjee A (*2021a*) Advances in sustainable development and management of environmental and natural resources: economic outlook and opinions volume I. CRC Press/Taylor and Francis Group/Apple Academic Press Inc, pp 1–437. https://doi.org/10.1201/9781003187455. Hardcover ISBN: 9781774910344

Prasad R, Jhariya MK, Banerjee A (*2021b*) Advances in sustainable development and management of environmental and natural resources: economic outlook and opinions volume II. CRC

Press/Taylor and Francis Group/Apple Academic Press Inc, pp 1–428. Hardcover ISBN: 9781774910368
Raj A, Jhariya MK, Banerjee A, Nema S, Bargali K (2023) Land and environmental management through forestry. Wiley-Scrivener, pp 1–372. https://onlinelibrary.wiley.com/doi/book/10.1002/9781119910527. ISBN: 9781119910404
Raj A, Jhariya MK, Banerjee A, Jha RK, Singh KP (2024) Agroforestry. Wiley-Scrivener. ISBN 978-1-394-23113-3. https://www.wiley.com/en-us/Agroforestry-p-9781394231133
Raj A, Jhariya MK, Meena RS (2025) Agroforestry for monetising carbon credits. Springer Nature. ISBN: 978-3-031-97150-1. https://link.springer.com/book/9783031971501
Rey Benayas JM, Bullock JM (2012) Restoration of biodiversity and ecosystem services on agricultural land. Ecosystems 15:883–899
Rosenstock TS, Dawson IK, Aynekulu E, Chomba S, Degrande A, Fornace K, Jamnadass R, Kimaro A, Kindt R, Lamanna C, Malesu M, Mausch K, McMullin S, Murage P, Namoi N, Njenga M, Nyoka I, Paez Valencia AM, Sola P, Shepherd K, Steward P (2019) A planetary health perspective on agroforestry in Sub-Saharan Africa. One Earth 1(3):330–344
Sanz MJ, de Vente J, Chotte JL, Bernoux M, Kust G, Rulz I, Almagro M, Alloza JA, Vallejo R, Castillo V, Hebel A, AkhtarSchuster M (2017) Sustainable land management contribution to successful land-based climate change adaptation and mitigation. United Nations Convention to Combat Desertification, Bonn. ISBN 978-92-95110-96-0
Sharma M, Kaushal R, Kaushik P, Ramakrishna S (2021) Carbon farming: prospects and challenges. Sustainability 13(19):11122
Sharma R, Mina U, Kumar BM (2022) Homegarden agroforestry systems in achievement of sustainable development goals. A review. Agron Sustain Dev 42(3):1–21
Sileshi GW, Mafongoya PL, Akinnifesi FK (2014) Agroforestry: fertilizer trees. Encycl Agric Food Syst 1:222–234
Singh B, Kaur N, Gill RIS, Dagar JC (2016) Agroforestry for crop diversification and carbon sequestration. Agroforestry Research Developments, p 99
Sistla SA, Roddy AB, Williams NE, Kramer DB, Stevens K, Allison SD (2016) Agroforestry practices promote biodiversity and natural resource diversity in Atlantic Nicaragua. PLoS One 11:e0162529
Sreenivas K, Sujatha G, Mitran T, Janaki KG, Ravisankar T, Rao PVN (2021) Decadal changes in land degradation status of India. Curr Sci 121(4):25
Suárez LR, Salazar JCS, Casanoves F, Bieng MAN (2021) Cacao agroforestry systems improve soil fertility: comparison of soil properties between forest, cacao agroforestry systems, and pasture in the Colombian Amazon. Agric Ecosyst Environ 314:107349
Toderich K, Shuyskaya E, Taha F, Matsuo N, Ismail S, Aralova D, Radjabov T (2013) Integrating agroforestry and pastures for soil salinity management in dryland ecosystems in Aral Sea basin. In: Shahid SA, Abdelfattah MA, Taha FK (eds) Developments in soil salinity assessment and reclamation-innovative thinking and use of marginal soil and water resources in irrigated agriculture. Springer, Dordrecht, pp 579–602
Torralba M, Fagerholm N, Burgess PJ, Moreno G, Plieninger T (2016) Do European agroforestry systems enhance biodiversity and ecosystem services? A meta-analysis. Agric Ecosyst Environ 230:150–161
UN (2017) World population prospects: the 2017 revision. The Population Division of the United Nations, Department of Economic and Social Affairs, United Nations Secretariat, New York. http://www.unfpa.org/world-population-trends
UN (2019) World population prospects 2019: highlights. The Population Division of the United Nations, Department of Economic and Social Affairs, United Nations Secretariat, New York. https://population.un.org/wpp/Publications/Files/WPP2019_Highlights.pdf. Accessed 19 June 2019
UNCCD (2012) United Nations convention to combat Desertifcation (UNCCD) in Bonn, Germany
UNEP (1991) Status of desertifcation and implementation of the United Nations plan of action to combat desertification. UNEP, Nairobi, p 77

van Noordwijk M (2018) Agroforestry as part of climate change response. IOP Conf Ser Earth Environ Sci 200:012002. http://sci-hub.tw/10.1088/1755-1315/200/1/012002
Van Noordwijk M (2019) Sustainable development through trees on farms: agroforestry in its fifth decade. World Agroforestry Centre (ICRAF), Bogor
Viswanath S, Nair PKR, Kaushik PK, Prakasam U (2000) Acacia nilotica trees in rice felds: a traditional agroforestry system in Central India. Agrofor Syst 50(2):157–177
Waldron A, Garrity D, Malhi Y, Girardin C, Miller DC, Seddon N (2017) Agroforestry can enhance food security while meeting other sustainable development goals. Trop Conserv Sci 10:1940082917720667
Wicke B, Smeets E, Dornburg V, Vashev B, Gaiser T, Turkenburg W, Faaij A (2011) The global technical and economic potential of bioenergy from salt-affected soils. Energy Environ Sci 4:2669–2681
Young A (1989) Agroforestry for soil conservation. CABI, Wallingford, p 276

Enhancing Biomass Production on Marginal Land: Avenue for Land Restoration and Bio-Based Economy

8

Vivek Verma, Nagaraja Tejo Prakash, and Sheikh Adil Edrisi

Abstract

Rapid industrialization, urbanization, globalization, and various anthropogenic activities have strained our finite natural resources due to the increasing energy demand. Therefore, renewable energy resources could be an alternative toward the future sustainable energy needs. In addition, global investment in renewable energy and fuels surged by 8.1%, reaching to 623 billion USD in 2023. Nations like the United States and other developed countries have reconfigured their trade and industrial policies. For instance, the United States initiated over 250 domestic clean energy-manufacturing projects, highlighting its commitment to a sustainable future. Simultaneously, the European Union also proposed the Net-Zero Industry Act, depicting an initiative toward a greener economy. To address the critical challenges of balancing food and energy production, ensuring environmental sustainability, and maintaining biodiversity is crucial. Therefore, the role of marginal and degraded land has become pivotal in this regard to produce biomass for bioenergy and avoiding conflict with food production. This strategy has an added benefit of reducing carbon footprint. Hence, it could be a viable

V. Verma
Department of Energy & Environment, Thapar Institute of Engineering & Technology, Patiala, Punjab, India

Thapar School of Liberal Arts & Sciences, Thapar Institute of Engineering & Technology, Patiala, Punjab, India

N. T. Prakash
Department of Energy & Environment, Thapar Institute of Engineering & Technology, Patiala, Punjab, India

S. A. Edrisi (✉)
Thapar School of Liberal Arts & Sciences, Thapar Institute of Engineering & Technology, Patiala, Punjab, India
e-mail: sheikhadil.edrisi@thapar.edu; sheikhadiledrisi@gmail.com

V. C. Pandey (ed.), *Plantation-Based Land Restoration*,
https://doi.org/10.1007/978-981-95-2860-8_8

option to cultivate energy crops by enriching biodiversity, improving soil quality, without affecting food security. Therefore, present chapter discusses the potential of marginal and degraded land to cultivate biomass for producing bioenergy, its management practices, and ecological economics perspective toward building the bio-based economy.

Keywords

Industrialization · Biodiversity · Renewable Energy · Biomass · Marginal land

8.1 Introduction

Energy is the driving force behind the world economy, and as economies develop, energy consumption tends to rise. It further follows a similar trend in the future due to a continuous increase in energy consumption. However, while energy demand grows, fossil fuel reserves are rapidly depleted. Further, it has been observed globally that approximately 19% of the energy demand is necessitated by renewable sources, which shows a significant inclination toward sustainability. It includes about 9% derived from biomass, while the remaining 10% comes from modern renewable sources, which include biofuels, wind, solar, geothermal energy, etc. (Edrisi and Abhilash 2016). Moreover, investment in fuel and renewable power increased by 8.1% in 2023 globally and reached around 623 billion USD (REN21 2023). By 2023, the surge in renewable power capacity reached an impressive 473 GW, making a remarkable 36% increase globally (REN21 2023). During the 2023 United Nations Climate Change Conference (UNCCC) held in Dubai, representatives from 130 nations made a strategic commitment toward addressing climate change by pledging to triple their renewable energy capability and bifold the rate of annual energy efficiency improvements by 2030 (REN21 2023). In the same year, other nations transformed their trade and industrial policies. At the same time, the United States spearheaded this movement by initiating over 250 clean energy manufacturing projects through the ground-breaking Inflation Reduction Act. Meanwhile, the European Union introduced the Net-Zero Industry Act and commenced the first stage of the Carbon Border Adjustment Mechanism. The renewable energy domain showcased the potential for economic growth with an 8% increase in employment in 2022, resulting in 13.7 million jobs (REN21 2023). Therefore, among various renewable resources, plant biomass stands out as the most effective alternative, along with hydropower, wind, and solar energy (Saqib et al. 2013). Furthermore, it is essential to understand that dependence on fossil energy leads to various problems related to health and the environment. Consequently, it is one of the significant contributors toward global warming and greenhouse gas (GHG) emissions. Therefore, energy scarcity, the need for cleaner energy, and climate change effects are primary factors that have stimulated renewable energy production on international agendas (Karp and Shield 2008).

Further, defining a piece of land as marginal depends on its economic returns and environmental outcomes. Eventually, it involves considering crop production costs, ecological services impacted by prevailing economic conditions, soil and climate factors, land use histories, and market valuations, which differ variably over time. The idea of utilizing marginal land for bioenergy crops is often undervalued, being regarded as less productive, low quality or degraded (Tilman et al. 2006). Later, it was found that more than half of recent studies that defined marginal land either explicitly or implicitly focused on the biophysical properties of the land, such as slope, soil pH, drainage, and soil salinization, which are crucial in determining crop productivity (Richards et al. 2014). Notably, the impact of these factors is anticipated as dissimilar for food and bioenergy crops, pointing toward untapped potential. Additionally, various studies have distinguished marginal land by examining current or historical land use, including land that has been left abandoned by agricultural activities (Stoof et al. 2015) or land, which are sensitive but retired from crop production (Emery et al. 2017). Embracing green technologies, for example, biofuels and bio-products, is regarded as the most effective way to decrease GHG emissions and global warming while fulfilling energy needs. Ultimately, it leads to necessary steps toward a sustainable future. To produce bioenergy, assessing land suitability for cultivating energy crops is vital. In 2007, China's government changed its biofuel production policy from a food-based to a non-food-based approach, resulting in no change in its arable land, and considered marginal land as an alternative (Qiu 2009). Biomass is regarded as a reliable carbon-neutral energy source since it absorbs CO_2 during growth and is subsequently released back into the atmosphere during combustion. Furthermore, there should also be a focus on the improving nutrient-use efficiency, which enables the biomass/bioenergy crop to reduce GHG emissions. Hence, such measures would help in achieving the net-zero targets (Edrisi et al. 2024; Perea-Moreno et al. 2017). As naturally occurring, non-fossil organic material, biomass can significantly reduce reliance on fossil fuels and serves as an excellent alternative with immense potential (Rozzi et al. 2020). Biomass resources derived from urban waste, forestry, agroforestry, and agriculture encompass a variety of diverse materials, including fine sawdust, crop residues, straw, organic manure, and household scraps (Brosowski et al. 2019).

The revised Renewable Energy Directive (RED) offers a way forward by limiting the contribution of renewable energy that comes from food-based biofuels, mandating that 3.6% of biofuels must be derived from nonfood "advanced" sources by 2030 (European Commission 2017). As one of the significant renewable energy sources, bioenergy garnered significant attention globally. Ultimately, there is a rising attention on utilizing marginal land for the energy feedstock. Further, spare land is mainly marginal and poorly suitable for traditional crops. However, it favors energy crops based on parameters, like economic, climatic, edaphic, environmental, and economic criteria (Levis and Kelly 2014).

The United Nations has considered 2021–2030 the Decade of Ecosystem Restoration (Herrick et al. 2019), highlighting the global need to rejuvenate our natural environments. Land restoration is recognized as a significant strategy in global environmental policy, which is ecosystem-based, as evidenced by the

Convention on Biological Diversity (CBD) after the recent development of the post-2020 Global Biodiversity Targets (Nicholson et al. 2021). Furthermore, restoring degraded land provides job opportunities because of soil improvement and helps mitigate and adapt to climate change. It was observed that bioenergy production from degraded lands solves various environmental issues and achieves multiple UN Sustainable Development Goals while mitigating land degradation (Keesstra et al. 2018). However, it is important to find innovative solutions to maximize land use for bioenergy production for a sustainable future.

Therefore, the world needs more land resources for biomass production to cater to a larger population. Simultaneously, ensuring food security, land availability to grow energy crops, and maintaining economic stability is a serious challenge. Consequently, no country can produce bioenergy from its cropland. In such a scenario, marginal land emerges as a significant alternative. Additionally, to increase bioenergy production globally, it is crucial to explore other avenues that can utilize marginal and degraded lands effectively. Therefore, this chapter explores the promising potential and practicality of utilizing marginal and degraded land for biomass energy production. It will further discuss its availability, bio-based economy, crop suitability, and other prospects. Additionally, the chapter will also cover the strategies for realizing the economic potential of marginal land for building bioeconomy and associated challenges and policy implications.

8.2 Marginal and Degraded Land: Current Status and Future Prospects

The concept of marginal land is often based on cost-benefit analysis and categorizing such areas as economically marginal. For instance, in China, the Ministry of Agriculture identifies "marginal land suitable for growing energy crops as wasteland." The classification of different land types, such as grassland, saline land, wasteland, reed swamp, bareland, tidal flat, marshland, and land, which are underutilized, is recognized as a potential land source and fit for biomass energy production (Yan et al. 2008). The term marginal land relates to biomass production and establishes a vital connection as a significant resource for bioenergy feedstock. Furthermore, the variations in the definitions of marginal land resulted in inconsistent assessments of marginal land (Zhuang et al. 2011). Current agricultural systems are primarily rooted in intensive farming practices and their modulations, with a strong emphasis on approaches such as improving productivity, mitigating environmental impacts, and enhancing resource-use efficiency (Sandhu 2021). However, it has been observed that sustainable intensification strategies often overlook significant dimensions of sustainability, especially the ecological and social aspects (Walthall et al. 2024). As per current trends in the agricultural settings and the burgeoning human population, intensification is urged to meet the global demand for food.

Degraded land is considered a significant threat to our ecosystems and communities that depend on them. The long-term decline in its functionality and services arose from disturbances that ecosystems were unable to recover on their own (UNEP 2007). Further, this degradation is primarily shaped by improper land use, resulting in a decline of natural land resources. Consequently, it results in reduced plant growth, which worsens the situation and may result in desertification (El-Beltagy 2000). Marginal lands are often categorized as unproductive or unsuitable for crop cultivation due to their undesired topology, inadequate soil properties, harsh climatic conditions, inadequate underground water, drought, and subsequently limited potential for profitability in traditional agriculture. However, it is crucial to understand that the extent of marginality is not uniform; it varies with intent and context. Marginal land encompasses a wide range of areas, such as contaminated lands and fallow agricultural land (Smith et al. 2013) resulting from unfavorable growing conditions. It also includes degraded land (Tilman et al. 2006) or former landfills used to dispose of urban waste (Nixon et al. 2001). Therefore, understanding and harnessing the potential of these marginal and degraded lands is critical for biomass production and encouraging programs like the Conservation Reserve Program to unlock their potential for sustainable use.

8.2.1 Potential of Marginal Land and Degraded Land

The marginal land concept initially relied on profitability in terms of the land's economic returns. However, very few studies have formally recognized marginal land from a financial perspective (Stoof et al. 2015; Richards et al. 2014; Kang et al. 2013). It is essential to consider biophysical characteristics and economic factors when estimating marginal land for food or bioenergy crop production. Focusing solely on productivity may overlook the economic viability of using the land for crops and any noneconomic benefits it may offer. Researchers at Michigan State University have suggested that marginal lands can be valuable for meeting alternative energy needs. It is being found that global marginal lands, especially in the US Midwest, could yield approximately 5.5 billion gallons of ethanol each year by growing cellulosic biomass from mixed plant species, subsequently mitigating GHG emissions (Gelfand et al. 2013).

As shown in Fig. 8.1, marginal lands are areas with low productivity and unsuitable for agriculture due to their limitations and essential ecosystem functions (Naveed et al. 2022). Throughout Europe, marginalization has led to a considerable decrease in arable land, pastures, and permanent meadows. The annual degradation

Fig. 8.1 Land uses showing a transitional phase of marginal land at the central focus. (Adapted from Naveed et al. 2022)

of land amounts to approximately ten million hectares (mha). Since 2008, the recovery of set-aside land has yet to be counterbalanced by using an innovative approach highlighting the potential of land use (Schröder et al. 2018). With such extensive areas of degraded land, there will be a significant opportunity to implement effective practices to utilize marginal land for biomass and bioenergy production toward meaningful land restoration. This restoration approach leads to enhanced bioenergy output as well as contributes to unlocking a range of valuable ecosystem services from the marginal land (Kidd et al. 2015).

Cities are densely populated and limiting renewable energy production. In such a scenario, marginal land can be a viable alternative for growing high-yield energy crops and reducing the challenges associated with long transportation and other issues, such as distant energy infrastructure. Studies suggest that improving food production will necessitate converting marginal lands into effective cropland systems and restoring degraded land and its ecosystem (Bigges 2007; Lal 2004). Interestingly, the significance of marginal lands should be considered for food production alongside the need for sustainability and land conservation. Furthermore, land plays a crucial role as a terrestrial sink for GHG. As a result, it is vital to implement sustainable land use practices to offset atmospheric carbon dioxide and reduce GHG emissions from land. Sustainable restoration of degraded lands could result in agricultural intensification, soil carbon sequestration, and biomass production.

In a study, the European Union has enough biomass potential to meet its future energy targets, but it still needs to make it easier to mobilize biomass plantations (Scarlat et al. 2013). The Intergovernmental Panel on Climate Change (IPCC) published a report based on renewable energy, highlighting that bioenergy potential could reach 50 EJ yr-1 by 2050 (Batidzirai et al. 2012) and maintaining sustainability criteria could increase it up to 500 EJ yr^{-1}. Furthermore, numerous studies have evaluated the sustainability of biomass and bioenergy under various scenarios in developed nations, but there is a lack of attention to this issue in developing nations. In the United States, it was conceived that conservation practices in larger farms consisting of marginal land could be profitable and sustainable through Conservation Effects Assessment Programs (CEAP) by national analysis (Richardson et al. 2008). The definition of marginal land has limitations because it overlooks the economic feasibility of converting current land use, such as food or feed production, to bioenergy crop production. The idea of marginal land is often based on a predetermined crop productivity threshold, which is applied consistently across different times and places without considering the potential for marginal land availability to change in response to developing conditions (Jiang et al. 2021).

8.2.2 Availability of Marginal Land and Degraded Land

Estimates have been made regarding the potential land available for biofuel production at national and global levels based on specific criteria. For instance, a study conducted in China found that 35–75 mha of marginal land, including steep hillsides, idle land, and saline land, is suitable for biofuel feedstock crops if it is utilized effectively (Li and Chan-Halbrendt 2009). Another study from China estimated that approximately 152 mha of marginal land is available, of which 10% to 60% could potentially be utilized for bioenergy production (Zhuang et al. 2011). Furthermore, Africa was estimated to have approximately 500 mha of marginal, underused, and unused land available in tropical regions for biofuel production (Dynes 2008). Additionally, there is an estimation of 385–472 mha of agricultural land available, which is left abandoned based on historical agriculture databases (Field et al. 2008). However, these estimates relied on inconsistent data sources based on land availability for bioenergy production and are under scrutiny due to their limited database.

According to Perlack et al. (2005), approximately 16 mha of non-cropland in the United States could produce billions of tons of perennial biomass as an energy feedstock, fulfilling the United States' transportation fuel requirements. In the Eastern United States, over 2 mha of land is no longer used for agriculture, presenting an opportunity for biomass production. Estimates suggest that this area could yield between 2.3 and 17.4 megagrams ha^{-1} of biomass from perennial grasses and between 4.5 and 15.5 megagrams ha^{-1} from woody crops of a short rotation period (Stoof et al. 2015). Similar estimations are yet to be discovered from other nations as well. Advanced technologies, such as web-based GIS approaches, have been used to determine the location and sustainability, and they can also be utilized for marginal land exploration in urban areas.

For example, a GIS framework has revealed a significant opportunity in Pittsburgh, Pennsylvania, where 2400 hectares of urban marginal land (UML) could be transformed to cultivate sunflowers for bioenergy production. Ultimately, this results in an impressive yield of 34,100 gallons of biodiesel derived from sunflower crops (Niblick et al. 2013). A geospatial evaluation of UML in the Patiala district of Punjab, India, suggests the availability of 990 ha for the forestry plantation coupled with bioenergy and carbon sequestration (Edrisi et al. 2025a). Furthermore, India boasts around 46.67 mha of wasteland, with 39.24 mha identified as suitable for bioenergy production at the national scale (NRSC 2011; Edrisi and Abhilash 2016). In contrast, another study estimated that India has a wasteland of 60 mha, of which 50% is suitable for *Jatropha-based* biodiesel (Ananthapadmanabhan 2006). Moreover, due to the limitations of the data and the complexity caused by spatial variations in socioeconomic conditions and technologies, it is challenging to obtain a precise estimate in a shorter time. In addition, land alongside highways and roads can be utilized to cultivate energy crops. The construction of roads and highways is an important key aspect, which is followed in any infrastructural development project globally. While developing these projects, new roads and highways are integrated into the transportation system each year, and planting trees along these routes is a common practice in various countries to mitigate pollution and biomass production. It will further lay the foundation for a bio-based economy to achieve sustainable development.

8.2.3 Bioenergy Production and Concept of Bio-Based Economy or Bioeconomy

The transformation toward a bio-based economy from a fossil-based economy plays a significant role in achieving various sustainable development goals. However, this transition requires new policies, innovative strategies, financial stability, a risk-taking approach, and sustainable ecological models. Managing local ecosystems for biomass production can improve stewardship among different stakeholders (Scheffer et al. 2015). In 2014 and 2015, the European Commission launched a circular economy concept to "close the loop," creating strong links between the bioeconomy and circular economy (Carus and Dammer 2018). This association is crucial, and now numerous studies focus on the bioeconomy, which aims to use renewable biological resources to produce sustainable energy, food, and industrial goods. The bioeconomy is viewed as a potential alternative to fossil resources to achieve sustainable development.

As we know, the bioeconomy is resource-based; agricultural biomass production plays a vital role in food and fodder production and acts as a raw material for several end-products in biorefineries. Once the biomass is prepared for further processing, biomass conversion into biofuel consists of several techniques that depend on end-product purity and environmental friendliness. As per the flowchart mentioned above (Fig. 8.2), pyrolysis is a process under which anaerobic decomposition of biomass occurs at a temperature range of 650 to 800 K. During this process, a vapor phase substance condenses into the mixture, which consists of various low molecular weight components such as aldehydes, esters, organic acids, sugars, alcohols, and aromatic compounds—ultimately, forming a mixture called as bio-oil and

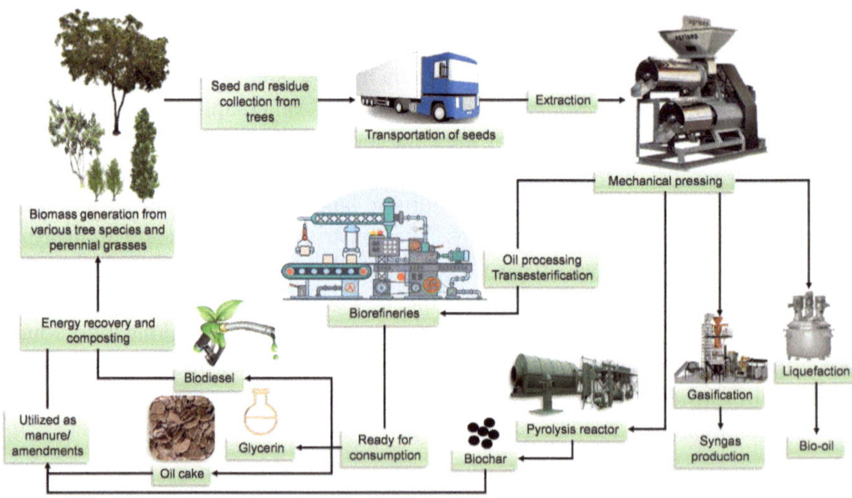

Fig. 8.2 Flowchart showing steps involved in cultivating and harnessing bioenergy feedstock under a circular process to approach a bio-based economy. (Adapted from Wang et al. 2023)

biochar. Further, this process is an effective way to decompose the lignocellulosic biomass (McKendry 2002). On the other hand, the process of gasification involves heating biomass to a temperature above 1000 K. This results in the formation of a gas mixture containing compounds, such as CO_2, CH_4, N_2, and others. Syngas, a combination of CO and H_2, can be easily obtained from the earlier-mentioned gas mixture (McKendry 2002).

Furthermore, liquefaction involves the catalytic thermal breakdown of biomass into minor components, which are then combined with bio-oils. This process requires a lower temperature than pyrolysis but higher pressures of 5–20 atmospheres. There is one more process called transesterification. During transesterification, triglycerides react, and the ester bond between the fatty acid chain and glycerol is hydrolyzed, followed by esterification with methanol. Later, this process produces fatty acid methyl esters, a significant biodiesel component (Hu et al. 2012).

Interestingly, bioeconomy refers to the transition of biological residues sustainably, eventually recycled, remanufactured, and reused (Carus and Dammer 2018). It has recently emerged and is more specifically known as a circular bio-based economy (Lange 2022). As compared to conventional agricultural products, the bio-based economy intends to utilize several forms of biomass that are acquired sustainably, consisting of residual biomass (Olofsson and Börjesson 2018) and agricultural wastes collected from various areas, which are degradable (Gontard et al. 2018). However, in terms such as "bioeconomy" and "bio-based economy," there are no perfect definitions to define them properly. More research is required to improve significantly and understand them in the future and make them more reliable regarding their socioeconomic credibility. Furthermore, energy crop suitability can be an essential criterion for achieving efficient biofuel production in terms of the bio-based economy.

8.2.4 Energy Crop Suitability and Other Prospects

Numerous plant species exhibit potential for use as bioenergy crops, depending upon land availability, climatic conditions, and suitability (Li et al. 2008). However, selected lignocellulosic crops, such as *Jatropha*, provide an opportunity to use as aboveground biomass for bioenergy feedstock instead of depending on harvested fruits. Grasses like *Miscanthus* and Switchgrass are ideal choices for cultivation on marginal lands due to their ability to thrive even in poor soils, require lesser nutrients, and have superior water use (Fargione et al. 2010; Sang and Zhu 2011). There is plenty of evidence in the literature that supports the global suitability of marginal and degraded lands for bioenergy crop cultivation. *Agave* spp., for instance, has the potential to produce up to 34 tons of dry biomass ha^{-1} yr^{-1} when used as a feedstock for biofuel in severely arid and degraded lands, according to research done by Davis et al. (2011). Furthermore, the potential of *Miscanthus* sp. was evaluated by Hastings et al. (2009) and Heaton et al. (2008) in the United States (US) and Europe, respectively. According to estimates, *Miscanthus* produces between 30 and 61 tons of harvested biomass per hectare annually in the United States (Heaton et al. 2008),

Table 8.1 Suitable tree species used in the tropics with annual biomass productivity (in Mg ha^{-1} yr^{-1}) depending upon bioenergy production (in GJ ha^{-1} yr^{-1}) based on forest

S. No.	Name of the tree species	Common name	Biomass and bioenergy yield	
			Mg ha^{-1} yr^{-1}	GJ ha^{-1} yr^{-1}
1.	*Agathis borneensis*	Kauri or Dammar	1.35	25.0
2.	*Aleurites moluccana*	Kemiri	4.65	86.0
3.	*Calamus caesius*	Rattan palm	2.25	42.0
4.	*Calliandra calothyrsus*	Powderpuff	15.0	277.5
5.	*Dyera polyphylla*	Jelutung	9.70	179.5
6.	*Fleroya ledermannii*	Abura	2.95	54.0
7.	*Gliricidia sepium*	Seema konna	7.00	129.5
8.	*Neolamarckia cadamba*	Kadamba	7.35	136.0
9.	*Sesbania bispinosa*	Daden or Sesban	12.5	231.5
10.	*Spondias mombin*	Amra	0.40	7.0

Source: Borchard et al. (2018)

whereas 40 tons of dry biomass per hectare are produced in Europe (Hastings et al. 2009). Further, in the assessment of *Panicum virgatum* (Switchgrass) biomass production across the United States, Wullschleger et al. (2010) discovered biomass yields of 8.7 and 12.9 tons ha^{-1} in upland and lowland regions, respectively.

Similarly, *Sorghum bicolor* has a significant potential to help China meet its 2050 biofuel targets, as noted by Nie et al. (2019). Corral et al. (2016) estimated that *Jatropha curcas* could produce more than 35.3 tons ha^{-1} of biodiesel on Spanish abandoned land. In addition, Kuhlman et al. (2013) investigated the possibility of using *Phragmites australis* to produce biogas, ethanol, and combustion in the Netherlands' marginal lands, with corresponding energy yields of 265.5 GJ ha^{-1}, 48 GJ ha^{-1}, and 52 GJ ha^{-1}. Apart from this, tree species such as *Azadirachta indica* and *Madhuca indica* are also utilized for biomass production in India. Still, their contribution to land restoration is comparatively less due to a long gestation period. There are a few tree species that are also suitable for bioenergy production as per available literature (Biswas et al. 2011; Atabani et al. 2013; Edrisi et al. 2023; Edrisi et al. 2025b; Abel et al. 2013) by incorporating information based on climate-adaptive agroforestry, silviculture and energy yield potential ha^{-1} yr^{-1} as mentioned in Table 8.1.

The International Energy Agency (IEA) estimates that India currently has 750 million tons of available biomass annually, with a surplus of 230 million metric tons. By 2040, this biomass is projected to yield 130 million tons of oil equivalent, thereby fulfilling about 15% of the nation's total energy demand (PwC report 2023). The Ministry of New and Renewable Energy (MNRE, India) committed to its vision by continuing with the National Bioenergy Programme to recover energy till 2025–2026; under this program, the ministry supports the Briquettes and Pellets manufacturing and promotes biomass (non-bagasse) based cogeneration in industrial sectors. Additionally, India is actively working toward creating a robust market for bioproducts to boost the economic viability of biomass pellets and briquettes. Around 230 manufacturers of biomass pellets and nearly 1030 manufacturers of briquettes were distributed across different states (PwC report 2023). Furthermore, these manufacturers are crucial suppliers to power plants and industries and pave the way for a sustainable and cleaner energy future in India (PwC report 2023). Various programs by governments are meant to enable sustainable utilization of bioenergy and fulfill the growing energy demand by implementing proper strategies to ace its economic potential and energy recovery.

8.3 Strategies for Assessing Economic Potential

Marginal lands are intended to provide land resources for food production. They comprise 1.3 billion ha of land, which is 36% of land globally, and can support approximately one-third of the global population if utilized reasonably (Wood et al. 2000). Identifying marginal lands is vital for implementing sustainable and innovative systems based on "smart-intensification" principles. These systems offer a variety of new products (such as organic and wild or semi-wild food, non-wood forest products, and ecosystem-based services) (Weltin et al. 2018; Schröder et al. 2018). Furthermore, marginal lands are the first step in unlocking their potential and transforming them into productive lands that can contribute to environmental, societal, and economic sustainability challenges (Schröder et al. 2018). Recent studies have shown that using marginal lands for energy crop production provides beneficial results (Rahman et al. 2014; Milbrandt et al. 2014). Later, it will help reevaluate less productive areas, align with European socioeconomic policies (EU 2018; FAO and IEA 2017), and support resilient land development. Therefore, it is essential to consider marginal lands' role in food production and the need for land conservation and sustainability. There are various strategies involved in marginal land assessment, such as:

- Application of phytoremediation techniques for land and biomass recovery
- Soil improvement on marginal land or degraded land
- Decision-making approach
- Site suitability
- Associated risk factors
- Life cycle assessment LCA
- Emergy-based ("energy memory" or "embodied energy") approach

The success of the phytoremediation technique relies on various distinct traits, such as contamination characteristics, hydrology, geology/hydrogeology, and site grade. Additionally, traits such as planned long- and short-term perspectives and land use are essential in choosing the appropriate phytoremediation technique. For example, recovery from biomass is often researched or trialed as part of managing contaminated land. Under phytoremediation, technology, such as phytoextraction, is used to a lesser extent as compared to phytostabilization for managing diffuse metal contamination across wider areas (Mench et al. 2010). Furthermore, in certain situations, marginal land may not have viable soil, so a series of "soil-forming" interventions are subjected to get the desired results (Golabi et al. 2007). Recycling organic matter into the soil can help create the right conditions for biomass production and maintain the soil during biomass production for several years or decades.

The restoration and development program should consist of a decision-making approach that clearly states its objectives regarding the marginal land and associated constraints related to it. For example, the ultimate aim should be a long-term, financially stable solution for the sustainable transformation of a degraded site to maximize land use for biomass production. Site suitability assessment aims to determine whether site conditions are viable for growing biomass crops or can be improved through specific interventions, by using soil remediation techniques. For instance, if considering establishing an on-site utilization facility, it is essential to evaluate the suitability and infrastructure of the site. Additionally, a risk-based approach is taken to assess the viability of marginal land, considering the potential identified in the initial stages by evaluating technology implications, conducting detailed diligence (e.g., financial constraints), and planning for implementation.

Furthermore, the life cycle assessment (LCA) findings highlight the potential of grasses cultivated on marginal land, mainly perennial species prevalent in the Mediterranean region. Eventually, it indicates that energy production in a joint heat and power plant can significantly reduce GHG emissions and reliance on nonrenewable energy resources. Therefore, *Miscanthus*, in particular, can save up to 13 tons of CO_2 equivalent ha^{-1} yr^{-1} and 230 GJ ha^{-1} yr^{-1}, respectively (Schmidt et al. 2015). Similarly, emergy, which represents "energy memory" or "embodied energy," is defined as available energy based on sustainability analysis (Brown and Ulgiati 2004). An analysis was done successfully on *Tectona grandis* and *Pongamia pinnata* and compared them with soybean and *Jatropha curcas*. Further, in a defined set of system boundaries, results revealed that for soybean total solar energy joules (sej) were 0.99E+16; for *Jatropha, Pongamia* sej were 1.01E+16; 1.33E+16 and for *Tectona* sej were 0.72E+16 based on bioenergy production alternatives (Edrisi et al. 2022).

In recent years, a growing consensus among regional and national organizations, governmental institutions, researchers, and experts has consistently highlighted the potential of biomass as an energy source on marginal lands. This approach aims to unlock the economic potential of underutilized land, such as marginal land. However, while competing demands for traditional food production on agricultural land, this initiative helps achieve a sustainable, low-carbon society (GBEP 2018; Dale et al. 2016; Blanco-Canqui 2016; E.C. 2017). Governments are implementing ambitious policies and developmental programs to cultivate dedicated bioenergy crops that can thrive on marginal lands by improving their yield, quality, and adaptability. The European Union (EU) is projected to increase its bioenergy-related cropland footprint to 14.3 mha by 2020 (De Schutter and Giljum 2014), driven by political goals. Globally, there is an expected increase in biofuels for the transportation sector from 2% to 10% by 2020 (US Congress 2007). In the United States, the Energy Independence and Security Act has mandated the utilization of 36 billion gallons of renewable fuels by 2022 (U.S. Congress 2007). Strategies, which include incentives or subsidy programs, are anticipated to facilitate the conversion of marginal lands into bioenergy production. Notably, one of the most significant benefits of biomass power and biofuels is their capacity to generate millions of jobs and stimulate substantial income growth, among the less privileged communities. Therefore, it is imperative to consider these strategies not solely from an energy perspective but also from a broader context of economic development (India 2002).

8.4 Challenges and Policy Implications

As discussed in earlier sections, marginal and degraded land has great potential for bioenergy production and offers various socioeconomic benefits, but it also comes up with challenges. The impact of using marginal land on sustainability, ecosystem services, and the environment has been a long-standing concern. The US Conservation Reserve Program (CRP) serves an important role in encouraging farmers to take farmland out of production in order to prevent soil loss and nonpoint source pollution (USDA 2006). To ensure long-term sustainability and minimize ecological hazards from consistent land use, the soil needs management and monitoring (Blanco-Canqui 2010). The degradation of soil and land is a critical challenge that demands urgent attention, as recommended by North American prairie grass systems analysis for lignocellulosic biofuel production (Gutierrez and Ponti 2009). Furthermore, reports suggest that producing bioenergy with land use changes may not effectively counterbalance carbon emissions (Searchinger et al. 2008; Righelato and Spracklen 2007).

A comprehensive systems study is required to address the complex challenges for sustainably utilizing the marginal and degraded lands (Gopalakrishnan et al. 2009; IBRDB 2008). Even though bioenergy is a spotless and renewable energy source, it does face numerous challenges. For example, in countries like China, Brazil, and the United States, the yearly demand for fuels is in billion liters, which can only be met by producing large-scale biomass from arable lands. Furthermore, it accounts for 10% of the total global land area. However, this approach is impractical; decreasing cropland leads to a surge in food prices (Runge and Senauer 2007) and triggers the increase of cropland in untouched areas, leading to GHG emissions and considerable deforestation (Searchinger et al. 2008). In these conditions, marginal lands will provide a pathway to obtain sustainable bioenergy by considering all factors to improve biomass productivity. Therefore, effective management practices are necessary to achieve the goals as mentioned earlier.

There are various limitations regarding marginal land utilization, such as elevation, slope, texture, depth, internal drainage, rockiness, remoteness, and fertility, which restrict profitability from the land. Furthermore, these characteristics affect the suitability for agricultural productivity through their impact on soil hydrology (Stoof et al. 2015). Most marginal land soils are affected due to seasonal variation in the atmosphere since they have shallow restrictive layers and relatively permeable soil surfaces on the top (Walter et al. 2000; Stoof et al. 2015). Moreover, soil on hill slopes with shallow restrictive layers and sandy, "excessively drained" soil has inadequate water-holding capacity (Walter et al. 2000). As a result, the soil has moisture deficits in summer and often limits its forage production.

Generally, unsuitable annual crops and poor prior management can lead to significant erosion, reduction in soil depth, native fertility, and organic matter, ultimately converting the soils to a marginal status. Marginal lands are very susceptible to natural processes and changing management scenarios as they are under a transitional state. Degradation of the land mainly occurs as a result of poor management of productive land. Another major component, which impacts the marginal land value, is the market-driven socioeconomic factors. The area of land used for cultivation will advance or recede depending on demand (Hollander 1895). The reduced land rent and increased market demands often change the profit scenario and ultimately lead the marginal lands into profitable ventures through production.

Recent approaches for distinguishing marginal lands are primarily empirical and qualitative, often containing subjective elements. However, these methods align with specific management targets for croplands that fluctuate across nations and vary spatiotemporally. The Food and Agriculture Organization within the United Nations defines marginal land as areas with significant limitations, which encompass wetlands, soil salinized, eroded, and lands that fall below acceptable cultivation standards (FAO 1993). As per USDA-NRCS, marginal land is assessed and utilized for farming suitability. With the current understanding, marginal lands can also be defined as physically not viable lands or have climate and soil restrictions, low production, fragile ecosystem services, and high environmental risk, hence being unrewarding in terms of economic returns. Furthermore, economic marginal land is not always the same as biophysical marginal land, according to Kang et al. (2013). However, crop productivity-based and biophysical measures do not always reflect the environmental implications associated with land use in the long term. Later, a hierarchical framework developed that recognizes extremely productive land becomes environmentally degraded due to continuous crop production and is changed into marginal land eventually (Kang et al. 2013).

Assessment and management of marginal land requires multiple concerns and needs. Later, it consists of a systemic approach that considers socioeconomic impacts and land functions. The important driving elements of marginal lands comprise food supply demand and bioenergy, the effects due to policies and incentives, ecological services, trade-offs related to the environment, and sustainability. Subsequently, improvement in soil and land quality is the need of the hour for further comprehensive management and quantitative assessment. Recent policies on biomass production are intended to achieve bioenergy production targets by channelizing ecosystem functions and sustainability, which is often at the expense of environmental effects such as climate change. It is critical to adjust land use policies to meet multifaceted goals with several trade-offs besides production, as highlighted by Acs et al. (2010), Strijker (2005), and Robertson et al. (2008). The following research domains are crucial regarding the assessment and management of marginal lands:

- Formulating and refining the criteria for marginal lands
- Networking of a land informatics framework
- Quantification of marginal land based on land use
- Interpretation of various land functions resulting from area development
- Integrating supportive policies to facilitate marginal land assessment

The formulation and refinement of criteria to define marginal land and its evaluation vary with time; however, it also depends on location and the precedence of the management targets. While classifying marginal land, it is more advantageous to consider a multiple-factor approach for robust decision-making. Further, a framework based on a land informatics system considers land functions and socioeconomic impacts, as depicted in Fig. 8.3. It will help to assess, manage, and monitor marginal land by using land databases and considering various land functions to classify land based on associated risks (Arshad et al. 2021). The integration of high-resolution spatial data enables the creation of two-way dynamic models, facilitating regular updates on marginal land conditions (Arshad et al. 2021). As a result, this innovation paves the way for more streamlined land use planning and management to become more accessible. Quantifying marginal land based on land use carefully assesses the marginal land potential for future biomass supply. Integrating bioenergy production on suitable marginal lands into the future might contribute to reaching renewable energy goals for 2050 (Gerwin et al. 2018). The importance of marginal lands is expected to grow in the coming decades, as bioenergy is anticipated to play a critical role in future energy supplies to meet GHG gas emissions reduction targets by 2050 (Gerwin et al. 2018).

Furthermore, the bioenergy market growth depends on supportive policies at local, state, and national levels (Stoof et al. 2015). However, earlier assessments of bioenergy potential require to be more accurate in terms of production value. They

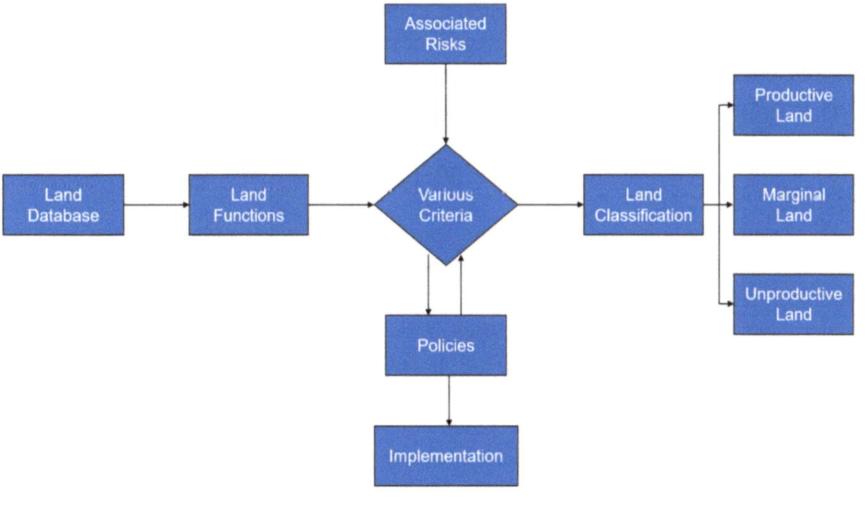

Fig. 8.3 Framework for Land Informatics system assessment. (Adapted from Arshad et al. 2021)

focused primarily on the recently available degraded land or abandoned agricultural lands, overlooking the longer-term abandonment issues experienced in a specific region (Stoof et al. 2015). Therefore, the responsibility for the area's development initiatives is left to states, localities, and individuals. Given the increasing development and use of opportunities at the state and national levels, it is imperative to introduce new supportive policies that encourage sustainable bioenergy production at the regional level. The growth of bioenergy production is obstructed by various policy-related challenges, which encompass research funding, regional GHG initiatives, biofuels policies, and national land conservation programs. Research initiatives are currently focused on dealing with the impacts and unique challenges of production on marginal land and degraded land, which are specific to each region. In contrast, bioenergy production has the potential to offer significant environmental benefits and be socio-economically stable. Therefore, it is important to address significant factors, which will expedite bioenergy production and ultimately contribute to enhancing livelihoods in rural areas and strengthening environmental quality on national and regional levels.

8.5 Conclusions and Future Recommendations

The concept of marginal land and degraded land utilization sparked a significant discourse regarding land management in various nations and academic disciplines. Furthermore, these lands are crucial in producing food and biomass, especially as land resources become increasingly limited. Therefore, investing in energy crops as a source of alternative fuel is a promising way to meet our growing energy demands. This approach addresses the challenge of diminished land resources and resolves the rising conflict between fuel and food production. Using marginal and degraded lands to grow perennial trees, shrubs, and grass species for dedicated energy production offers a significant enhancement in energy and food security. Furthermore, this strategy will reduce dependency on edible oils and other crops while delivering various environmental benefits. Additionally, to restore marginal lands, it is necessary to have a clear understanding of their various functions, economic, and social impacts. As a result, it allows landowners to upgrade their living standards by the effective use of barren lands into a profitable venture and develop a bio-based economy.

Moreover, there is also a challenge for the landowners to identify their specific end-users like industrial partners, biorefineries, which have a pivotal role in closing the loop of linear economic models under the fossil-based economy as discussed under Sect. 8.2.3. These biorefineries have various industrial mechanisms, like pyrolysis, liquefaction, transesterification, and others, that help transform the lignocellulosic biomass into biofuel and bioenergy. Furthermore, such processes and techniques need a thorough assessment for developing an efficient and effective bioeconomic model that includes a decision-making approach, site suitability analysis, assessment of risk factors, LCA, and emergy analyses. However, a single-directional approach would not be enough to fulfill the need for the ever-increasing fuel demands. Therefore, a comprehensive multi-dimensional approach is needed that integrates the utilization of GIS, modelling, and simulation to study the marginal and degraded l and potential for a thorough understanding of large-scale biomass production and its implications on the environment. Subsequently, by prioritizing purpose-driven research, we will be able to ensure long-term, sustainable, and cost-effective development.

Acknowledgments The authors would like to express gratitude to the Director, Dean, and Head TSLAS, TIET, Patiala, India, for providing institutional support and encouragement. SAE would like to thank Thapar ACT, Thapar Seed Fund, and SERB-SRG for the funding opportunities.

References

Abel S, Couwenberg J, Dahms T, Joosten H (2013) The Database of Potential Paludiculture Plants (DPPP) and results for Western Pomerania. Plant Divers Evol 130:219–228

Acs S, Hanley N, Dallimer M, Gastonvb KJ, Roberston P, Wilson P, Armsworth PR (2010) The effect of decoupling on marginal agricultural systems: implication for farm incomes, land use and upland ecology. Land Use Policy 27:550–563. https://doi.org/10.1016/j.landusepol.2009.07.009

Ananthapadmanabhan G (2006) Bio-fuels vs. Foods Implications for Indian Farmers; National Workshop, Hyderabad, December 8th and 9th. http://www.csa-india.org/kiaworkshop/KIA_TOTAL/KIA_PAPERS/8.ananth-biofuels%20vs%20food-implica-tions.pdf

Arshad MN, Donnison I, Rowe R (2021) Marginal Lands: Concept, classification criteria, and management. Supergen Bioenergy Hub report no. 02/2021. Published in Birmingham, England. Available from: https://www.supergen-bioenergy.net/wp-content/uploads/2021/09/Marginal-Land-Report.pdf

Atabani AE, Silitonga AS, Ong HC, Mahlia TMI, Masjuki HH, Badruddin IA, Fayaz H (2013) Non-edible vegetable oils: a critical evaluation of oil extraction, fatty acid compositions, biodiesel production, characteristics, engine performance and emissions production. Renew Sust Energ Rev 18:211–245

Batidzirai B, Smeets E, Faaij A (2012) Harmonising bioenergy resource potentials; methodological lessons from review of state of the art bioenergy potential assessments. Renew Sustain Energy Rev 16:6598–6630

Bigges A (2007) The landscape and marginal cropping lands-inherent and induced. Trop Grassl 41:133–138

Biswas B, Scott PT, Gresshoff PM (2011) Tree legumes as feedstock for sustainable biofuel production: opportunities and challenges. J Plant Physiol 168:1877–1884

Blanco-Canqui H (2010) Energy crops and their implications on soil and environment. Agron J 102:403–419. https://doi.org/10.2134/agronj2009.0333

Blanco-Canqui H (2016) Growing dedicated energy crops on marginal lands and ecosystem services. Soil Sci Soc Am J 80:845. https://doi.org/10.2136/sssaj2016.03.0080

Borchard N, Bulusu M, Hartwig AM, Ulrich M, Lee SM, Baral H (2018) Screening potential bioenergy production of tree species in degraded and marginal land in the tropics. Forests 9:594. https://doi.org/10.3390/f9100594

Brosowski A, Krause T, Mantau U, Mahro B, Noke A, Richter F (2019) How to measure the impact of biogenic residues, wastes and by-products: development of a national resource monitoring based on the example of Germany. Biomass Bioenergy 127:105275

Carus M, Dammer L (2018) The circular bioeconomy—concepts, opportunities, and limitations. Ind Biotechnol 14(2):83–91

Corral S, De Lara DRM, Salguero MT, Mendoza CCJ, De La Nuez DL, Santos MD, Pena FD (2016) Assessing Jatropha crop production alternatives in abandoned agricultural arid soils using MCA and GIS. Sustainability 8:505. https://doi.org/10.3390/su8060505

Dale VH, Kline KL, Buford MA, Volk TA, Tattersall Smith C, Stupak I (2016) Incorporating bioenergy into sustainable landscape designs. Renew Sust Energ Rev 56:1158–1171. https://doi.org/10.1016/j.rser.2015.12.038

Davis SC, Dohleman FG, Long SP (2011) The global potential for agave as a biofuel feedstock. GCB Bioenergy 3:68e78. https://doi.org/10.1111/j.1757-1707.2010.01077.x

De Schutter L, Giljum SA (2014) Calculation of the EU bioenergy land footprint. Discussion paper on land use related to EU bioenergy targets for 2020 and an outlook for 2030. Vienna

Dynes M (2008) Growing up. Africa Investor 31

Edrisi SA, Abhilash PC (2016) Exploring marginal and degraded lands for biomass and bioenergy production: an Indian scenario. Renew Sustain Energy Rev 54:1537–1551

Edrisi SA, Sahiba SA, Chen B, Abhilash PC (2022) Emergy-based sustainability analysis of bioenergy production from marginal and degraded lands of India. Ecol Model 466:109903

Edrisi SA, Singh A, Dubey PK, Abhilash PC (2023) *Tectona grandis* L.f. mediated restoration of marginal lands in Eastern Uttar Pradesh, North India. Land Degrad Dev 34(4):929–42. https://doi.org/10.1002/ldr.4506

Edrisi SA, Chaurasiya R, Verma V (2024) Sustainable Plant Nutrition and Soil Carbon Sequestration Greenhouse Gas Emissions and Plant Soil Nutrition: Towards Nutrient-Use Efficiency for Achieving Net-Zero Emissions. In: Edrisi SA, Dubey PK, Rao CS, Baral H, Chaturvedi RK, Abhilash PC (eds) Sustainable Plant Nutrition and Soil Carbon Sequestration (2024). Sustainable Plant Nutrition in a Changing WorldSpringer Nature, Switzerland, Cham 65–85

Edrisi SA, Verma V, Kumar D (2025a) Geospatial evaluation of urban marginal lands for the forestry development in Northern India. Environment Development and Sustainability. https://doi.org/10.1007/s10668-025-06787-z

Edrisi SA, Tripathi V, Prakash NT, Abhilash PC (2025b) Climate-adaptive agroforestry: restoring marginal lands for fuel, food, and nutrition security. Environmental Sustainability. 8(1) 127-134. https://doi.org/10.1007/s42398-025-00344-0

El-Beltagy A (2000) Land degradation: a global and regional problem; in: conference proceeding for the international conference United Nations and global governance in the new millennium, 19–21 January 2000, Tokyo. http://www.unu.edu/millennium/environment.html

Emery I, Mueller S, Qin Z, Dunn JB (2017) Evaluating the potential of marginal land for cellulosic feedstock production and carbon sequestration in the United States. Environ Sci Technol Food Industry 51:733–741

European Commission, Proposal for a Directive of the European Parliament and of the Council on the Promotion of the Use of Energy from Renewable Sources (Recast), Brussels (2017). https://eur-lex.europa.eu/legal-content/EN/TXT/?uri=CELEX:52016PC0767R%2801%29

European Union (2018) Directive 2018/2001 of the European Parliament end of the Council of 11 December 2018 on the promotion of the use of energy from renewable sources

FAO and IEA (2017) HOW2GUIDE for Bioenergy, Roadmap Development and Implementation. https://www.iea.org/publications/freepublications/publication/How2GuideforBioenergyRoadmapDevelopmentandImplementation.pdf

Fargione JE, Plevin RJ, Hill JD (2010) The ecological impact of biofuels. Annu Rev Ecol Evol Syst 41:351–377

Field CB, Campbell JE, Lobell DB (2008) Biomass energy: the scale of the potential resource. Trends Ecol Evol 23(2):65–72

Food and Agricultural Organization of United Nations (FAO) (1993) FESLM: an international framework for evaluating sustainable land management. World Resources Report 73. FAO, Rome, Italy

Gelfand I, Sahajpal R, Zhang X, Izaurralde RC, Gross KL, Robertson GP (2013) Sustainable bioenergy production from marginal lands in the US Midwest. Nature 493:514–517

Gerwin W, Repmann F, Galatsidas S, Vlachaki D, Gounaris N, Baumgarten W et al (2018) Assessment and quantification of marginal lands for biomass production in Europe using soil-quality indicators. Soil 4(4):267–290

Golabi MH, Denney MJ, Iyekar C (2007) Value of composted organic wastes as an alternative tosynthetic fertilizers for soil quality improvement and increased yield. Compost Sci Util 15:267–271

Gontard N, Sonesson U, Birkved M, Majone M, Bolzonella D, Celli A et al (2018) A research challenge vision regarding management of agricultural waste in a circular bio-based economy. Crit Rev Environ Sci Technol 48(6):614–654. https://doi.org/10.1080/10643389.2018.1471957

Gopalakrishnan G, Negri MC, Wang M, Wu M, Snyder SW, Lafreniere L (2009) Biofuels, land, and water: a systems approach to sustainability. Environ Sci Technol 43:6094–6100. https://doi.org/10.1021/es900801u

Gutierrez AP, Ponti L (2009) Bioeconomic sustainability of cellulosic biofuel production on marginal lands. Bull Sci Technol Soc 29:213–225. https://doi.org/10.1177/0270467609333729

Hastings A, Clifton-Brown J, Wattenbach M, Mitchell CP, Stampfl P, Smith P (2009) Future energy potential of Miscanthus in Europe. GCB Bioenergy 1:180–196. https://doi.org/10.1111/j.1757-1707.2009.01012.x

Heaton EA, Dohleman FG, Long SP (2008) Meeting US biofuel goals with less land: the potential of Miscanthus. Glob Chang Biol 14:2000–2014. https://doi.org/10.1111/j.1365-2486.2008.01662.x

Herrick JE, Abrahamse T, Abhilash PC, Ali SH, Alvarez-Torres P, Barau AS et al (2019) Land restoration for achieving the sustainable development goals: an international resource panel think piece. United Nations Environment Programme

Hollander JH (1895) The concept of marginal rent. Q J Econ 9:175–187

Hu J, Yu F, Lu Y (2012) Application of Fischere Tropsch synthesis in biomass to liquid conversion. Catalysts 2:303e26. https://doi.org/10.3390/catal2020303

India (2002) India Vision 2020. Planning Commission, Government of India. December. Available at http://planningcommission.nic.in/plans/planrel/pl_vsn2020.pdf

Interagency Biomass Research and Development Board (IBRDB) (2008) Increasing feedstock production for biofuels: economic drivers, environmental implications, and the role of research, Washington D.C.

Jiang C, Guan K, Khanna M, Chen L, Peng J (2021) Assessing marginal land availability based on land use change information in the contiguous United States. Environ Sci Technol 55(15):10794–10804

Kang S, Post W, Wang D, Nichols J, Bandaru V, West T (2013) Hierarchical marginal land assessment for land use planning. Land Use Policy 30:106–113

Karp A, Shield I (2008) Bioenergy from plants and the sustainable yield challenge. New Phytol 179:15–32

Keesstra S, Mol G, De Leeuw J, Okx J, De Cleen M, Visser S (2018) Soil-related sustainable development goals: four concepts to make land degradation neutrality and restoration work. Land 7(4):133

Kidd P, Mench M, Álvarez-López V, Bert V, Dimitriou I, Friesl-Hanl W, Neu S (2015) Agronomic practices for improving gentle remediation of trace element-contaminated soils. Int J Phytoremediation 17(11):1005–1037

Kuhlman T, Diogo V, Koomen E (2013) Exploring the potential of reed as a bio energy crop in The Netherlands. Biomass Bioenergy 55:41–52. https://doi.org/10.1016/j.biombioe.2012.06.024

Lal R (2004) Soil carbon sequestration to mitigate climate change. Geoderma 123:1–22. https://doi.org/10.1016/j.geoderma.2004.01.032

Lange L (2022) Christian Patermann and the bioeconomy: a testimony on his 80th anniversary: important next steps for speeding up and unlocking the full potential of a circular biobased economy. EFB Bioecon J 2:100037

Levis S, Kelly M (2014) Mapping the potential for biofuel production on marginal lands: differences in definitions, data and models across scales. ISPRS Int J Geo Inf 3:430–459. https://doi.org/10.3390/ijgi3020430

Li S, Chan-Halbrendt C (2009) Ethanol production in China: potential land technologies. Appl Energy 86:S162–S169
Li X et al (2008) Major energy plants and their potential for bioenergy development in China. Environ Manag 46:579–589
McKendry P (2002) Energy production from biomass (part 1): overview of biomass. Bioresour Technol 83(1):37–46
Mench M, Lepp N, Bert V, Schwitzgubel J, Gawronski SW, Schroeder P, Vangronsveld J (2010) Successes and limitations of phytotechnologies at field scale: outcomes, assessment and outlook fromCOST action 859. J Soils Sediments 10:1039–1070
Milbrandt AR, Heimiller DM, Perry AD, Field CB (2014) Renewable energy potential on marginal lands in the United States. Renew Sust Energ Rev 29:473–481
MNRE, India. Schemes | Ministry of New and Renewable Energy | India (mnre.gov.in)
Naveed MA, Donnison I, Rowe R (2022) Marginal lands: concept, classification criteria and management. https://doi.org/10.13140/RG.2.2.11996.46721
Niblick B, Monnell JD, Zhao X, Landis AE (2013) Using geographic information systems to assess potential biofuel crop production on urban marginal lands. Appl Energy 103:234–242
Nicholson E, Watermeyer KE, Rowland JA, Sato CF, Stevenson SL, Andrade A et al (2021) Scientific foundations for an ecosystem goal, milestones and indicators for the post-2020 global biodiversity framework. Nat Ecol Evol 5(10):1338–1349
Nie Y, Cai W, Wang C, Huang G, Ding Q, Yu L, Li H, Ji D (2019) Assessment of the potential and distribution of an energy crop at 1-km resolution from 2010 to 2100 in China – the case of sweet Sorghum. Appl Energy 239:395–407. https://doi.org/10.1016/j.apenergy.2019.01.214
Nixon D, Stephens W, Tyrrel S, Brierley E (2001) The potential for short rotation energy forestry on restored landfill caps. Bioresour Technol 77:237–245
NRSC, Wastelands Atlas of India: hange Analysis Based on Temporal Satellite Data of 2005-06 and 2008-09. Department of Land Resources, Ministry of Rural Development, Government of India. 2011. Available at: ⟨http://www.dolr.nic.in⟩
Olofsson J, Börjesson P (2018) Residual biomass as resource–life-cycle environmental impact of wastes in circular resource systems. J Clean Prod 196:997–1006
Perea-Moreno A-J, García-Cruz A, Novas N, Manzano-Agugliaro F (2017) Rooftop analysis for solar flat plate collector assessment to achieving sustainability energy. J Clean Prod 148:545–554
Perlack RD, Wright LL, Turhollow AF, Graham RL, Stokes BJ, Erbach DC (2005) Biomass as feedstock for a bioenergy and bioproducts industry: the technical feasibility of a billionton annual supply. DTIC Document
PwC report, India (2023) Fuelling India's future with bioenergy (pwc.in)
Qiu J (2009) China's climate target: is it achievable? Nature 462:550–551. https://doi.org/10.1038/462550a
Rahman MM, Mostafiz SB, Paatero JV, Lahdelma R (2014) Extension of energy crops on surplus agricultural lands: a potentially viable option in developing countries while fossil fuel reserves are diminishing. Renew Sust Energ Rev 29:108–119
Renewables (2023) Global Status Report-2023 http://www.ren21.net
Richards BK, Stoof CR, Cary IJ, Woodbury PB (2014) Reporting on marginal lands for bioenergy feedstock production: a modest proposal. Bioenergy Res 7(3):1060–1062. https://doi.org/10.1007/s12155-014-9408-x
Richardson CW, Bucks DA, Sadler EJ (2008) The conservation effects assessment project benchmark watersheds: synthesis of preliminary findings. J Soil Water Conserv 63:590–604
Righelato R, Spracklen DV (2007) Carbon mitigation by biofuels or by saving and restoring forests? Science 317:902. https://doi.org/10.1126/science.1141361
Robertson GP, Dale VH, Doering OC, Hamburg SP, Melillo JM, Wander et al (2008) Sustainable biofuels redux. Science 322:49–50. https://doi.org/10.1126/1161525
Rozzi E, Minuto FD, Lanzini A, Leone P (2020) Green synthetic fuels: renewable routes for the conversion of non-fossil feedstock into gaseous fuels and their end uses. Energies 13:420
Runge CF, Senauer B (2007) How biofuels could starve the poor. Foreign Aff:41–53

Sandhu H (2021) Bottom-up transformation of agriculture and food systems. Sustainability 13:2171
Sang T, Zhu W (2011) China's bioenergy potential. GCB Bioenergy 3:79–90
Saqib A, Tabbssum MR, Rashid U, Ibrahim M, Gill SS, M.A. (2013) Mehmood marine macro algae Ulva: a potential feed-stock for bioethanol and biogas production Asian. J Agric Biol 1:155–163
Scarlat N, Dallemand J-F, Banja M (2013) Possible impact of 2020 bioenergy targets on European Union land use. A scenario-based assessment from national renewable energy action plans proposals. Renew Sustain Energy Rev 18:595–606
Scheffer M, Barrett S, Carpenter SR, Folke C, Green AJ, Holmgren M, Hughes TP, Kosten S, van de Leemput IA, Nepstad DC, van Nes EH, Peeters ET, Walker B (2015) Creating a safe operating space for iconic ecosystems. Science 347:1317–1319
Schmidt T, Fernando AL, Monti A, Rettenmaier N (2015) Life cycle assessment of bioenergy and bio-based products from perennial grasses cultivated on marginal land in the Mediterranean region. Bioenergy Res 8:1548–1561
Schröder P, Beckers B, Daniels S, Gnädinger F, Maestri E, Marmiroli N, Mench M, Millán R, Obermeier MM, Oustriere N, Persson T (2018) Intensify production, transform biomass to energy and novel goods and protect soils in Europe—A vision how to mobilize marginal lands. Sci Total Environ 616:1101–1123
Searchinger T, Heimlich R, Houghton RA, Dong F, Elobeid A, Fabiosa J et al (2008). Use of U.S. Croplands for Biofuels Increases Greenhouse Gases Through Emissions from Land-Use Change Science 319:1238–1240. https://doi.org/10.1126/science.1151861
Smith SL, Thelen KD, MacDonald SJ (2013) Yield and quality analyses of bioenergy crops grown on a regulatory brownfield. Biomass Bioenergy 49:123–130
Stoof CR, Richards BK, Woodbury PB, Fabio ES, Brumbach AR, Cherney J, Das S, Geohring L, Hansen J, Hornesky J, Mayton H, Mason C, Ruestow G, Smart LB, Volk TA, Steenhuis TS (2015) Untapped potential: opportunities and challenges for sustainable bioenergy production from marginal lands in the Northeast USA. Bioenergy Res 8(2):482–501
Strijker D (2005) Marginal lands in Europe – causes of decline. Basic Appl Ecol 6:99–106. https://doi.org/10.1016/j.baae.2005.01.001
The Global Bioenergy Partnership (GBEP) (2018) ⟨http://www.globalbioenergy.org⟩
Tilman D, Hill J, Lehman C (2006) Carbon- negative biofuels from low-input high- diversity grassland biomass. Science 314(5805):1598–1600
UNEP (United Nations Environment Programme) (2007) Global Environment Outlook (GEO-4); Nairobi http://unep.org/geo/geo4/media/
United States Congress (2007) Energy Independence and security act of 2007. Public Law No: 110–140
United States Department of Agriculture (USDA) (2006) Agricultural conservation reserve protection program (CRP). USDA Farm Service Agency, Washington, DC
Walter MT, Walter MF, Brooks ES, Steenhuis TS, Boll J, Weiler K (2000) Hydrologically sensitive areas: variable source area hydrol ogy implications for water quality risk assessment. J Soil Water Conserv 55(3):277–284
Walthall B, Vicente-Vicente JL, Friedrich J, Piorr A, López-García D (2024) Complementing or co-opting? Applying an integrative framework to assess the transformative capacity of approaches that make use of the term agroecology. Environ Sci Pol 156:103748
Wang J, Fu J, Zhao Z, Bing L, Xi F, Wang F et al (2023) Benefit analysis of multi- approach biomass energy utilization toward carbon neutrality. Innovation 4(3):100423
Weltin M, Zasada I, Piorr A, Debolini M, Geniaux G, Moreno Perez O, Scherer L, Tudela Marco L, Schulp CJE (2018) Conceptualising fields of action for sustainable intensification– a systematic literature review and application to regional case studies. Agric Ecosyst Environ 257:68–80. https://doi.org/10.1016/j.agee.2018.01.023
Wood S, Sebastian K, Scherr SJ (2000) Pilot analysis of global ecosystems: agroecosystems. International Food Policy Research Institute and World Resources Institute, Washington, DC

Wullschleger SD, Davis EB, Borsuk ME, Gunderson CA, Lynd LR (2010) Biomass production in switchgrass across the United States: database description and determinants of yield. Agron J 102:1158. https://doi.org/10.2134/agronj2010.0087

Yan L, Zhang L, Wang S, Hu L (2008) Potential yield of bioethanol from energy crops and their regional distribution in China. Trans CSAE 24:213–216

Zhuang D, Jiang D, Liu L, Huang Y (2011) Assessment of bioenergy potential on marginal land in China. Renew Sustain Energy Rev 15:1050–1056. https://doi.org/10.1016/j.rser.2010.11.041

Plantation: A Sustainable Restoration Approach

9

Manoj Kumar Jhariya, Arnab Banerjee, Abhishek Raj,
Vimal Chandra Pandey, S. S. Bargali,
Santosh Kumar Verma, and Taher Mechergui

Abstract

Land degradation is one of the critical issues in front of the globe. The scientific community is trying to explore the restoration solution for it, and so far, a lot of work has been done for a specific region as well as for a broad spectrum. Among various techniques and applications, afforestation is regarded as a potential approach to land restoration. As per the reported works till 2018, approximately 300 million hectares of degraded lands were aimed for plantation globally. Such approaches involving tree plantation on marginal lands of about 174 million

M. K. Jhariya
Department of Farm Forestry, Sant Gahira Guru Vishwavidyalaya,
Ambikapur, Chhattisgarh, India

A. Banerjee
Department of Environmental Science, Sant Gahira Guru Vishwavidyalaya,
Ambikapur, Chhattisgarh, India

A. Raj
Pt. Deendayal Upadhyay College of Horticulture and Forestry, Dr. Rajendra Prasad Central Agriculture University, Samastipur, Bihar, India

V. C. Pandey (✉)
CSIR – National Botanical Research Institute,
Lucknow, Uttar Pradesh, India

Society for Green Development, Lucknow, Uttar Pradesh, India

S. S. Bargali
Department of Botany, D.S.B. Campus, Kumaun University, Nainital, Uttarakhand, India

S. K. Verma
Department of Civil Engineering, G H Raisoni University, Amravati, Maharastra, India

T. Mechergui
Faculte des Sciences de Bizerte (FSB), Zarzouna, Tunisie

V. C. Pandey (ed.), *Plantation-Based Land Restoration*,
https://doi.org/10.1007/978-981-95-2860-8_9

hectares are able to sequester 35 gigatons of carbon dioxide (CO_2) till 2050. Such activities would also provide additional benefits of about US dollar (USD) three trillion. Among developing countries, most of the countries reflected altered land use patterns, rapid industrialization, urbanization, and population rise that have continued the process of land degradation. This degradation leads toward unproductiveness, limits to production of particular crops or growth of vegetation, low groundwater table and water availability, soil loss, biodiversity loss, etc. Thus, it is urgent to move toward the restoration of degraded landscapes as well as halt the degradation process through combating measures. Plantation forestry has now become an accepted approach for increasing the production of forest-based products as well as reducing the pressure on the intact forest lost by deforestation. India has a wide availability of degraded land and forests as potential sites for plantation. This is properly managed with the correct choice of species can restore the degraded land, enhance the carbon reserve, and supply wood and non-wood forest products on a sustainable basis.

Keywords

Afforestation · Ecology · Forestry · Plantation · Restoration · Sustainability

9.1 Introduction

Forest and land degradation have emerged as major problems across the globe. This prevailing scenario is due to population bursts, the transformation of forest lands into agricultural lands, deforestation, and unscientific practices of agricultural production (FAO 2010). Intensive and unsustainable farming practices, along with the non-judicious application of agrochemicals, accelerate the degradation process of land and associated resources and the ecosystem as a whole (Meena et al. 2020; Jhariya et al. 2019; Raj et al. 2023). This further causes the enhancement in the acreage of cultivable wastelands.

The increment in the consumptive use of natural resources has led the scientific community to search for alternatives or best management perspectives for moving toward a sustainable world through harmony with nature and natural resources (Jhariya et al. 2022). Biofuel plantation has a significant potential to address the issue of energy crisis as well as energy dependency, along with the restoration of the land ecosystem. It has been reported that genetically modified biofuel crops help to prevent a land degradation process along with land sustainability on the other hand (Chatterton et al. 2017). In the case of mining areas, land degradation is a serious issue, and restoration of such lands is a challenging task. Restoration of such land requires strategy formulation for plantation and selection of suitable species (Deal et al. 2012). In the present context, the vegetation appears to be the most potential carbon (C) sink, which accumulates the C and transfers it into the soil ecosystem for the gradual buildup of the soil C pool (Samal et al. 2022; Banerjee et al. 2023).

From a restoration perspective, the plantation of trees for timber or biomass yield is the most promising strategy to regain the quality of degraded land (Perrow and Devy 2002). Plantation, on the other hand, also addresses the issue of greenhouse gas (GHG) emission reduction due to deforestation activities (FAO 2020). C sequestration by trees seems to be the most potential climate change mitigation strategy along with ecological restoration of degraded lands (Yadav et al. 2022). Various plantation systems can be fruitful in this perspective by accumulating atmospheric carbon dioxide (CO_2) as well as restoring degraded ecosystems.

The term land degradation can be described as any undesirable change in the land ecosystem, hampering the prosperity and well-being of human civilization. Some major factors of land degradation include mining, deforestation, invasion, along intensive agriculture practices (Awasthi et al. 2022a, b; Thakrey et al. 2022). On a global basis, more than one billion hectares of forest area appear to be depleted, and proper initiative needs to be taken for their restoration. In this connection, a plantation is a suitable option for the ecological restoration of the land ecosystem (Vincent et al. 2021). Plantation activities improve soil health by adding organic matter and the gradual buildup of the soil C pool. From a sustainability perspective, maintaining the economic and biological dimensions of plantations is a big challenge (Mansourian 2020). Further, in the landscape restoration process, integration of community participation would add more value (FAO 2021). Specific characteristics, such as indigenous resistance toward natural hazards, biological nitrogen fixation ability, fast-growing rate, and climate-resilient species selection, along optimum seed dispersal mechanism, are favorable for the plantation. In this aspect, community-based plantation activities such as joint forest management and social forestry are fruitful approaches. The present chapter deals with the role of the plantation approaches for the restoration of the landscape with an eco-friendly or nature-based approach on a sustainable basis.

9.2 Plantation Coverage in the World

Global forest covers 31%, which is 4.06 billion hectares of the total land area. This is not covered equally among the peoples in the world and makes availability of 0.52-hectare (ha) land per person. Among the forest types, the maximum coverage is shared by tropical forests at 45%, which is followed by boreal, temperate, and subtropical forests (FAO 2021). More than half of the world's forest areas (54%) are covered by only five countries, viz., the Russian Federation, China, Brazil, the USS, and Canada, which is an interesting fact. However, plantation areas and their coverage are also estimated and recorded in the world. Plantation comprised one or two species, even-aged with regular spacing, which is intensively managed for productive purposes. Similarly, "other planted forests" is a key term that is used for the protection of soil-water and ecosystem restoration. Other planted forests, similar to natural forests, i.e., not managed intensively and covered 55% of the total planted forests. As per the figure, 131 million (m) ha (45%) of areas are covered by plantations in the world. The maximum plantation area is covered by South America,

Table 9.1 Countries' goals for promoting plantations and forest areas in the world (United Nations Department of Economic and Social Affairs 2021)

Country	Goal for promoting plantations and forest areas
Australia	Set a goal for promoting plantation by planting 20 million trees by the year 2030.
India	Targeted a plantation goal by adding 200,00 hectares of forests and tree cover annually
Sri Lanka	Targeted a plantation goal by enhancing forest cover to 32% of the country
Japan	Preserved approx. 25 million hectares of forest areas in the country
New Zealand	Set a goal for promoting plantation by adding one billion trees in the country in the next 10 years (2018 to 2028)
Bulgaria	Set a rigid goal for promoting plantation by retaining or preserving forested areas
Madagascar	Targeted a plantation goal by restoring 4 m ha of forests in the year 2030
Thailand	Targeted a plantation goal by enhancing forest cover to 55% of the country
Nigeria	Targeted a plantation goal by enhancing forest cover from 6% to 32% of the country in the year 2030
Suriname	Targeted a plantation goal by enhancing forest cover to 93% of the country
Kenya	Targeted a plantation goal by enhancing forest cover to 10% of the country
Côte d'Ivoire	Targeted a plantation goal by enhancing forest cover to 20% of the country by the year 2030
Brazil	Set a goal for enhancing the forest area by 2 m ha of plantation in the country
Jamaica	Targeted a plantation goal by enhancing forest cover and checking deforestation activities
Myanmar	Targeted a plantation goal by enhancing forest cover to 30% of the country
Philippines	Set a goal for enhancing the plantation through the restoration of approx. 1.50 m ha of degraded forest area in the country
Panama	Targeted a plantation goal through the restoration of one million hectares of degraded forest area between 2015 and 2035

whereas Europe has the lowest area of plantation. As per FAO (Food and Agriculture Organization) (2020), approx. 44% of the total plantation forests are shared by introduced species in the world. Many countries set up a goal for enhancing plantations and forest areas for a better ecosystem and land restoration. In the Indian subcontinent, 19 m ha of land area is under the plantation forestry alone among 70 m ha in the Asia-Oceania region (FAO 2020). Therefore, the goal for promoting plantations and forest areas is depicted in Table 9.1 (United Nations Department of Economic and Social Affairs 2021).

Plantation forestry performs several practices in the form of revenue generation, livelihood development, and green-belt development for combating pollution; it promotes sustainable forest management practices (SFM) across the region. Hence, plantation can be promoted through various approaches that fulfill the target of urban greening, green infrastructure, home-gardens, agroforestry, social forestry, farm forestry, wood lots, block plantation, restoration purposes, etc. All these activities can be implemented by using them as environmental indicators and should come under the purview of national policy for effective plantation and an increase in plantation area (Jhariya et al. 2024; Yadav et al. 2024).

9.3 Plantation-Based Ecosystem Services

Plantation delivered multifarious ecosystem services along with better land management and restoration (MEA 2005). Provisioning, regulating, supporting, and cultural services are four main categories of ecosystem services developed by the Millennium Ecosystem Assessment (MEA). Availability of timber, fuelwood, and non-timber forest products (NTFPs), including food, fiber, medicine, and other genetic resources covered under provisioning services through plantation. Air purification, water regulation, watershed management, soil fertility maintenance, erosion control, waste treatment, pest and disease regulation, climate security, and pollination are regulating services. Supporting services include soil formations, nutrient cycling, and primary production, whereas aesthetic recreation, ecotourism, spiritual, and religious values are identified as cultural services (MEA 2005). These diversified services through plantation make a healthy ecosystem and environment. However, the delivery of ecosystem services increases day by day due to the expansion of plantation areas (Szulecka et al. 2016). Integration of MPTs (multipurpose tree/timber species) in plantations can promote species diversification that further intensifies ecosystem services, which maintains soil, food, and climate security (Kumar et al. 2015; Banerjee et al. 2024). In earlier times, many industrial markets were incapable of capturing forest-based non-product benefits (Stephens and Grist 2014). Market-based instruments used by many land users and natural resource managers help in maintaining ecosystem services (Pirard 2012). This approach surely maintains plantation-based services and offers financial incentives for achieving many conservation goals (Deal et al. 2012). Thus, plantation is a nature-based solutions that ensure the future sustainable development goals (SGDs) (Stephens and Grist 2014). Plantation-based ecosystem services are depicted in Table 9.2 (United Nations Department of Economic and Social Affairs 2021).

9.4 Approaches to Land Restoration

The scenario of land degradation on a global basis is depicted in Fig. 9.1 (FAO 2021). Restoration of land is a process that aims to restore the degraded conditions of soils as well as provide additional economic benefits, help in combating climate change and natural hazards, and improve soil quality, production, and yields. From an economic perspective, restoring the land ecosystem gives a good output. According to UNEP (United Nations Environment Program), the GDP (gross domestic product) of the globe is very much dependent upon nature, and approximately USD 30 is invested in this purpose. Across the globe, land restoration practices have become the heart of UNCCD's (United Nations Convention to Combat Desertification) mission toward achieving land degradation neutrality (LDN) by 2030, as well as building up climate resiliency (World Wildlife Fund for Nature; WWF 2022). Global programs such as Initiative 20x20 in Caribbean islands were initiated to restore 20 million hectares of degraded lands by 2020. Similarly, in

Table 9.2 Plantation-based ecosystem services in the world (United Nations Department of Economic and Social Affairs 2021)

Country	Program	Plantation-based activities	Ecosystem services
China	"A green miracle" program of the Saihanba Mechanized Forest Farm	This program ensures plantation activities by enhancing forest cover from 11% to 80%. Also, added 10.12 million cubic meters of the forest stock volume	Plantation activities ensure 7.69 million tons of carbon storage along with 545,000 tons of pure oxygen and 137 m cubic meters of clean water to nearby region of Tianjin and Beijing
Ghana	Modified Taungya System (MTS) program	This program selected degraded and wasteland areas for plantation activities through ecorestoration and provides a variety of ecosystem services	Strengthen food and nutritional security for forest-dependent peoples
Thailand	Forest restoration program	Royal Forest Department has created broad buffer zones (1.60 m ha) in between national forest reserves (NFR) which helps in the restoration of degraded land	Strengthen food and nutritional security, along with NTFPs (non-timber forest products) provision to forest-dependent peoples
Australia	Sustainable Forest Management (SFM)-based program for Tiwi Islands	Plantations promoted by local communities	Provide food, shelters, and many important medicinal plants, which maintain socioeconomic, spiritual, and cultural services in the region
Australia	"Growing a Better Australia: A Billion Trees for Jobs and Growth"	Targeted a plantation goal by adding a billion trees in the next decade	Provide food, shelter, and many important medicinal plants to peoples. Also boost industrial growth and supports scientific research activities in this region
New Zealand	"One billion trees program"	Targeted a plantation goal by adding one billion trees in the year 2028, which strengthens the forest cover, along with ecorestoration of degraded land in the country	Provides countless ecosystem services, which maintain socioeconomic, spiritual, and cultural services in the region
Senegal	Program for managing trans-boundary forests in the country	The country signed a MoU (memorandum of understanding) with the Gambia for the sustainable management of trans-boundary forests	Enhance storage, stabilization, and flux of carbon and ensure food and job security to local peoples
Côte d'Ivoire	Cocoa forest initiatives based eco-friendly program	The program minimizes deforestation and enhances plantations on degraded and wasteland	Strengthen the people's livelihood and socio-economic status, which also helps monitoring, management, and protection of forests

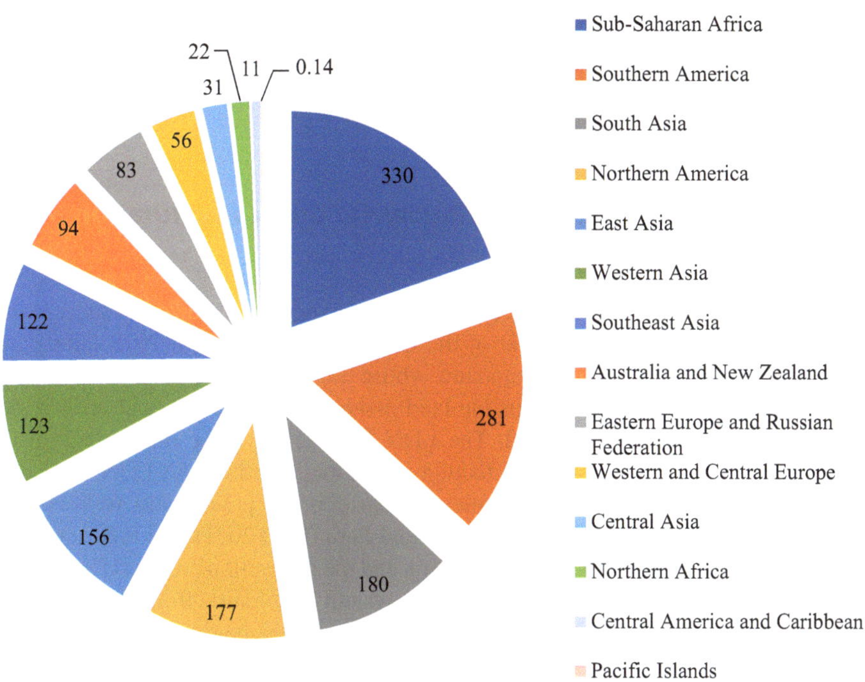

Fig. 9.1 Human-induced land degradation (m ha) in the world (FAO 2021)

Africa, AFR (African region) 100 was initiated for the restoration of 100 million hectares till 2030 (FAO 2021).

For productive land ecosystems, achieving the target of zero net losses has become a major initiative across the globe (Banerjee et al. 2021). In this connection, various technologies in the form of crop rotation, tree plantation, agroforestry, watershed management, and organic farming are the most suitable approaches and nature-based solutions for land degradation. According to one estimate, 115 countries across the world have shown commitment toward restoring approximately 150 m ha of degenerated land ecosystems (FAO 2021).

In the agriculture sector, approximately 250 m of land are targeted for restoration purposes. Agriculture is a major sector that provides a significant amount of food for the global population, and hence, proper care should be taken in order to maintain the agroecosystem quality and sustainability (Kumar et al. 2015; Haj-Amor et al. 2022). Ever-growing human population has necessitated the addition of 20% of the natural ecosystem into farmland by 2050 in order to feed the growing human population. It was found that restoration of 150 m ha of degenerated wasteland could provide economic benefits up to USD 85 billion, followed by USD 30–40 billion on an annual basis for the farming community. This, in turn, would ensure food security for up to 200 million people globally (FAO 2021). Topsoil is the key factor in the restoration of degraded land as it is the most fertile area for soil nutrients and microorganisms (Mansourian 2020; Manral et al. 2022). It has been reported that approximately USD 1 trillion can be gained by preventing the loss of topsoil for the upcoming 50 years in the African subcontinent. On the other hand, no action in this regard would create an economic loss of up to USD 2 trillion (FAO 2020).

9.5 Land Management Initiatives for the Eco-restoration of Degraded Lands

In the process of land restoration, proper financial support is the key to success. In this connection, the establishment of the LDN fund was done in 2018 jointly by UNCCD and Mirova Finance Company, whose major objective was to utilize and generate the fund through sustainable land management practices and restoration approaches throughout the world. The LDN fund seems to be highly fruitful in terms of sustainable hazelnut production in Bhutan. Further, the LDN fund has shown significant promise in implementing agroforestry practices as well as sustainable production of coffee in Colombia and Peru (FAO 2021; WWF 2022).

Considering the issue of land degradation and the restoration process UN (United Nations) has declared the UN Decade on Ecosystem Restoration from 2021 to 2030. The major motto behind such a program is to restore the degenerated ecosystems across the globe. Secondarily, it would help to sustain the ecosystem on a long-term basis and provide additional economic benefits (FAO 2020). From a forest land perspective, it has been reported that about 2% of forest area is affected by forest fire globally, and more than half of the forest area of India is prone to forest fire (Jhariya and Singh 2021a, b). On an annual basis, USD 6.3 trillion of economic loss is associated with the process of land degradation. Further, according to the Bonn challenge, economic benefits of about USD 7–9 trillion annually can be achieved by sustainable land restoration initiatives. On an annual basis, 12 billion ha of land are degraded which is associated with an annual GDP loss of 10% through loss of ecosystem services (FAO 2021). This is a very serious issue as the existence of humanity largely depends upon the land ecosystem in the form of forests, agricultural land, etc., which also provides diverse ecosystem services (Jhariya et al. 2021a, b).

In the future, land degradation will become a serious issue and the most cost-intensive gesture to follow. The major cause behind land degradation is unsustainable practices in agriculture and animal husbandry that led to the loss of ecosystem services along with biological resources (Chatterton et al. 2017). It also hampers food security as well as the supply of freshwater resources. The problem is that ignorance in terms of causes of degradation would aggravate the problem more, and thus land degradation would become a continuous process (Mansourian 2020). Hence, formulation of a proper strategy on a case-to-case basis would be the successful option of combating the land degradation process (WWF 2022). Further land degradation does not mean only the removal of vegetal cover but also the loss of ecosystem services; hence, proper determination of the problem and development of climate resiliency is required. Natural selection happens to be the key process for the successful implementation of plantations to combat the land degradation process. At the present time, restoration project has been implemented across the world which provide benefits to the indigenous population of the ecosystem, which in turn would work efficiently for the restoration of degraded ecosystem. Moreover, we must thrust over nature-based solutions and infrastructure along with the conservation of natural assets for all-around sustainability (Chatterton et al. 2017). An integrated approach by providing common platforms to all the different stakeholders, policymakers, researchers, and academicians toward addressing the issue is necessary for sustainability purposes for various land restoration projects (WWF 2022).

9.6 Forest Land Restoration

The forest ecosystem has its inherent importance by providing various ecosystem services. Hence, global initiatives were taken to increase the forest area by the restoration of degraded lands. The major aim of the scheme was to bring back millions of hectares of degraded lands under vegetal cover. In this connection, the leading initiatives were taken during 2017 under the UN strategic plan for forests. The main objective was to increase the forest cover by 3% by 2030, which is equivalent to 120 m ha of land on a global basis. Further, 61 countries across the globe have committed to increasing the forest cover by about 350 m ha under the scheme of the Bonn Challenge (UNCCD 2022).

Policymaking and its effective implementation require effective guidance and technical support from the domain of science and technology. Screening of suitable habitats for plantation and reforestation is the first step toward the land restoration process. Further, the selection of species should address the issue of food security and the prevailing ecologic condition of the region, followed by recognizing the economic and sociocultural values of the local community stakeholders (Banerjee et al. 2023). Recent advancement includes the preparation of maps and documentation with areas having potential for restoration. But, the conversion percentage of such type of lands is yet to be determined properly. According to the report of FAO (2020), the major contribution to restoration approaches includes plantation forestry, which is approximately of about 290 m ha. This much land area encompasses 7% of forest covers on a global basis and gives 50% of the timber output.

Species selection criteria are a big issue in restoration approaches. For instance, whether one needs to consider the species on the basis of their ecosystem services or on the basis of climate resiliency is a big challenge. In upcoming times, there will be a huge hurdle of climatic extremes and hence the plantation scheme should have the capability to resist them. It is an evident fact that no single species has all these attributes, and hence mixed plantations would be effective for reforestation purposes as well as to combat the major problem of land degradation (WWF 2022).

To date, the plantation that has been done across the globe is based upon monoculture practices, giving prior importance to their economic output. For example, species, such as eucalyptus, spruce, pines, fir, poplar, shisham, teak, mangium, etc., were recommended for plantation purpose. Such approaches do not give proper information regarding the climate resiliency of species. Through research, it was observed that up to 20-year-old tree species in terms of forest development, but their growth phase is critical for the successful establishment of the forest. The major focus of the work includes screening species with high nutrient utilization efficiency, followed by higher growth rates. Species are also screened for their resistance to climatic extremes (FAO 2020; UNCCD 2022).

9.7 Approaches for Forest Land Restoration

Forest land restoration is a big issue that requires a systematic approach to an area. The first step is the involvement of local people. Proper motivation and encouragement of adopting FLR (forest land restoration) programs by the local stakeholders are the key factors toward achieving positive outputs from plantation activities (Kumar et al. 2015). The most important thing for the local people is how much revenue is generated from such activities. It's an important point of attraction for the involvement of the people. PFM (participatory forest management) is the role model in this connection for the protection of forests owned by the state government. The income from such a scheme is dependent upon timber and non-timber forest species. However, factors such as rules and regulations and proper channels of marketing often reduce the benefit to a quite extent (Mansourian 2020). Awareness generation in terms of extension activities in the forest area is also another important part for successful FLR. This includes capacity development, enhancing the communication skills of local community stakeholders, as well as the involvement of academic personnel, followed by a significant amount of livelihood generation. Suitable nature-based solutions such as FLR, watershed management have also proved to be fruitful in terms of combating climatic extremes and negative impacts of the changing climate (WWF 2022). Reforms in the legal framework and promotion of effective plantation practices is required for successful land restoration programs along with the involvement of local community people. Maintaining suitable ecological conditions through the use of indigenous species and emphasizing natural regeneration can be an effective strategy toward FLR.

Subsequently, factors such as uncontrolled grazing activity, reducing dependency of households toward energy consumption, need to be done properly for effective maintenance of vegetation cover (Mansourian 2020). From an economic perspective effective market mechanism for the products obtained from planted tree species is required for promoting the local inhabitants to participate in the program actively. Various schemes such as silvopastoral, agroforestry, social forestry, community forestry, extension forestry, urban forestry, and farm forestry can be highly effective for FLR programs and their management (FAO 2021). Proper policy in terms of land tenure system, recognition of forest rights in the form of forest ownership would be highly fruitful toward effective implementation of FLR. Socio-economic development is also another important part of the successful implementation of FLR programs. In this connection, communal plantation was found to be an effective strategy for capacity building of the farming community, as

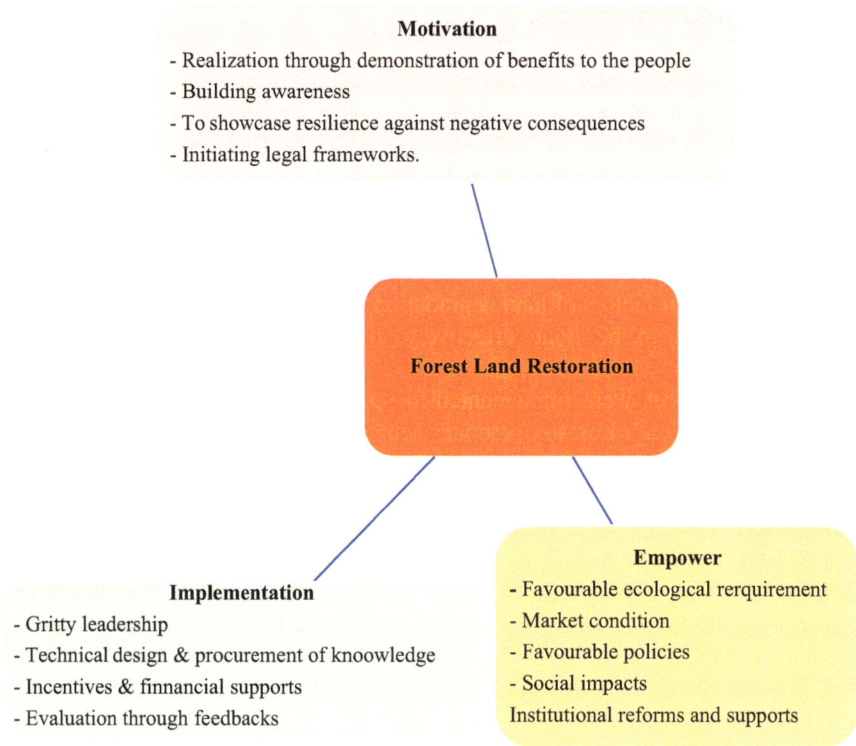

Fig. 9.2 Steps toward intervention of forest land restoration (FLR)

evident from various works across the world. Clear-cut guidelines at the institutional level dealing with FLR programs are also required to increase the effectivity of the restoration process (Fig. 9.2).

At the international level, the scope of implementation needs to be broadened by fulfilling the targets of the treaties, conventions at the global level. For example, the Bonn Challenge was adopted by various countries at their national level in order to fulfill nations' land degradation neutrality target along with nationally determined contributions from combating desertification and changing climate (UNCCD 2022; WWF 2022). Further international collaboration among nongovernment organizations (NGOs) and government organizations often leads to successful implementation of FLR programs by promoting local community people to adapt greener approaches toward environmental sustainability (Chatterton et al. 2017). Designing and formulation of plans for the successful implementation of FLR programs can also be achieved by integrating such restoration approaches within national action plans, following suitable guidelines. Proper funding and financial incentives in terms of additional income from practices such as agroforestry, community forestry, etc. along with improved market infrastructure would also help in this process. Continuous improvement requires effective feedback mechanism of the stakeholders who are benefitted from the scheme is also required (Mansourian et al. 2021).

9.8 Land Restoration

Restoration of land involves processes such as regeneration of degraded land, combating desertification, etc., with a focus on revenue and livelihood generation for local community stakeholders (Mansourian et al. 2021). The process involves selection of suitable species for plantation, promotion of native forest cover and fodder plantation, focusing on useful plant species, promoting natural regeneration by community participation, followed by the generation of an e-database for good germplasm and gene pool buildup through rural livelihoods. In this process, the reasons and causal factors of land degradation, followed by an effective process of restoration, need to be done effectively to get positive output (FAO 2021; WWF 2022).

Various practices were implemented to restore the degenerated and degraded land. For example, agroforestry practices help to prevent soil erosion and conservation of soil (Raj et al. 2020). Climate-resilient agriculture practices help in the development of the soil C pool, as reported by various countries across the globe. Restoration of the riparian ecosystem can help to regenerate the hydrological cycle. Forest land protection through plantation, afforestation, and reforestation activities help to reduce the atmospheric CO_2 load. Further, ecorestoration of wetlands also enhances multiple ecosystem services such as enhanced water quality and improvement in the habitat condition of wildlife. Ecological restoration of mined wastelands also helps in the rehabilitation of mined areas as well as improvement in soil quality and health (UNCCD 2022).

9.9 Plantation Activities for Climate Change Mitigation

Plantation on degraded and wasteland improves biodiversity, which intensifies multifarious ecosystem services, including climate change mitigation. Reducing GHGs with greater C storage is other prime services provided by plantation in degraded land. Climate-resilient plantation management stabilizes and reduces excessive GHGs by capturing atmospheric C and storing it into plant parts as biomass through the C sequestration process. This could help in reducing C footprints and enhance biomass energy with greater C management. However, biomass can be derived from the storage of carbon C in forests, which is a good alternative to fossil fuels (FAO 2016). Plantation management for biomass production is also needed for low fossil fuel inputs as compared to other land use practices (St Clair et al. 2008). Expansion of plantations improves standing tree volume and its growth rates, which also provide many wood-derived products that replace consumption of fossil fuels and reduce global problems of climate change (McEwan et al. 2020). Thus, plantation enhances storage, stabilization, and flux of C, which maintains greater environmental sustainability (Parks et al. 1997). Reducing deforestation and maximizing plantation in degraded and wasteland improves C status in forests and soils that helps in climate change mitigation for a sustainable world (Michetti and Rosa 2012).

9.10 Future Thrust

FLR through plantation or afforestation can be the best option due to its cost-effectiveness, eco-environmental friendliness, as well as a nature-based approach for the reclamation and restoration of the degraded forest areas, as well as wastelands. To ensure the successful FLR implementation on a sustainable basis requires proper measures and comprehensive management planning toward land restoration and management (Chatterton et al. 2017). From a future perspective, capacity building and development of institutional infrastructure through cross-sectoral collaboration are required for improving the effectivity of FLR program (FAO 2021). Economic incentives or mechanisms in the form of microfinance, loan, etc., can be effective for FLR programs. In this connection, promotion of marketing of NTFPs and mixed plantation can be good for sufficient revenue generations. Other schemes, such as green accounting of resources, payment for ecosystem services, and certification of sustainable land management practices, can also improve the output of FLR programs. The funding mechanism if FLR programs needs to be distributed at both macro and micro levels. Facilitating livelihood security practices such as short rotation forestry, going for plantation giving significant economic output, would also provide long-term sustainability of the FLR programs (WWF 2022). At present times, FLR programs need to be reoriented around land use and land cover planning, followed by biodiversity conservation, address livelihood security, and should provide diverse opportunities for income generation. The development of a business model for FLR should also involve the private sector. A greater share of decision-making among the local community stakeholders would enhance the community participations in FLR programs (UNCCD 2022).

9.11 Conclusions

Ecological restoration of degraded lands is the need of the hour. In the present context, the growing human population has promoted unsustainable land use practices that lead to severe land degradation. For higher productivity, yield-intensive agricultural practices have led to land degradation, alteration of land use, and loss of soil quality. As a consequence, the area of degraded land across the globe is increasing day by day. To combat such mega events, different forms of plantation practices for diverse sectors with having sustainable approach need to be formulated for the betterment of land ecosystem. Further, capacity building, development of institutional framework, holistic approach, and integration of livelihood generation and security would be essential for the overall sustainability of land ecosystem. Considering the issues of climate change, food security, and crisis, proper policies such as agroforestry, plantation forestry, social forestry, farm forestry, and climate-resilient agricultural practices are the need of the hour and hence need to be implemented at the grassroot level with utmost care about the benefit of the stakeholders of various spheres of society.

References

Awasthi P, Bargali K, Bargali SS, Jhariya MK (2022a) Structure and Functioning of *Coriaria nepalensis* Wall dominated Shrublands in degraded hills of Kumaun Himalaya. I. Dry Matter Dynamics. Land Degrad Dev 33(9):1474–1494. https://doi.org/10.1002/ldr.4235

Awasthi P, Bargail K, Bargali SS, Khatri K, Jhariya MK (2022b) Nutrient Partitioning and Dynamics in *Coriaria nepalensis* Wall Dominated Shrublands of Degraded Hills of Kumaun Himalaya. Front For Glob Change 5:913127. https://doi.org/10.3389/ffgc.2022.913127

Banerjee A, Meena RS, Jhariya MK, Yadav DK (2021) Agroecological footprints management for sustainable food system. Springer Nature Singapore. eBook ISBN: 978-981-15-9496-0, Hardcover: 978-981-15-9495-3. p 514. https://doi.org/10.1007/978-981-15-9496-0

Banerjee A, Jhariya MK, Bargali SS, Palit D (2023) Ecorestoration for sustainability. Wiley-Scrivener, USA, pp 1–441. ISBN: 9781119879718. https://onlinelibrary.wiley.com/doi/book/10.1002/9781119879954

Banerjee A, Jhariya MK, Raj A, Mechergui T (2024) Ecosystem management: climate change and sustainability. Wiley-Scrivener, USA, pp 1–669. ISBN 978-1-394-23121-8. https://onlinelibrary.wiley.com/doi/book/10.1002/9781394231249

Chatterton P, Ledecq T, Dudley N (eds) (2017) Landscape Elements: steps to achieving sustainable landscape management. WWF, Vienna

Deal RL, Cochran B, LaRocco G (2012) Bundling of ecosystem services to increase forestland value and enhance sustainable forest management. Forest Policy Econ 17:69–76. https://doi.org/10.1016/j.forpol.2011.12.007

FAO (2010) Food and Agriculture Organization of the United Nations, The State of the World's Land and Water Resources for Food and Agriculture. Rome

FAO (2016) Global forest resources assessment 2015. How are the world's forests changing? Second Edition, Food and Agricultural Organisation of the United Nations, Rome, Italy, p 56. http://www.fao.org/3/a-i4793e.pdf

FAO (2020) Global Forest Resources Assessment 2020 – Key findings. Rome. https://doi.org/10.4060/ca8753en

FAO (2021) The state of the world's land and water resources for food and agriculture – systems at breaking point. Synthesis report 2021. Rome. https://doi.org/10.4060/cb7654en

Haj-Amor Z, Araya T, Kim DG, Bouri S, Lee J, Ghiloufi W, Yang Y, Kang H, Jhariya MK, Banerjee A, Lal R (2022) Soil salinity and its associated effects on soil microorganisms, greenhouse gas emissions, crop yield, biodiversity and desertification: a review. Sci Total Environ 843:156946. https://doi.org/10.1016/j.scitotenv.2022.156946

Jhariya MK, Singh L (2021a) Herbaceous diversity and biomass under different fire regimes in a seasonally dry forest ecosystem. Environ Develop Sustain 23(5):6800–6818. https://doi.org/10.1007/s10668-020-00892-x

Jhariya MK, Singh L (2021b) Effect of fire severity on soil properties in a seasonally dry forest ecosystem of Central India. Int J Environ Sci Technol 18(12):3967–3978. https://doi.org/10.1007/s13762-020-03062-8

Jhariya MK, Banerjee A, Meena RS, Yadav DK (2019) Sustainable agriculture, forest and environmental management. Springer Nature Singapore, p 606. https://doi.org/10.1007/978-981-13-6830-1

Jhariya MK, Meena RS, Banerjee A (2021a) Ecological intensification of natural resources for sustainable agriculture. Springer Nature Singapore, p 655. https://doi.org/10.1007/978-981-33-4203-3

Jhariya MK, Banerjee A, Meena RS, Kumar S, Raj A (2021b) Sustainable intensification for agroecosystem services and management. Springer Nature Singapore, p 870. https://doi.org/10.1007/978-981-16-3207-5

Jhariya MK, Meena RS, Banerjee A, Meena SN (2022) Natural resources conservation and advances for sustainability. Elsevier, Academic Press. ISBN: 9780128229767. https://doi.org/10.1016/C2019-0-03763-6

Jhariya MK, Raj A, Banerjee A, Bargali SS, Bargali K, Pandey VC (2024) Plantation forestry for ecorestoration: a working premises. In: Pandey VC (ed) Biodiversity and ecosystem services on post-industrial land. Wiley, pp 171–187. ISBN:9781394187386

Kumar C, Begeladze S, Calmon M, Saint-Laurent C (eds) (2015) Enhancing food security through forest landscape restoration: lessons from Burkina Faso, Brazil, Guatemala, Viet Nam, Ghana, Ethiopia and Philippines. IUCN, Gland, Switzerland, pp 5–217

Manral V, Bargali K, Bargali SS, Jhariya MK, Padalia K (2022) Relationships between soil and microbial biomass properties and annual flux of nutrients in Central Himalayan forests, India. Land Degrad Dev 33:2014–2025. https://doi.org/10.1002/ldr.4283

Mansourian S (2020) Enabling factors to scale up forest landscape restoration: the roles of governance and economics. WWF & IUFRO, Berlin

Mansourian S, Diederichsen A, Vallauri D (2021) Twenty Years Later: lessons learnt from seven forest landscape restoration initiatives worldwide. WWF France, FLR Field series, Experiences in Forest Landscape Restoration, Paris, 64 pages

McEwan A, Marchi E, Spinelli R, Brink M (2020) Past, present and future of industrial plantation forestry and implication on future timber harvesting technology. J For Res 31(2):339–351

Meena RS, Kumar S, Datta R, Lal R, Vijaykumar V, Brtnicky M, Sharma MP, Yadav GS, Jhariya MK, Jangir CK, Pathan SI, Dokulilova T, Pecina V, Marfo TD (2020) Impact of agrochemicals on soil microbiota and management: a review. Land 9(2):34. https://doi.org/10.3390/land9020034

Michetti M, Rosa R (2012) Afforestation and timber management compliance strategies in climate policy. A computable general equilibrium analysis. Ecol Econ 77:139–148. https://doi.org/10.1016/j.ecolecon.2012.02.020

Millennium Ecosystem Assessment [MEA] (2005) Ecosystems and human well-being synthesis. Island Press, Washington D.C., USA, p 155. https://www.millenniumassessment.org/documents/document.356.aspx.pdf

Parks PJ, Hall DO, Kristrom B, Masera OR, Moulton RJ, Plantinga AJ, Swisher JN, Winjum KK (1997) An economic approach to planting trees for carbon storage. In: Sedjo RA, Sampson RN, Wisniewski J (eds) Economics of carbon sequestration in forestry, 1st Eds. CRC Press, Boca Raton, pp S9–S21. ISBN 9780849311581

Perrow MR, Davy AJ (2002) Handbook of Ecological Restoration. Vol. 1. Principles of Restoration. Cambridge University Press, Cambridge

Pirard R (2012) Market-based instruments for biodiversity and ecosystem services: a lexicon. Environ Sci Pol 19-20:59–68. https://doi.org/10.1016/j.envsci.2012.02.001

Raj A, Jhariya MK, Yadav DK, Banerjee A (2020) Climate change and agroforestry systems: adaptation and mitigation strategies. Apple Academic Press Inc., CRC Press- A Taylor and Francis Group, US & Canada, p 383. ISBN: 9781771888226. https://doi.org/10.1201/9780429286759

Raj A, Jhariya MK, Devi A, Banerjee A, Poonam, Jaiswal SK (2023) Agrochemicals and pollinator diversity: a socio-ecological synthesis. Springer, pp 137–159. https://doi.org/10.1007/978-981-99-3439-3_5

Samal B, Singh L, Jhariya MK (2022) Carbon storage, mitigation and sequestration potential of *Haldina cordifolia* and *Mitragyna parvifolia* in Tropical Dry Deciduous Environment of Chhattisgarh, India. Ecol Eng 175:106490. https://doi.org/10.1016/j.ecoleng.2021.106490

St Clair S, Hillier J, Smith P (2008) Estimating the pre-harvest greenhouse gas costs of energy crop production. Biomass Bioenergy 32:442–452. https://doi.org/10.1016/j.biombioe.2007.11.001

Stephens ML, Grist P (2014) Market failure for plantations: past experience and emerging trends for delivering wood production and ecosystem services in Australia. Int For Rev 16:205–215

Szulecka J, Obidzinski K, Dermawan A (2016) Corporate-society engagement in plantation forestry in Indonesia: evolving approaches and their implications. Forest Policy Econ 62:19–29. https://doi.org/10.1016/j.forpol.2015.10.016

Thakrey M, Singh L, Jhariya MK, Tomar A, Singh AK, Toppo S (2022) Impact of disturbance on biomass, carbon, nitrogen storage in vegetation and soil properties in tropical dry deciduous forest in Chhattisgarh, India. Land Degrad Dev 33:1810–1820. https://doi.org/10.1002/ldr.4263

UNCCD (2022) Global Land Outlook 2. UN Convention to Combat Desertification, Bonn
United Nations Department of Economic and Social Affairs (2021) Global forest goals report 2021. (Sales No.: E.21.IV.3) New York: United Nations. https://www.un.org/esa/forests/wp-content/uploads/2021/04/Global-Forest-Goals-Report-2021.pdf
Vincent JR, Curran SR, Ashton MS (2021) Forest Restoration in Low- and Middle-Income Countries. Annu Rev Environ Resour 46:4.1–4.29. https://doi.org/10.1146/annurev-environ-012220-020159
WWF (2022) Towards nature positive – the role of restoration for land degradation neutrality. WWF, Gland, Switzerland, pp 1–13
Yadav VS, Gupta SR, Yadav SS, Meena RS, Lal R, Sheoran NS, Jhariya MK (2022) Carbon sequestration potential and CO_2 fluxes in a tropical forest ecosystem. Ecol Eng 176:106541. https://doi.org/10.1016/j.ecoleng.2022.106541
Yadav SK, Banerjee A, Jhariya MK, Raj A, Bargali SS, Pandey VC (2024) Bauxite mine restoration and management. In: Pandey VC (ed) Biodiversity and ecosystem services on post-industrial land. Wiley, pp 241–261. ISBN:9781394187386

Land Restoration and Livelihood Improvement: A Prescription for Sustainable Development?

10

Prachi Upadhyay, Neetu Arya, and Joystu Dutta

Abstract

The issue of land degradation is growing quickly and becoming a significant environmental challenge. It is a burning issue. Targeted restoration techniques are required to lessen the effects of land degradation, which present serious obstacles to socioeconomic development and environmental sustainability. This chapter explores innovative technologies and strategies for land restoration, while analyzing key drivers of land degradation, including deforestation, unsustainable agricultural practices, and climate change. It also emphasizes the link between land restoration efforts and sustainable livelihoods. The ecological and socioeconomic advantages of important topics, including soil protection, reforestation, agroforestry, and water management, are highlighted. The chapter also looks at how these restoration initiatives improve livelihoods by boosting agricultural output, generating jobs, and guaranteeing food security. Successful case studies of land restoration projects that highlight scalable solutions and their revolutionary effects on communities are examined in order to offer useful insights. The combination of these components emphasizes how important land restoration is to promoting sustainable livelihoods and environmental resilience.

P. Upadhyay · N. Arya
Indian Institute of Forest Management, Bhopal, Madhya Pradesh, India

Forest Research Institute Deemed to be University, Dehradun, India

J. Dutta (✉)
Department of Environmental Science, Sant Gahira Guru University Sarguja, Ambikapur, Chhattisgarh, India

IUCN Commission on Ecosystem Management (CEM), Gland, Switzerland

V. C. Pandey (ed.), *Plantation-Based Land Restoration*,
https://doi.org/10.1007/978-981-95-2860-8_10

Keywords

Land degradation · Restoration techniques · Socioeconomic development · Sustainable livelihoods · Environmental resilience

10.1 Introduction

Forests have been cleared for various uses ever since people discovered how to manage fire and control flora and fauna. Over the past 10,000 years, human activities have transformed nearly half of the Earth's ice-free land surface (Lambin et al. 2003). This transformation has taken a toll on the environment, primarily due to the increasing energy demands and the pressures of meeting those needs (Areendran et al. 2013). Land cover and land use undergo continuous transformation. Understanding these changes is crucial for effective planning, monitoring, evaluation, and conservation of natural resources. Human-driven activities like pollution, deforestation, soil degradation, and climate change—especially with rising population pressure—have significantly harmed ecosystems (Thakur et al. 2022). The Earth's system processes are currently being impacted by land use change, which was once a local human practice (Turner et al. 2007; Foley et al. 2005; Le Quéré et al. 2016; Alkama and Cescatti 2016; Song et al. 2018). The carbon and water cycle, surface energy balance, and diversity. Overuse of resources has resulted from mining, industrialization, and deforestation (Areendran et al. 2013; Thakur et al. 2022a).

UNEP (1992) defines land degradation as the rate at which soil quality deteriorates, leading to a reduction in land productivity. It lowers the region's total productivity and encompasses several types of soil degradation as well as the deterioration of water sources. Given its effects on the environment, agriculture, food security, and human livelihood, it is regarded as a significant barrier to sustainable development (UNCCD 2015). As noted by Nachtergaele et al. (2010), it may also be defined as a decline in the land's ability to sustain biological, chemical, and economic functions for those who depend on it. It exposes the land to other difficulties and causes problems like vegetation loss and soil erosion. Particularly in tropical nations, erosion—particularly from precipitation—is a serious problem (EPA 2003). Top soil displacement, silting in water bodies, and the formation of gullies and ravines are the main features of this type of erosion (Zhao et al. 2013; Abdul-Wadood Moomen and Dewan 2016). Economic and noneconomic components of livelihood have been the subject of numerous research on land degradation brought on by mining and other human activities (Eswaran et al. 2001; Akabzaa and Darimani 2001; Abdul-Wadood Moomen and Dewan 2016). Many developing nations are researching the effects of mining on land degradation (EPA 2003). However, the majority of the research that has been done is qualitative. Only vegetational disturbances are included in quantitative research (Abdul-Wadood Moomen and Dewan 2016).

Restoring the natural landscape for plant and animal species is known as land restoration. This promotes economic growth, protects ecosystems, averts natural disasters, and raises soil production overall. Thus, it can be claimed that land restoration contributes to environmental preservation, resilience building, and population support worldwide. The economic benefits of land restoration are enormous. A UNEP report highlights that approximately 50% of the global GDP relies on nature, and for every dollar invested in restoration efforts, there is an estimated return of USD 30 in benefits. The United Nations Convention to Combat Desertification (UNCCD) places land restoration at the core of its goal to achieve land degradation neutrality by the year 2030. Key recommended practices for halting land degradation and promoting recovery include tree plantation, crop rotation, water conservation techniques, and the application of organic fertilizers. At present, around 115 countries have committed to restoring nearly one billion hectares of degraded land. This initiative includes the rehabilitation of about 250 million hectares of agricultural land, which is expected to enhance economic value and strengthen global food security (UNCCD n.d.).

In many low-income nations, land degradation is often aggravated by the expansion of human settlements into ecologically sensitive zones, such as hillsides and wetlands, primarily driven by growing population pressures (FAO 2011). The over-harvesting of trees and shrubs for fuelwood and animal feed further escalates the risk. Additionally, inadequate land ownership systems and weak governance have resulted in vast stretches of severely degraded land being neglected—particularly high-carbon peatlands in Southeast Asia—thereby encouraging unsustainable practices like short-term logging, pulpwood extraction, and palm oil cultivation. According to Chirwa (2014), extensive human activity has led to severe deforestation and environmental degradation in the natural forests of Tanzania's Shinyanga region over recent decades. In Uganda, land degradation is responsible for an economic loss estimated between 4% and 12% of the country's GDP (Bolwig 2002), with the most severely impacted areas being the mountainous and agriculturally productive zones. Soil erosion is a major physical factor contributing to this decline. The districts of Mbale, Rakai, and Kotido, known for cattle grazing, are among the hardest hit, showing degradation levels of 75–80%, while in the highland regions of Kabale and Kisoro in the southwest, the rate of agricultural land degradation reaches as high as 85–90% (Bolwig 2002).

The chapter further explores various methods of land restoration, focusing on innovative approaches such as soil conservation, reforestation, agroforestry, and integrated water management. These strategies are not only aimed at mitigating land degradation but also at enhancing ecological balance and delivering socioeconomic benefits. Particular attention is given to the key factors that must be considered during restoration activities to ensure their success and sustainability. These include community participation, proper planning, and alignment with local environmental conditions. Additionally, the chapter highlights how these restoration efforts contribute significantly to livelihoods by improving agricultural productivity, creating employment opportunities, and securing food supplies for affected communities. Successful case studies of land restoration initiatives are presented to illustrate practical, scalable solutions and their transformative impacts on both ecosystems and communities. Through an integrated synthesis of these elements, the chapter emphasizes the critical role of land restoration in fostering environmental resilience and enhancing sustainable livelihoods, particularly in regions where the impacts of land degradation are most acute.

10.2 Factors Contributing to Land Degradation

Land degradation stems from a range of intricate and varied factors (see Table 10.1). A single driver can lead to different outcomes depending on how it interacts with other immediate or underlying conditions. The effects of direct drivers—such as soil composition, terrain, and climate—are generally well understood. For instance, steep slopes are more likely to experience erosion from water runoff (Wischmeier 1976; Voortman et al. 2000), and soils rich in silt are naturally more susceptible to degradation (Bonilla and Johnson 2012). Over time, various Sustainable Land Management (SLM) practices have been introduced to combat soil and land degradation (Liniger and Schwilch 2002; Liniger and Critchley 2007). However, there is ongoing debate surrounding the root causes of land degradation (von Braun et al. 2013; Nkonya et al. 2011) and the challenges associated with the adoption of SLM practices by land users (see Pender et al. 2009, for Central Asia). Key issues often discussed include whether poverty is a leading driver of land degradation (Way 2006), whether greater access to markets accelerates degradation or supports sustainable practices, and whether population growth leads to environmental harm (Grepperud 1996) or encourages improved land management (Tiffen et al. 1994). These are summarized in Mirzabaev et al. (2015). These fundamental forces and the theoretical underpinnings of their cause-and-effect mechanisms are expounded upon in Table 10.1. Depending on the datasets used, the methodology used, the dates taken into consideration, and the locations researched, the conclusions drawn have been rather varied and frequently contradictory (Mirzabaev et al. 2015).

Table 10.1 Immediate and root causes of land degradation and their possible cause-and-effect relationships (selected examples)

Drivers	Types	Example of causality	References
Topography	Immediate and natural	Steep terrains are particularly at risk of intense soil erosion caused by water flow	Wischmeier (1976), Voortman et al. (2000)
Land cover transformation	Immediate, both natural and human-induced	Shifting rangelands into irrigated agricultural fields can increase soil salinity; deforestation also contributes to degradation	Gao and Liu (2010), Lu et al. (2007)
Climate factors	Immediate and natural	Arid and hot climates are naturally susceptible to wildfires, which in turn can trigger soil degradation. Heavy rainfall may lead to flooding and erosion. Erratic or infrequent rainfall, such as in monsoon regions, promotes salinity and erosion	Safriel and Adeel (2005), Barrow (1991)
Soil erodibility	Immediate and natural	Soils with a high silt content are inherently more vulnerable to erosion processes.	Bonilla and Johnson (2012)
Population density	Indirect	Effects vary by context. In some cases, higher population density leads to better land use practices; in others, it accelerates land degradation.	Bai et al. (2008), Tiffen et al. (1994)
Access to market	Indirect	Results are mixed. On one hand, proximity to markets can motivate farmers to adopt sustainable practices; on the other, higher opportunity costs can reduce their willingness to engage in labor-intensive land conservation	Pender et al. (2006)
Land tenure security	Indirect	Ambiguous impact. While insecure tenure may discourage sustainable land use, in other cases, farmers continue to invest in conservation despite insecure rights	Kabubo-Mariara (2007), Brasselle et al. (2002)
Poverty	Indirect	Poor households often depend heavily on land, motivating them to conserve it—provided that market systems allow for efficient resource use	Nkonya et al. (2008)

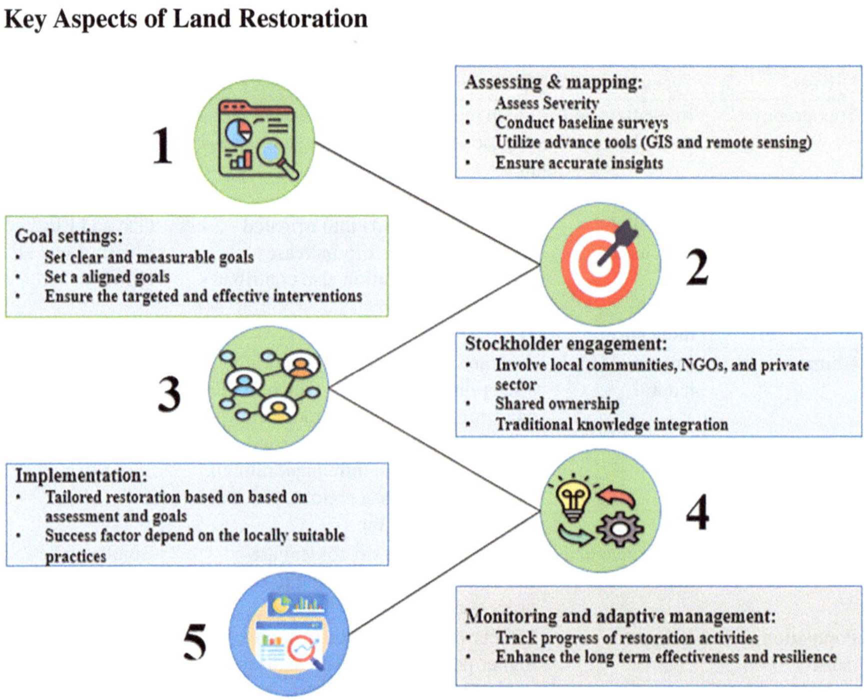

Evaluating and mapping the scale of land degradation and ecosystem decline is a vital initial step toward successful land restoration efforts. Conducting baseline assessments helps in pinpointing critical zones that require immediate attention and in measuring the extent of damage. Technological innovations such as remote sensing and Geographic Information Systems (GIS) have significantly enhanced this process by offering accurate, real-time monitoring and detailed visualization of degraded areas. As highlighted by Bastin et al. (2019) and Weng and Alavalapati (2020), these tools improve the precision of restoration planning and help in tailoring strategies to specific site conditions. Establishing clear restoration goals is equally important, as it allows for the formulation of measurable targets that align with broader environmental and socioeconomic outcomes, such as enhancing crop productivity, improving soil quality, or boosting ecosystem services. Clearly defined objectives also inform the selection of the most appropriate restoration methods, ensuring interventions are both targeted and effective (Chazdon et al. 2017).

Involving local communities, governmental organizations, nongovernmental organizations, and the commercial sector is a crucial component of effective rehabilitation. According to Reed et al. (2009), this collaborative method promotes shared ownership, incorporates traditional ecological knowledge, and aligns restoration efforts with the socioeconomic structure of the area. The next step is implementation, which uses site-specific strategies such as agroforestry systems, soil amendments, reforestation, and water management. Effective execution requires stakeholder capacity-building and locally appropriate techniques (Laestadius et al.

2015). Lastly, adaptive management and monitoring guarantee the long-term viability of restoration initiatives. Indicators like plant cover, soil health, and biodiversity indices are used to gauge progress, and adaptive management techniques improve methods in response to monitoring results. Resilience and long-term ecosystem restoration success are guaranteed by this dynamic framework (Holl and Aide 2011).

10.3 Methods of Land Restoration

10.3.1 Reforestation: Reviving Degraded Forest Ecosystems

Reforestation refers to the act of replanting trees in regions that were previously forested but have been degraded due to factors such as deforestation, agriculture, or urban development. This approach not only helps revive biodiversity but also enhances soil health, regulates water cycles, and captures atmospheric carbon dioxide—playing a crucial role in mitigating climate change. A prominent example is India's Green India Mission, which emphasizes large-scale tree planting to combat land degradation and sustain forest-based livelihoods (MoEFCC 2018). Furthermore, successful reforestation efforts often rely on active community participation by integrating indigenous ecological wisdom with modern restoration practices to promote the long-term health of these ecosystems.

10.3.2 Afforestation: Creating New Forests for Sustainable Development

Restoring degraded land often involves reforestation, which refers to planting trees in areas that were once barren or lacked forest cover. This approach plays a vital role in semiarid and degraded regions, as it helps prevent soil erosion, boosts groundwater recharge, and promotes local biodiversity. A prominent example is the large-scale tree planting efforts in China's Loess Plateau, where previously unproductive land was transformed into thriving ecosystems, resulting in better crop production and reduced soil degradation (Lü et al. 2012). Furthermore, afforestation initiatives aligned with global efforts such as the United Nations Decade on Ecosystem Restoration and the Bonn Challenge aim to rejuvenate millions of hectares of degraded landscapes worldwide by 2030.

10.3.3 Agroforestry: A Sustainable Land Restoration Technique

Agroforestry is a cutting-edge method of land restoration that blends forestry practices with agricultural crops to create a harmonious blend of ecological sustainability and agricultural productivity. This approach creates a multipurpose environment that supports both ecological balance and financial gains by utilizing the cooperative efforts of trees, crops, and cattle. The ability of agroforestry to improve soil health is one of its main advantages. Tree roots create a favorable environment for agricultural growth by reducing soil erosion, enhancing soil structure, and increasing water retention capacity. Additionally, organic materials from tree litter, like leaves and bark, provide nutrients to the soil, boosting microbial activity and soil fertility in general (Jose 2009).

Additionally, agroforestry is crucial to the preservation of biodiversity. By creating many ecosystems in one location, this technique supports a wide variety of plant and animal species. Birds, insects, and small animals find refuge and sustenance in the tree canopies, and soil organisms essential to ecosystem function are supported by the understory grasses and crops. By drawing in natural predators, this diversity not only aids in pest management but also enhances pollination services, which are vital to agricultural productivity. Because it stores atmospheric carbon dioxide in soil and tree biomass, agroforestry has a significant environmental impact. In addition to enhancing local lives by increasing access to food, fruits, and timber, this slows down climate change. Agroforestry systems can also act as buffer zones, reducing the impacts of deforestation on natural forests. It is particularly crucial to use agroforestry techniques in regions where land degradation is occurring. It provides a sustainable substitute for monocropping systems, which often lead to habitat loss and soil depletion. Agroforestry projects have been successfully implemented in nations including Brazil, Kenya, and India, proving their capacity to restore degraded lands and raise rural incomes (McNeely and Schroth 2006).

10.3.4 Land Restoration Technique: Soil Amendments

Soil amendment refers to the application of organic or inorganic substances aimed at enhancing soil quality, increasing fertility, and promoting overall ecosystem well-being. This method plays a vital role in rehabilitating degraded soils affected by activities such as agriculture, deforestation, industrialization, and urban development. Among the different approaches, incorporating organic materials, biochar, and mycorrhizal fungi emerges as an environmentally sustainable and beneficial option.

10.3.4.1 Application of Organic Matter

Key contributors to soil restoration include organic inputs such as compost, animal manure, and crop residues. These materials provide essential nutrients to plants, enhance the soil's water-holding capacity, and improve its physical structure. Incorporating organic matter also promotes better air circulation within the soil, minimizes erosion, and boosts microbial activity—factors that are vital for effective nutrient cycling. Moreover, these organic inputs serve as carbon reservoirs, helping to capture atmospheric CO_2 and mitigate climate change. For example, research by Lal (2015) indicates that applying compost to dry, degraded soils can significantly improve crop productivity and increase soil organic carbon levels.

10.3.4.2 Biochar Application

Biochar, a carbon-rich material produced by pyrolyzing organic biomass in low oxygen conditions, is another beneficial soil additive. By increasing cation exchange capacity, enhancing nutrient retention, and reducing soil acidity, it increases soil fertility. Biochar is suitable for sandy and degraded soils because it increases water retention and reduces nitrogen loss. Its porous structure enhances soil health by giving beneficial soil microbes a place to live. Additionally, the use of biochar has been widely acknowledged for its capacity to sequester carbon over the long term, which is consistent with global climate goals. According to studies, biochar can lower soil greenhouse gas emissions, especially nitrous oxide (Lehmann and Joseph 2011).

10.3.4.3 Mycorrhizal Fungi Application

A symbiotic relationship between fungi and plant roots is encouraged by the use of mycorrhizal fungi in soil amendments. By extending the root system with fungal hyphae that may acquire nutrients outside the root zone, these fungi enhance the absorption of nutrients and water. By aggregating soil particles, mycorrhizal fungi also enhance soil structure, lowering erosion and boosting plant resistance to environmental stresses. Their use is particularly beneficial for repairing damaged soils and reestablishing native plants. For instance, it has been shown that arbuscular mycorrhizal fungi significantly increase plant growth and phosphate uptake in nutrient-poor soils (Smith and Read 2008).

10.3.5 Wetland Restoration: Reestablishing Hydrology and Native Vegetation

Restoring wetlands that have been damaged by natural disasters, climate change, or human activities is the goal of wetland restoration, a crucial ecological process. Restoring the wetland's native flora and natural hydrological conditions is the major goal of wetland restoration. These elements are essential for preserving resilience, ecosystem services, and biodiversity. Because the water table, flow, and seasonal inundation patterns affect the distribution of aquatic fauna and the mix of plant species, hydrology is essential to wetland functioning. Improving water management, reestablishing natural water flow, and tackling issues including altered drainage patterns, invasive species, and excessive nutrient loading are common goals of restoration efforts (Zedler and Kercher 2005). Since native vegetation provides ecological services, including carbon sequestration, water filtering, and habitat provision, it is a crucial part of wetland restoration. The reintroduction of native plant species promotes ecological balance since they are well suited to the hydrological conditions of the marsh. Utilizing native plants is essential for improving wetland biodiversity, stabilizing the soil, and reducing erosion. To guarantee that the vegetation reflects the wetland's natural ecological history, restoration projects frequently involve meticulous plant selection, transplantation, and, in some situations, sowing (National Research Council 1992).

10.3.6 Mangrove Restoration for Coastal Land Stabilization and Biodiversity Enhancement

Mangrove restoration is a crucial land restoration tactic, particularly in coastal areas where mangroves are vital to the preservation of ecosystems. Coastal forests, especially mangroves, play a vital role in supporting both natural ecosystems and human communities. They help anchor the soil, reduce shoreline erosion, and promote biodiversity. Kathiresan and Bingham (2001) emphasize that restoring mangrove species contributes significantly to maintaining nutrient levels, preventing sediment loss, and improving soil integrity—factors crucial for a healthy coastal habitat. Furthermore, the intricate root systems of mangroves, such as prop roots and pneumatophores, serve as natural barriers by absorbing the energy of waves, storm surges, and other coastal forces (Murray et al. 2014).

Additionally, restoring mangroves is essential to boosting biodiversity along the shore. The health of local fisheries and marine ecosystems depends on the breeding, feeding, and rearing grounds that mangrove forests offer to a variety of species, such as fish, crabs, and mollusks (Lewis 2005). The restoration of mangroves creates a rich and varied environment by serving as home for a variety of aquatic animals, terrestrial organisms, and bird species. Additionally, these locations offer vital ecological services that help mitigate climate change, like carbon sequestration (Alongi 2008). Planting native mangrove species in deteriorated coastal areas is the standard practice for mangrove rehabilitation. Because different mangrove species are adapted to particular climatic conditions, such as salinity and tidal variations, species selection is crucial. For long-term success, restoration initiatives—like eliminating invasive species, reviving natural water systems, and implementing continuous monitoring and management—need to be carefully designed (McLeod and Salm 2009). In addition, involving local communities and stakeholders is crucial, as their traditional knowledge and sustainable approaches can significantly enhance the resilience of ecosystems (Bayraktarov et al. 2016).

10.3.7 Land Restoration Techniques: Terracing and Contour Farming to Reduce Soil Erosion in Hilly Areas

One of the most common methods of restoring land in mountainous areas is terracing. It involves constructing horizontal, stepped platforms along a hill's slope. By dividing the water flow, these terraces reduce the likelihood of soil erosion and decrease the water's descent down the slope. A steep slope can be effectively terraced into a number of level areas that are more suitable for farming. Additionally, terraces aid in soil water retention, enhancing the land's fertility and sustainability for farming. In regions like Southeast Asia, the Andes, and the Himalayas, where erosion is a persistent issue due to steep terrain and severe rainfall, this technology has been widely employed. In 1995, Pimentel et al. worked on contour farming, which entails planting crops or plowing in accordance with the land's contours rather than in straight lines and is another technique related to soil erosion control.

This tactic lowers water flow and the erosive impact of precipitation by utilizing the topography of the soil. By creating natural barriers, planting crops along contours slows water flow and lets it soak into the soil instead of washing away and losing topsoil. The best results from contour farming come from combining it with other soil conservation techniques like mulching or cover crops, which shield the soil's surface from erosion caused by wind and water (Erenstein 2003).

10.3.8 Bioremediation: Use of Microbes and Plants to Detoxify Contaminated Soils

Using living organisms, usually bacteria and plants, bioremediation is a long-term and efficient land restoration technique that detoxifies or eliminates pollutants from soil and water. This technique breaks down contaminants using the natural metabolic processes of microorganisms or plants, rendering them completely nontoxic or less harmful. Microbial bioremediation and phytoremediation have drawn the most interest among the many bioremediation methods. Utilizing specific microorganisms, including bacteria, fungi, and algae, microbial bioremediation breaks down, transforms, or immobilizes environmental pollutants. These bacteria have developed mechanisms to effectively transform organic and inorganic contaminants into less hazardous substances by using them as sources of energy or nourishment. Common microorganisms used in bioremediation to break down hydrocarbons, heavy metals, and organic contaminants like pesticides and solvents include *Pseudomonas* species, *Bacillus* species, and *Mycobacterium* species. Because of its effectiveness and affordability in eliminating various forms of soil contamination, microbial bioremediation is commonly chosen, particularly in industrial and agricultural contexts (Ghosal et al. 2015). The use of plants to absorb, accumulate, and detoxify contaminants from soil through processes like phytoextraction, phytodegradation, and phytostabilization is known as phytoremediation. The ability of *Brassica juncea, Populus* species, and *Salix* species to absorb heavy metals, pesticides, and organic pollutants through their root systems and retain or alter them in their tissues is well-known. Because plants can be used to restore damaged landscapes, improve soil fertility, and increase biodiversity, phytoremediation offers the dual benefits of being aesthetically pleasing and environmentally benign (Ali et al. 2013).

10.4 Livelihood Improvement Strategies

Livelihood refers to the means through which individuals secure the essentials of life, encompassing their capabilities, assets, income, and activities. It represents a dynamic interplay of resources—natural, physical, human, and financial—along with skills that enable survival and contribute to overall well-being (UN-REDD n.d.). The concept of livelihood is not uniform; it varies across nations, influenced by economic conditions, environmental factors, and social structures. These contextual elements shape how livelihoods are perceived and pursued in diverse regions (Zhen et al. 2022). The ongoing degradation of natural resources has emerged as a major global concern, significantly endangering livelihoods across the world. Currently, this issue has already affected more than 25% of the Earth's land surface, impacting the lives of around 3.2 billion people (Olsson et al. 2019; Mirzabaev et al. 2019). Although land comprises just 30% of the planet's surface, it is home to 85% of all recognized species and includes diverse ecosystems such as forests, grasslands, and agricultural lands. These ecosystems are essential, offering crucial services and resources that support life. However, increasing land degradation and the transformation of natural ecosystems are escalating the risks to both biodiversity and human well-being. As the global population continues to rise, so does the pressure on land resources and the demand for related goods. At the same time, a widening gap between people and their environmental responsibilities threatens to undermine efforts to address land degradation and related ecological issues. This growing disconnect not only weakens environmental conservation but also puts at risk the livelihoods of millions who rely directly on these natural systems (UNCCD 2023).

Land degradation results from a multifaceted set of drivers that are generally classified into social, political, cultural, and economic dimensions (Olsson et al. 2019). Tackling this issue is particularly challenging, as it predominantly impacts populations in low-income regions of developing countries. Beyond damaging terrestrial environments, land degradation also affects nearby aquatic ecosystems—both freshwater and marine—and poses serious risks to the health and livelihoods of adjacent communities (Altieri et al. 2017). The situation is further intensified by the growing impacts of climate change, which not only worsen land degradation but are also fueled by it, creating a self-reinforcing cycle. Therefore, it is crucial to align land restoration efforts with wider strategies aimed at combating climate change (Arneth et al. 2021). The consequences of land degradation are particularly severe for poor and marginalized communities who rely heavily, both directly and indirectly, on the environment for their livelihoods. Recognizing this, global initiatives such as the UN Decade on Ecosystem Restoration (2021–2030) provide a crucial opportunity to intensify efforts toward combating environmental degradation. However, restoration is often resource-intensive, requiring substantial financial investment. Ensuring adequate funding, coupled with the formulation of inclusive policies and active stakeholder engagement, is critical to addressing the needs and aspirations of affected communities.

Tailoring livelihood opportunities to the local context—considering factors such as location, type, and extent of degradation—can help align community preferences with ecosystem restoration efforts. By fostering a balance between trade-offs and the ecosystem services provided by restored environments, sustainable outcomes can be achieved (Lengefeld et al. 2022). Open dialogues about restoration processes, policies, and payment mechanisms can further bolster financial restoration efforts while simultaneously supporting community development and livelihood generation (Aronson et al. 2010; Lengefeld et al. 2022). Land degradation poses significant challenges that demand integrated and coordinated efforts across ecological, economic, and social dimensions to foster resilience and sustainable development. The deterioration of land quality, including soil erosion, nutrient depletion, and reduced agricultural productivity, profoundly impacts livelihoods, particularly in agriculture-dependent communities of developing nations. These challenges exacerbate the vulnerabilities of communities already grappling with climate change, poverty, food insecurity, and low incomes (Birhanu et al. 2024).

Land degradation has far-reaching consequences that affect all three core dimensions of sustainable development—social, environmental, and economic—emphasizing the critical need for effective restoration strategies. Encouragingly, global efforts have begun to yield positive outcomes, with approximately 16% of degraded land worldwide showing improvement. Nevertheless, the journey ahead remains challenging, demanding enhanced and sustained actions to preserve productive land for both present and future generations (Davis and Bessonova 2014).

Restoration initiatives in different regions have delivered significant results, illustrating the powerful role ecological restoration can play in supporting both the environment and communities. For example, in Guinea, the use of short-rotation coppice systems has led to economic gains of $11–$20 per person per day. In Jerusalem, notable improvements in soil health were observed—soil moisture increased from 34% to 89%, while erosion rates dropped by 94%. These achievements demonstrate how land restoration not only revives ecosystems but also strengthens the resilience and livelihoods of those who depend on them (Zhen et al. 2022). People's choices in managing natural resources are often shaped by their need to secure basic needs and a stable livelihood. This connection underscores the vital role of ecosystem restoration in enhancing quality of life. Restoration policies significantly influence aspects of livelihood such as human well-being, natural resources, and community ties. Furthermore, they play a crucial role in ensuring the continued supply of essential services like clean air, energy sources, and spaces for recreation.

Achieving successful restoration requires individuals to adapt their behaviors to align with sustainability goals. For example, people may need to reduce cropping and grazing areas, alter fuel consumption patterns, or diversify their income sources beyond farming. While these changes are essential for ecological recovery, they directly impact individuals who have relied on these resources for generations. Understanding and addressing these behavioral adjustments are key to fostering a harmonious balance between restoration goals and community livelihoods (Zhen et al. 2022). To achieve successful restoration, it is crucial to encourage individuals to adapt their behaviors in ways that align with sustainability objectives, so some success stories are the following:

10.4.1 China's Loess Plateau

It is a well-known example of land restoration throughout the world. The World Bank-backed Loess Plateau Watershed Rehabilitation Project significantly restored the plateau, which was once thought to be among the most degraded environments as a result of centuries of overgrazing and deforestation. The project, which was started in the 1990s, included terracing, extensive reforestation, and the adoption of sustainable farming methods. This led to the restoration of more than 1.2 million hectares of damaged land, which increased plant cover and greatly decreased soil erosion. Increased agricultural output, better water supplies, and a reduction in poverty all benefited the neighborhood. In addition to demonstrating the efficacy of landscape-level restoration, this project highlighted how community involvement may be incorporated into sustainable land management (World Bank 2007).

10.4.2 Rwanda's Gishwati Forest

The Gishwati-Mukura Landscape Restoration Project restored Rwanda's Gishwati Forest, which had been badly damaged by deforestation and agricultural growth. This program, which was backed by the Rwandan government and international groups like as the International Union for Conservation of Nature (IUCN), aimed to preserve biodiversity, reforest degraded areas, and provide eco-friendly livelihoods to local communities. Since it was put into place, the area's forest cover has grown, improving water quality, reducing erosion, and allowing endangered species like the golden monkey to reappear. This project serves as an example of how restoration initiatives can concurrently address ecological and socioeconomic issues (IUCN 2020).

10.4.3 Aravalli Hills Restoration in Haryana and Rajasthan

The oldest mountain range in India, the Aravalli Hills, suffered greatly from urbanization, deforestation, and mining. Mining prohibitions and reforestation campaigns have resulted from recent initiatives by state governments, environmental groups, and the Indian Supreme Court. Raising awareness and encouraging afforestation efforts have been greatly aided by nongovernmental organizations (NGOs), such as the Aravalli Bachao Citizens Group. The region's desertification trends have decreased, natural flora has returned, and groundwater recharge has increased as a result of these efforts, demonstrating the possibility of community-led restoration projects (Citizen-led Initiative: Aravalli Bachao 2023).

10.4.4 Miyawaki Forests in Urban Areas

Akira Miyawaki, a Japanese botanist, invented the Miyawaki approach, which has been used to repair urban environments across India. This technique has been employed, for example, in Chennai and Bengaluru to establish dense urban forests on underutilized or degraded land. These mini-forests have demonstrated milder microclimates, improved air quality, and a sharp increase in biodiversity. Because it can restore ecosystems in constrained areas, the method has become popular for sustainable urban development (Miyawaki 1999).

10.4.5 Odisha's Mangrove Restoration in Bhitarkanika

One outstanding example of mangrove restoration is the Bhitarkanika National Park in Odisha. Large-scale replanting operations were started by the state government in 1999 after Cyclone Paradip severely damaged mangroves, with assistance from NGOs such as Wetlands International South Asia. In addition to restoring more than 12,000 hectares of mangrove forests, this endeavor increased fishing production and served as a buffer against storms and coastal erosion, improving the livelihoods of nearby residents. The importance of mangroves for ecological and climate resilience is demonstrated by the Bhitarkanika instance (Wetlands International South Asia 2022).

10.5 Conclusion

Land degradation has emerged as a critical global challenge, impacting ecosystems, landscapes, and biodiversity. With a finite amount of land available and a rapidly growing population, restoring degraded land to maintain its productivity has become increasingly important. Land restoration refers to the process of bringing degraded ecosystems back to a state that supports native plant and animal life. This process not only enhances soil fertility but also fosters ecological balance and contributes to economic development. Various techniques are being employed to rehabilitate degraded land, including reforestation, agroforestry, and soil improvement practices. The effectiveness of these methods varies depending on regional conditions and the extent of degradation. Since land serves as a fundamental resource for human survival, especially for vulnerable and marginalized communities in developing nations, its deterioration significantly impacts their livelihoods and well-being. In many regions, successful land restoration efforts have gone hand-in-hand with livelihood improvement programs. Actively involving local communities in these initiatives not only deepens their understanding of environmental sustainability but also creates new employment and income opportunities. This integrated approach contributes to long-term socioeconomic stability and improves overall quality of life.

References

Ali H, Khan E, Sajad MA (2013) Phytoremediation of heavy metals—concepts and applications. Chemosphere 91(7):869–881. https://doi.org/10.1016/j.chemosphere.2013.01.075

Alongi DM (2008) Mangrove forests: resilience, protection from storms, and carbon sequestration. Oceanography 21(4):3–18

Altieri AH, Harrison SB, Seemann J, Collin R, Diaz RJ, Knowlton N (2017) Tropical dead zones and mass mortalities on coral reefs. PNAS 114:3660–3665

Abdul-Wadood Moomen A, Dewan A (2016) Assessing the spatial relationships between mining and land degradation: evidence from Ghana. Int J Min Reclam Environ 31(7):505–518. https://doi.org/10.1080/17480930.2016.1188253

Akabzaa TM, Darimani A (2001) Impact of mining sector investment in Ghana: a study of the Tarkwa mining region (Draft Report). Structural Adjustment Participatory Review Initiative (SAPRI)

Arneth A, Olsson L, Cowie A, Erb KH, Hurlbert M, Kurz WA, Mirzabaev A, Rounsevell MDA (2021) Restoring degraded lands. Annu Rev Environ Resour 46:569–599. https://doi.org/10.1146/annurev-environ-012320-054809

Alkama R, Cescatti A (2016) Biophysical climate impacts of recent changes in global forest cover. Science 351(6269):600–604. https://doi.org/10.1126/science.aac8083

Aronson J, Blignaut JN, Milton SJ, Le Maitre D, Esler KJ, Limouzin A, Fontaine C, De Wit MP, Prinsloo P, Van Der Elst L (2010) Are socioeconomic benefits of restoration adequately quantified? A meta analysis of recent papers (2000–2008) in restoration ecology and 12 other scientific journals. Restor Ecol 18:143–154

Areendran G, Rao P, Raj K, Mazumdar S, Puri K (2013) Land use/land cover change dynamics analysis in mining areas of Singrauli district in Madhya Pradesh, India. Trop Ecol 54:239–250

Bai Z, Dent D, Olsson L, Schaepman M (2008) Proxy global assessment of land degradation. Soil Use Manag 24(3):223–234

Barrow C (1991) Land degradation: development and breakdown of terrestrial environments. Cambridge University Press, Cambridge, UK

Bastin JF et al (2019) The global tree restoration potential. Science 365(6448):76–79

Bayraktarov E et al (2016) The effectiveness of mangrove restoration. PLoS One 11(2):e0158679

Birhanu BZ, Desta G, Cofie O, Tilahun SA, Mabhaudhi T (2024) Restoring degraded landscapes and sustaining livelihoods: sustainability assessment (cum-review) of integrated landscape management in sub-Saharan Africa. Front Clim 6:1338259. https://doi.org/10.3389/fclim.2024.1338259

Bolwig S (2002) Land use change and soil degradation in the Southwestern highlands of Uganda. A Contribution to the Strategic Criteria for Rural Investments in Productivity (SCRIP) Program of the USAID Uganda Mission. The International Food Policy Research Institute, Washington, DC

Bonilla CA, Johnson OI (2012) Soil erodibility mapping and its correlation with soil properties in Central Chile. Geoderma 189:116–123

Brasselle F, Brasselle A, Gaspart F, Platteau JP (2002) Land tenure security and investment incentives: puzzling evidence from Burkina Faso. J Dev Econ 67:373–418

Chazdon RL et al (2017) A policy-driven framework for forest and landscape restoration. Front Ecol Environ 15(9):483–490

Chirwa PW (2014) Restoration practices in degraded landscapes of Eastern Africa. African Forest Forum, Working Paper Series 2(11):55

Citizen-led Initiative: Aravalli Bachao (2023) Restoring Haryana's Hills

Davis, M., M. Palm, E. Mattsson and E. Bessonova (2014). Sharing the Land: Restoring Degraded Ecosystems and Improving Livelihoods Through Agroforestry. Swedish International Agricultural Network Initiative (SIANI) and Forest, Climate & Livelihood Research Network (Focali), "Forests, Landscapes and Food Security" theme brief.

Erenstein O (2003) Contour farming in the semi-arid regions of India. Land Degrad Dev 14(5):463–472

Eswaran H, Lal R, Reich PF (2001) Land degradation: an overview. In: Bridges EM, Hannam ID, Oldeman LR, Pening de Vries FWT, Scherr SJ, Sompatpanit S (eds) Responses to land degradation (Proc. 2nd Int. Conf. on land degradation and desertification, Khon Kaen, Thailand). Oxford Press, New Delhi, pp 20–35

FAO (2011) The state of the world's land and water resources for food and agriculture. Managing systems at risk. UN FAO, Rome, p 50

Foley JA, DeFries R, Asner GP, Barford C, Bonan G, Carpenter SR, Chapin FS, Coe MT, Daily GC, Gibbs HK, Helkowski JH, Holloway T, Howard EA, Kucharik CJ, Monfreda C, Patz JA, Prentice IC, Ramankutty N, Snyder PK (2005) Global consequences of land use. Science 309(5734):570–574. https://doi.org/10.1126/science.1111772

Gao J, Liu Y (2010) Determination of land degradation causes in Tongyu County, Northeast China via land cover change detection. Int J Appl Earth Obs Geoinf 12:9–16
Ghosal S, Maiti SK, Bose P (2015) Role of microorganisms in bioremediation of contaminated soil. Biorem J 19(1):1–14. https://doi.org/10.1080/10889868.2015.1118322
Grepperud S (1996) Population pressure and land degradation: the case of Ethiopia. J Environ Econ Manag 30(1):18–33
Holl KD, Aide TM (2011) When and where to actively restore ecosystems? For Ecol Manag 261(10):1558–1563
International Union for Conservation of Nature (IUCN) (2020) Gishwati-Mukura: a Model for Landscape Restoration in Rwanda. Retrieved from [IUCN Resources]
Jose S (2009) Agroforestry for ecosystem services and environmental benefits: an overview. Agrofor Syst 76(1):1–10
Kabubo-Mariara J (2007) Land conservation and tenure security in Kenya: Boserup's hypothesis revisited. Ecol Econ 64:25–35
Kathiresan K, Bingham BL (2001) Biology of mangroves and mangrove ecosystems. Adv Mar Biol 40:81–251
Laestadius L et al (2015) Strategies for scaling up forest landscape restoration in the tropics. Conserv Lett 8(1):14–21
Lal R (2015) Restoring soil quality to mitigate soil degradation. Sustainability 7(5):5875–5895
Lambin EF, Geist HJ, Lepers E (2003) Dynamics of land-use and land-cover change in tropical regions. Annu Rev Environ Resour 28(1):205–241. https://doi.org/10.1146/annurev.energy.28.050302.105459
Lehmann J, Joseph S (2011) Biochar for environmental management: science and technology. Earthscan
Lengefeld E, Stringer LC, Nedungadi P (2022) Livelihood security policy can support ecosystem restoration. Restor Ecol 30:e13621. ISSN 1526-100X
Lewis RR (2005) Ecological engineering for successful management and restoration of mangrove forests. Ecol Eng 24(4):403–418
Le Quéré C, Andrew RM, Canadell JG, Sitch S, Korsbakken JI, Peters GP, Manning AC, Boden TA, Tans PP, Houghton RA, Keeling RF, Alin S, Andrews OD, Anthoni P, Barbero L, Bopp L, Chevallier F, Chini LP, Ciais P et al (2016) Global carbon budget 2016. Earth Syst Sci Data 8(2):605–649. https://doi.org/10.5194/essd-8-605-2016
Liniger H, Critchley W (2007) Where the land is greener. Technical Center for Agricultural and Rural Cooperation, Wageningen
Liniger H, Schwilch G (2002) Enhanced decision-making based on local knowledge: the WOCAT method of sustainable soil and water management. Mt Res Dev 22(1):14–18
Lu D, Batistella M, Mausel P, Moran E (2007) Mapping and monitoring land degradation risks in the Western Brazilian Amazon using multitemporal landsat TM/ETM+ images. Land Degrad Dev 18:41–54
Lü Y, Fu B, Feng X, Zeng Y, Liu Y, Chang R, Wu B (2012) A policy-driven large scale ecological restoration: quantifying ecosystem services changes in the Loess Plateau of China. PLoS One 7(2):e31782
McLeod E, Salm RV (2009) Managing mangroves for resilience to climate change. IUCN, Gland
McNeely JA, Schroth G (2006) Agroforestry and biodiversity conservation—traditional practices, present dynamics, and lessons for the future. Biodivers Conserv 15(2):549–554
Ministry of Environment, Forest and Climate Change (MoEFCC) (2018) Green India Mission: Securing Ecosystems and Livelihoods. Government of India
Mirzabaev A, Nkonya E, von Braun J (2015) Economics of sustainable land management. Curr Opin Environ Sustain 15:9–19
Mirzabaev A, Wu J, Evans J, García-Oliva F, Hussein IAG et al (2019) Desertification. In: Shukla PR, Skea J, Buendia EC, Masson-Delmotte V, Pörtner HO et al (eds) Climate change and land: an IPCC special report on climate change, desertification, land degradation, sustainable land management, food security, and green house gas fluxes in terrestrial ecosystems. Intergov. PanelClim.Change, Geneva, pp 249–343

Miyawaki A (1999) Creating native forests by the quickest method. Springer
Murray BC et al (2014) Green infrastructure and the role of coastal ecosystems in protecting the United States from storm surge damage. J Coast Res 30(5):1339–1347
National Research Council (1992) Restoration of aquatic ecosystems: science, technology, and public policy. National Academy Press
Nkonya E, Pender J, Kaizzi K, Kato E, Mugarura S, Ssali H, Muwonge J (2008) Linkages between land management, land degradation, and poverty in Sub-Saharan Africa: the case of Uganda. IFPRI Research Report 159. Washington D.C., USA
Nachtergaele F, van Velthuizen H, Verelst L, Batjes NH, Dijkshoorn JA, van Engelen VWP, Fischer G, Jones A, Montanarella L (2010) Harmonized World Soil Database (Proceedings of the 19th world congress of soil science, Brisbane, Australia). International Union of Soil Sciences
Nkonya E, Gerber N, Baumgartner P, von Braun J, De Pinto A, Graw V et al (2011) The economics of desertification, land degradation, and drought—toward an integrated global assessment. ZEF-Discussion Papers on Development Policy No. 150. Center for Development Research (ZEF), Bonn
Olsson L, Barbosa H, Bhadwal S, Cowie A, Delusca K et al (2019) Land degradation. In: Shukla PR, Skea J, Buendia EC, Masson-Delmotte V, Pörtner HO et al (eds) Climate change and land: an IPCC special report on climate change, desertification, land degradation, sustainable land management, food security, and green house gas fluxes in terrestrial ecosystems. Intergov. PanelClim.Change, Geneva, pp 345–436
Pender J, Nkonya E, Jagger P, Sserunkuuma D, Ssali H (2006) Strategies to increase agricultural productivity and reduce land degradation in Uganda: an econometric analysis. In: Pender J, Ehui S (eds) Strategies for sustainable land management in the East African highlands. International Food Policy Research Institute, Washington, DC, pp 165–190
Pender J, Mirzabaev A, Kato E (2009) Economic analysis of sustainable land management options in Central Asia. Final report for the ADB.IFPRI/ICARDA, 168. Washington DC, USA
Pimentel D et al (1995) Environmental and economic costs of soil erosion and conservation benefits. Science 267(5201):1117–1123
Reed MS et al (2009) Who's in and why? A typology of stakeholder analysis methods for natural resource management. J Environ Manag 90(5):1933–1949
Safriel UN, Adeel Z (2005) Dryland systems. In: Hassan R, Scholes R, Ash N (eds) Ecosystems and human well-being: current state and trends, vol 1. Island Press, Washington, DC, pp 623–662
Smith SE, Read DJ (2008) Mycorrhizal Symbiosis. Academic Press
Song X-P, Hansen MC, Stehman SV, Potapov PV, Tyukavina A, Vermote EF, Townshend JR (2018) Global land change from 1982 to 2016. Nature 560(7720):639–643. https://doi.org/10.1038/s41586-018-0411-9
Thakur TK, Dutta J, Upadhyay P, Patel DK, Thakur A, Kumar M, Kumar A (2022) Assessment of land degradation and restoration in coal mines of central India: a time series analysis. Ecol Eng 175:106493. ISSN 0925-8574, https://doi.org/10.1016/j.ecoleng.2021.106493
Thakur TK, Dutta J, Bijalwan A, Swamy SL (2022a) Evaluation of decadal land degradation dynamics in old coal mine areas of Central India. Land Degrad Dev 33(16):3209–3230. https://doi.org/10.1002/ldr.4383
Tiffen M, Mortimore M, Gichuki F (1994) More people, less erosion: environmental recovery in Kenya. Wiley and Sons, London, UK
Turner BL II, Lambin EF, Reenberg A (2007) The emergence of land change science for global environmental change and sustainability. Proc Natl Acad Sci USA 104(52):20666–20671. https://doi.org/10.1073/pnas.0704119104
UNCCD (2023) Land Restoration to Safeguard Nature and Livelihoods. https://www.unccd.int/resources/publications/land-restoration-safeguard-nature-and-livelihoods-unccd-and-cbd-working
UNREDD (n.d.). https://www.un-redd.org/glossary/livelihood
United Nations Environment Programme (1992) World atlas of desertification. UNEP
United Nations Convention to Combat Desertification (2015) Combating desertification/land degradation and drought for poverty reduction and sustainable development: the contribution

of science, technology, traditional knowledge and practices: book of abstracts/UNCCD 3rd Scientific Conference, 9–12 March 2015, Cancún, Mexico. UNCCD
U.S. Environmental Protection Agency (2003) EPA's report on the environment (2003 draft report) (EPA/600/R-03/050). Author, Washington, DC
von Braun J, Gerber N, Mirzabaev A, Nkonya E (2013) The economics of land degradation. ZEF working papers 109. Bonn, Germany
Voortman RL, Sonneveld BG, Keyzer MA (2000) African land ecology: opportunities and constraints for agricultural development. Center for International Development Working Paper 37. Harvard University, Cambridge, MA
Way SA (2006) Examining the links between poverty and land degradation: from blaming the poor toward recognizing the rights of the poor. In: Johnson P, Mayrand K, Paquin M (eds) Governing global desertification: linking environmental degradation, poverty, and participation. Ashgate, Burlington, VT, pp 27–41
Weng YC, Alavalapati JRR (2020) GIS applications for sustainable development and environmental management. Springer
Wetlands International South Asia (2022) Mangrove restoration in Bhitarkanika: lessons and achievements
Wischmeier WH (1976) Use and misuse of the universal soil loss equation. J Soil Water Conserv 13(1):5–9
World Bank (2007) Restoring China's Loess Plateau: turning back the clock on environmental degradation. Retrieved from [World Bank Publications]
Zedler JB, Kercher S (2005) Wetland resources: status, trends, and restoration priorities. Ecol Appl 15(3):722–734
Zhao et al (2013) Soil erosion and its response to the changes of precipitation and vegetation cover on the Loess Plateau. J Geogr Sci 23(6):1091–1106. https://doi.org/10.1007/s11442-013-1065-z
Zhen L, Xu Z, Cheng W (2022) Ecological protection and livelihood improvement in ecologically vulnerable regions. J Resour Ecol 13(5):759–762

Agroforestry-Based Degraded Land Restoration: An Indian Perspective

11

Narayan Lal, Nisha Sahu, Khushboo Rani, Dhiraj Kumar, Bharat Prakash Meena, Sanjib Kumar Behera, and Vikas Ramteke

Abstract

Land degradation is one of the primary issues in India that leads to the deterioration of the land's quality for agricultural production and environmental protection. In India, there are approximately 147.75 million hectares (Mha) of degraded land, of which 93.68 Mha are caused by water erosion, 16.03 Mha by acidification, 14.29 Mha by waterlogging, 9.48 Mha by wind erosion, 5.89 Mha by salinity, and 8.38 Mha by other factors such as ice caps and barren mountains. Different chemical, physical, and biological processes, like shifting agriculture, deforestation, overgrazing, steep slope farming, excessive use of chemical fertilizers, and depletion of forest resources, can cause land degradation. These processes can be caused directly or indirectly by human actions. Land can still be used while preventing degradation thanks to agroforestry. Through the various service and production roles that woody perennials play in agricultural and livestock systems, agroforestry, the intentional integration of perennial plants has the potential to reduce the problem of land degradation. Agroforestry can reduce salt, boost soil fertility, decrease acidity, alkalinity, and desertification, among other things. In the end, it improves soil health, maintaining the land's suitability for the long-term, sustainable production of a variety of goods. By restoring and boosting land productivity, the agroforestry tactic includes relay-cropping, the use of multipurpose tree species (MPTs), terracing and contour cultivation, alley and strip cropping are suitable to meet the needs of low-resource farmers. Agroforestry has the potential to reduce land degradation when farmers and the community are fully engaged in the planning, development, and execution stages

N. Lal (✉) · N. Sahu · K. Rani · D. Kumar · B. P. Meena · S. K. Behera
ICAR-Indian Institute of Soil Science, Bhopal, Madhya Pradesh, India

V. Ramteke
Indira Gandhi Krishi Vishwavidyalaya, Raipur, Chhattisgarh, India

V. C. Pandey (ed.), *Plantation-Based Land Restoration*,
https://doi.org/10.1007/978-981-95-2860-8_11

of the project. The design of an agroforestry system must incorporate a variety of stakeholders in order to efficiently support sustainable land production, improve biodiversity, and improve ecosystem services at the plot and landscape scales. Additionally, best practices for diversifying agroforestry systems must be identified, and soil properties and land use in degraded landscapes must be better understood.

Keywords

Land degradation · Agroforestry · Diversification · Mitigation

11.1 Introduction

The biological, economic, and quality of land are all lost as a result of changes to its chemical, physical, and biological characteristics is known as land degradation. There are several factors (Fig. 11.1) that can lead to land degradation, which might include soil compactness, a decrease in soil pH value, destruction of forests, aridity, weathering, degradation of organic carbon, and salt accumulation (Turner et al. 2016). Land degradation is an issue that impacts numerous kinds of ecosystems worldwide. Land degradation already impacts approximately 40% of the planet's terrain, and it diminishes another 12 million hectares (Mha) of land annually (Kapur 2016). Land degradation is becoming a global problem. It threatens not just the health of people but also the potability of the water and the essential components of ecological systems, which serve as habitat to all forms of life. For instance,

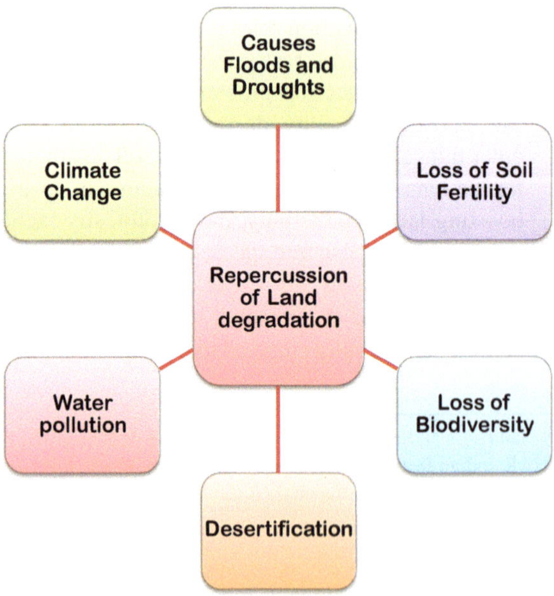

Fig. 11.1 Potential threats from land degradation

biodiversity and climate change are closely related to it, as are other significant global concerns. Land degradation brought on by unjustifiable land use/land cover change is impacting the provision of ecosystem services from healthy ecosystems, and this is continuing to undermine the landscape's potential for multifunctionality. Of those affected by this land degradation, 3.2 billion are poor and underprivileged people, mostly living in rural regions with limited options for adaptation (Sena and Ebi 2021). More than 42% of the world's underprivileged inhabitants reside in degraded regions, and an estimated 1.5 billion individuals worldwide relied on these lands for their sustenance in 2007 (Nachtergaele et al. 2010). Globally, a projected 27,000 biospecies and 24 billion tons of fertile soil are lost every year. As a result, it is critical to bring an end to soil deterioration in order to enhance living conditions for those who live there while also ensuring food, water, and environmental security.

Around USD 40 billion is lost to degradation yearly on a global scale. Its negative effects on the availability of food, ecosystems, and agricultural production mean that it will remain an important challenge in the coming decades. It has grown to be an important ecological problem (Scherr 1999). Approximately 18.1 million km^2 of degraded land exists worldwide, of which 92% is the result of ineffective management and 38% is the result of excessive livestock grazing (Hamdy and Aly 2014). More than 1.5 billion humans live in a world where 20% of cultivated land, 30% of flora and 10% of pastures are affected by land degradation. There are several possible human-caused for it, including hazardous land use practices and global warming (Bai et al. 2008).

A viable farming system and environment depend on the restoration of land used for agriculture. The ever-growing population is placing enormous strain on land, which in turn is driving up demand for food, fiber, and housing. Agricultural land degradation is caused by a multitude of ecological and man-made factors. The primary sources of jeopardy brought on by soil erosion. Contraction, acidification, and salt accumulation are other elements that deteriorate arable lands. Aggressive farming, forest clearing, excessive grazing, and poor soil health management are the main causes of erosion on farms. These damaged lands are restored by a combination of ecological and farming practices, which include agroforestry practices, crop rotation, reducing tillage, using cover crops, establishing vegetative buffer zones, and no-till farming practices. The soil is protected against erosion by biological techniques like crop residues applied with compost, buffers, and conditioners used in direct contact with the soil surface.

The goal of restoration is to start and speed up land recovery. Depending on the degrading factor and the environment in question, several restoration techniques are used. Forest ecosystems can be restored by replanting in natural or planted areas (Gastauer et al. 2020). Rangelands can be kept from eroding and becoming decertified by reseeding to produce vegetation cover (Stock et al. 2020). While wetland restoration seeks to restore the natural hydrology of rivers and lagoons, conservation tillage, nutrient replenishment, and organic replenishment can repair deficient soils (Santini et al. 2019).

Land degradation can be mitigated and its recovery aided by the use of agroforestry as a sustainable system. Unlike monocrop systems, a variety of plant species are employed to create a biodiverse agricultural composition. Trees are a necessary component of agroforestry systems, which are generally made up of a variety of crops, herbaceous plants, and forage species. With several positive effects on the environment and the economy, this composition promotes biodiversity and regional ecosystem services and functions, resulting in sustainable rural development. However, because trees occupy ground and reduce the amount of area available for food production, farmers may view the planting of trees unfavorably (Leakey et al. 2015).

The technique of cultivating trees alongside crops is known as agroforestry, is becoming more recognized for its ability to offset the detrimental effects of soil degradation and greenhouse gas emissions (Verchot et al. 2007). According to reports, this land use strategy promotes biodiversity conservation, improves air and water quality, increases soil nutrient status, and has the capacity to sequester carbon (Dollinger and Jose 2018; Jose 2009). The effectiveness of agroforestry systems in mitigating soil erosion, preserving ideal soil organic matter and other physicochemical qualities, and ultimately increasing soil biodiversity makes them particularly significant for improving soil quality and fertility (Dollinger and Jose 2018; Marsden et al. 2019). This method has the potential to improve ecosystem performance while also enhancing the products and services acquired from agricultural and forest ecosystems (Gold and Garrett 2009). Because agroforestry enhances soil carbon accumulation, it may also be a significant factor in reducing greenhouse gas buildup in the atmosphere on a worldwide scale (IPCC 2000). Agroforestry is generally favored for food provisioning over forest plantations because it allows for the introduction of a greater variety of edible species into the system. Thus, small-scale farmers can practice agroforestry may be an effective means of reducing poverty and safeguarding natural ecosystems (Nath et al. 2005).

Numerous studies have demonstrated that the socioeconomic, monetary, and environmental aspects of the land, as well as the decrease of poverty, are the primary motivators for a community's desire to adopt agroforestry (Knowler and Bradshaw 2007). The choice of whether or not to adopt agroforestry, however, is more difficult since it entails a number of choices, such as choosing the best combination of products, which necessitates a thorough evaluation of input consumption and time (Oostendorp and Zaal 2012). We still don't fully grasp the adoption process or the extent to which community agroforestry is being adopted (Jara-Rojas et al. 2020). While many earlier studies (Nkamleu and Manyong 2005; Jera and Ajayi 2008) have treated the adoption of agroforestry as a binary option, other factors—namely, social capital and network variables—also play a significant role in determining whether or not potential adopter farmers choose to use agroforestry as a viable cultivating technique (Jara-Rojas et al. 2020). The major objective of a book chapter is to focus on presenting how agroforestry can be an effective tool in restoring degraded lands, and this chapter will provide a comprehensive understanding of how agroforestry can be a powerful tool in addressing land degradation and ensuring sustainable development in India.

11.2 Global Scenario of Land Degradation

Land degradation is a significant global challenge that currently affects about two billion hectares, or nearly 25% of the Earth's land area. This degradation is primarily driven by unsustainable agricultural practices, deforestation, urbanization, and climate change. The Food and Agriculture Organization (FAO) estimates that around 1.3 billion people depend on degraded land for their livelihoods, impacting food security and economic stability (FAO 2021). Recent data indicate that land degradation could reduce global agricultural productivity by up to 30% by 2050 if current trends continue (Intergovernmental Science-Policy Platform on Biodiversity and Ecosystem Services 2021). Furthermore, the UN Environment Programme (UNEP) reported that land degradation currently results in annual economic losses exceeding $10 trillion due to decreased ecosystem services, which include soil fertility and water regulation (UNEP 2022). To combat land degradation, initiatives such as the United Nations Decade on Ecosystem Restoration (2021–2030) aim to restore 350 Mha of degraded land globally. Effective management strategies, policy reforms, and community involvement are crucial for reversing land degradation and promoting sustainable land use practices to secure a healthier environment for future generations.

11.3 Global Scenario of Agroforestry

Agroforestry systems encompass a wide range of socio-ecological contexts and biophysical environments, from urban farmland and backyard gardens to vegetation plantation farms and animal husbandry or pastoral vitality systems. The various agroforestry systems followed in different countries are presented in Table 11.1. Since the scope of its practice has never been measured, opinions regarding its significance are greatly divided. Agroforestry offers an opportunity to enhance agricultural productivity, reduce the effects of global warming, and achieve resilience in the agricultural sector. An estimated 1.2 billion people worldwide depend on agroforestry farming systems; it is extensively utilized in tropical and developing nations (World Bank 2004). More than 1.2 billion individuals worldwide engage in agroforestry on over 1 billion ha of land, according to the UN Food and Agriculture Organization. It is urgently necessary to compensate for the harm caused by improper agricultural practices and other variables, as recent FAO (Food and Agriculture Organization) reports reveal that almost one-third of the planet's arable land is already damaged (FAO 2022). According to the FAO and ICRAF (2019), agroforestry refers to land-use systems and technologies where agricultural or animal crops are grown on the same plot in specified spatial and temporal arrangements alongside woody perennial plants.

An estimated 160 Mha worldwide are covered by agroforestry systems, with roughly 78% of that area located in tropical regions and 22% in temperate regions. However, these figures are approximations due to the absence of appropriate protocols for defining the boundaries of areas containing erratic tree stands mixed with

Table 11.1 Agroforestry systems followed in various countries

Country	Area under agroforestry (Mha)	Types of Agroforestry Systems	Reference
China	37.1	Agroforestry in farming systems	Nair et al. (2009b)
India	28.2	Home gardens, agroforestry systems	Kumar (2006b)
Indonesia	20.4	Cocoa and coffee agroforestry	Julian (2016)
Brazil	15.5	Silvopastoral, agro-silvopastoral	Nair et al. (2009a)
Mexico	16.0	Shade coffee, silvopastoral systems	Perfecto et al. (2005)
Ethiopia	14.0	Traditional agroforestry systems	Abebe et al. (2019)
Nigeria	12.5	Cocoa and cashew agroforestry	Amerino et al. (2024)
Kenya	10.0	Alley cropping, agroforestry parks	Karanja et al. (2018)
Peru	10.0	Agroforestry with native species	Parodi et al. (2022)
Tanzania	8.0	Agroforestry with fruit trees	Mziray et al. (2020)
Philippines	7.5	Rubber agroforestry, coconut-based systems	Tabora (1991)
Vietnam	6.1	Home gardens, mixed cropping systems	Le et al. (2017)
Colombia	5.5	Coffee agroforestry systems	De Leijster et al. (2021)
Australia	3.4	Silvopastoral systems	Ainsworth et al. (2012)
Bangladesh	1.5	Homestead agroforestry	Miah and Hussain (2010)

understory crops (Nair et al. 2021). According to Lal et al. (2018), there are about 600 Mha of tree intercropping systems worldwide. According to Zoomer et al. (2016), 43% of all farmland now have more than 10% tree cover, an improvement of 2% over the previous year. According to Nair et al. (2009a), 307 Mha of the 823 Mha that include silvopastoral and agroforestry systems globally are devoted to agroforestry. But their estimate comes from multiplying the FAO's estimate of total agricultural area by an imagined 20% agroforestry coverage. There is no objectively measured evidence on which to base the 20% value. Dixon (1995) provides another commonly cited estimate of the worldwide agroforestry extent, speculating that agro-silvopastoral and agroforestry systems span between 585 and 1215 Mha across Africa, Asia, and the Americas. But rather than being inhabited by them, this is an approximation of the space they determine to be technically appropriate for these systems. 45.4 Mha of land is used for agroforestry, according to the FAO (2020) most recent Global Forest Resources Assessment. Reisner et al. (2007) projected that European silvoarable systems encompass 65.2 Mha, whereas Kumar (2006a) estimated that household gardens in South and Southeast Asia occupy 8.0 Mha.

More than 900 million people engage in agroforestry, which covers almost 100 Mha, or 43% of the world's agricultural land, according to a working paper from the World Agroforestry Centre (Zoomer et al. 2016). Agroforestry is also common in Asia, where it has long been essential to indigenous lives (Cannell 1987). Eighty-seven percent of all agricultural land in Southeast Asia, 55% in East Asia, 27% in South Asia, and 26% in Northern and Central Asia is covered by agroforestry (Zoomer et al. 2016). The most recent report from the IPCC Working Group III (IPCC 2022), found that the differences in accounting techniques for human terrestrial CO_2 fluxes amount to 5.5 $GtCO_2$ annually. This discrepancy results from different national greenhouse gas inventories' depictions of anthropogenic ecosystem shifts in comparison to global simulation studies.

11.4 Environmental Services of Agroforestry

Soil is dynamic and continually undergoes development through various processes, which include both formation and degradation. Agroforestry acts as a boon for proper utilization of nutrients and enhances soil health (Dollinger and Jose 2018). Agroforestry systems play a vital role in providing environmental services, as about 20% of the world's population in rural and urban areas of developing countries depends directly or indirectly on agroforestry products. Agroforestry is a type of agriculture that overcomes the gap between agriculture and forestry so that climatic, economic, and personal benefits can be achieved. The integration of trees, agricultural crops, and/or animals into an agroforestry system contributes to the solution of many environmental problems through better use of resources. The role of agroforestry in the agricultural sector is multifunctional. The multifunctional role like providing ecosystem services, environmental benefits, and economic commodities ecosystems has been focused on by both the Millennium Ecosystem Assessment (2005) and the International Assessment of Agricultural Science and Technology for Development (2008). It improves soil health, air health, biological health, and most importantly sequester carbon (C) (Williams-Guilleen et al. 2008; Nair et al. 2009b; Mosquera-Losada et al. 2020). Agroforestry has a huge scope and numerous benefits for the environment. There are four major broad categories beneficial to the environment: soil improvement, C sequestration, biodiversity conservation, and quality of air and water.

11.4.1 Soil Improvement Through Suitable Selection of Trees

Due to agroforestry structural stability of soil increased, deep tree roots enhance water infiltration and improve water storage by increasing the number of soil pores. Macropores work as channels and allow air and moisture to move into the soil; thus, risk of soil erosion is reduced and tree roots and trunks act as a barrier to reduce runoff and sediment, improving the overall soil quality. The two important points of consideration are the type of tree species and the establishment methods in agroforestry. Many tree species are characterized as "wonder trees" because of their multiple environmental role and products. *Moringa oleifera* (Lam) can be a good choice in arid and semiarid zones where water is a limiting growth factor (Boumenjel et al. 2021). This tree species can tolerate water-deficient conditions due to deep root growth and enhances soil quality. Similarly, poplars absorbed water from the deeper soil layers even if the soil was dry on the top layers. The nitrogen release from poplar leaf litter was equivalent to 7 kg N ha^{-1} yr^{-1} (Thevathasan and Gordon 2004). In agroforestry, trees not only provide 5Fs (food, fodder, fuel, fertilizer, and fiber) but also enhance soil quality, filters air, reduce erosion and most importantly they maintain C balance. For farmers this system is very much helpful especially during diverse climatic condition in terms of survival in rural areas. If main crop fails, they have other sources for sustaining their livelihood. Hasnol Othman et al. (2012) revealed that leguminous cover crops (LCC) have beneficial effects on soil by fixing and providing nitrogen (N) to the main crop, improve growth, as well as lessen competition from harmful weeds.

11.4.2 C Sequestration

In India, C sequestration potential has been estimated to be 25 t C ha^{-1} over 96 million ha in agroforestry (Sathaye and Ravindranath 1998). One of the studies conducted under AICRP-AF shows that in subtropical sub-montane low hill areas, agroforestry can enhance C sequestration by 50% for both arable and nonarable lands. Similarly, researchers revealed that plantations and agroforestry have a huge potential in sequestrating the C produced by heavy industries (Ajit Dhyani et al. 2014). Agroforestry is important for two reasons: (1) the tree component captures atmospheric C through photosynthesis and stores it underground, behaving as a C sink, and (2) agroforestry reduces deforestation of temperate and tropical forests related to clearing for agriculture. In an agroforestry system, the amount of C sequesters increases compared to a single crop or barren region (Sharrow and Ismail 2004; Kirby and Potvin 2007). In addition, the significant amount of C stored in aboveground biomass, agroforestry systems can also store C belowground (Fig. 11.2). This sequestered C can be sold in C credit markets. The total organic C concentration and pools are higher in soils under agroforestry and increase with tree age. Bateni et al. (2021) found that olive groves have the capacity to sequester more C than in monocrop, barren, and plantation ecosystems. In general, agroforestry systems in drier climates, like hot, arid, and eroded zones, have less C sequestration potential than those in wet and fertile zones, and temperate zones with low temperatures have less potential than hot and tropical zones (Nair et al. 2009b). Hence, researchers recognized the significance of agroforestry system to overcome global warming due to its combination of plant species (Handa et al. 2015).

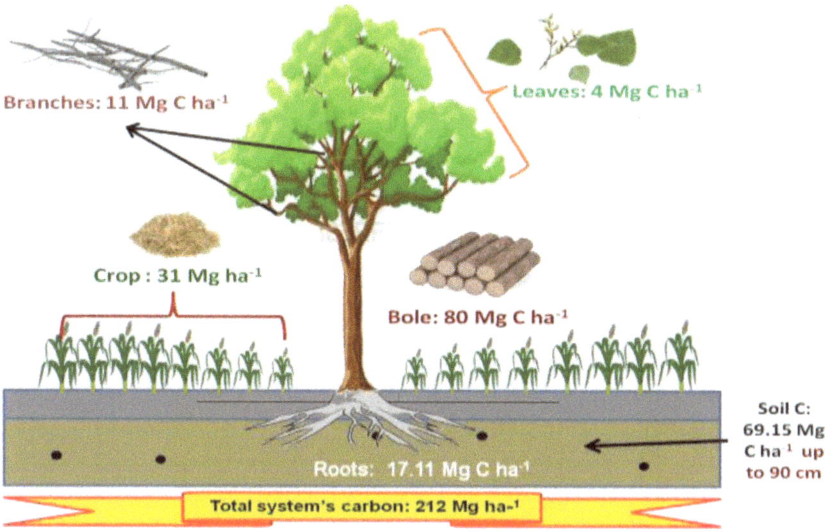

Fig. 11.2 C storage in an agroforestry system. (Source: Chavan et al. 2023)

11.4.3 Enhance Biodiversity

In cold temperate climates and hot tropical areas, agroforestry acts as a major enhancer of biodiversity and increases microbial diversity both above- and below-ground to a significant extent. This has been revealed by many researchers also (McNeely 2004; Harvey et al. 2006).

The larger abundance of individuals and species diversity was found in the multiple plantations than in the pure plantations. Birds and bats had dispersed the majority of the regenerating individuals. The combination of different tree species in this system has variation in shading and litter accumulation situations, which affects the germination of individuals.

11.4.4 Water and Air Quality

Agroforestry plays a pertinent role in increasing the quality of air of the surrounding environment. Trees acts as windbreaks and shelterbelts to protect agricultural crops and livestock. In addition, prevent soil erosion, protect crops, provide wildlife habitat, wind velocity is reduced and thereby limiting wind erosion and particulate matter in the air, and reduce smell from dung and urine of the livestock caused by wind quality. Handayani and Prawito (2011) stated that agroforestry systems have 80% higher conservation value of water than monoculture systems. The quality of water was improved by controlling pollutants, encouraging infiltration, increasing nutrient uptake and storage, sediment deposition, increasing transpiration, and promoting denitrification in the shallow subsurface.

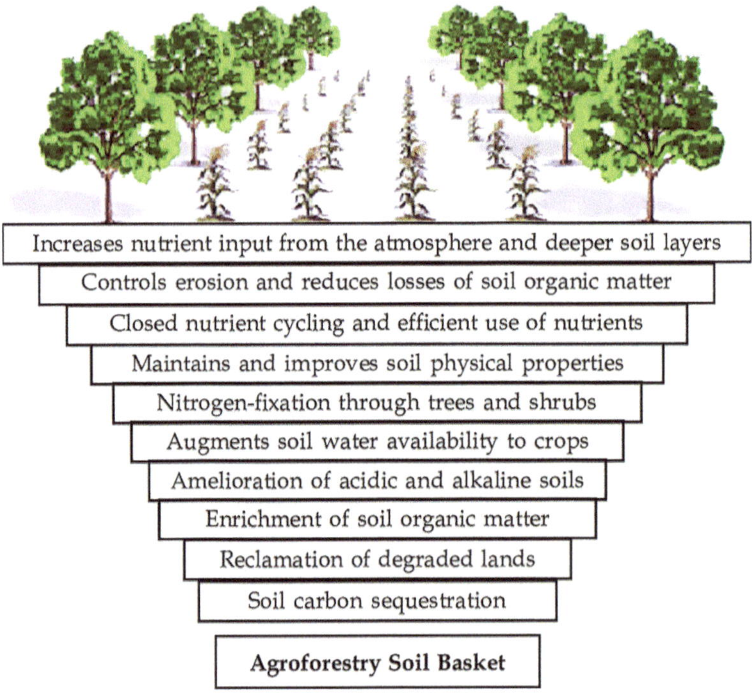

Fig. 11.3 Benefits of agroforestry to the environmental system. (Source: Uthappa et al. 2015)

The other environmental benefits **(Fig. 11.3)** *are***:**

- Effective utilization of natural resources.
- The greenhouse effect is countered by the constitution of an efficient system for C sequestration, by integrating the stock maintenance of organic material in the soil layer.
- Provides protection to soil and water.
- Enhance biodiversity.
- Due to the deep roots of trees, soil erosion and runoff are reduced.
- Biological activity and soil organic matter can be maintained at satisfactory levels for soil fertility.
- Through organic matter maintenance and the impact of tree roots, physical properties of soil can be maintained.
- The soil problems like acidification and salination can be monitored, and crop plants help to reclaim polluted soils.
- Result in a reduction in insect pests and related diseases.
- Economic risk is reduced due to the production of a number of products.
- Improve water infiltration into the soil and reduce surface runoff.
- Helps to maintain a conducive temperature for plants and animals, mainly during warmer days.
- Reductions in the velocity of wind protect crops from damage and lodging.

Thus, now it's time to adopt and aware the alarming population about the agroforestry system on a large scale, which has huge advantages to make the production system economically sustainable, environmentally friendly, and socially reasonable. A better understanding of the integration of agricultural crops, plant materials, and animal species in agroforestry systems should be encouraged for wider adoption, especially among rural communities.

11.5 Causes of Land/Soil Degradation

The impact of soil degradation can be understood from the extent of deterioration of almost all soil properties and soil functions. Broadly, soil degradation can be categorized into (1) physical, (2) chemical, (3) biological, and (4) ecological soil degradation. Changes in these dominant aspects of soil as a consequence of anthropogenic and natural factors affect the ability of soil to function (Fig. 11.4).

11.5.1 Natural Factors

High torrential and irregular rainfall, different types of soil, topography, earthquakes, volcanic eruptions, and steep terrain are the main natural causes driving land degradation (Abdi et al. 2013). A raindrop's powerful impact can immediately strike the surface of the land, degrading and separating the soil structure, which is worsened on bare soil. Land degradation is mainly are often caused by distribution frequency and intensity of rain fall, which disrupts the nature of soil properties and

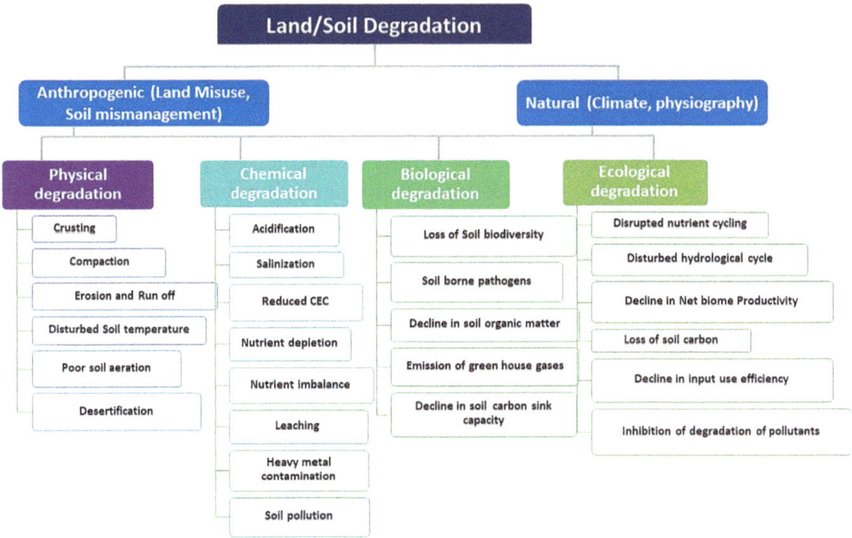

Fig. 11.4 Classification of soil degradation. (Adapted from Lal 2015)

ecosystem services (Sklenicka 2016). A high slope is more likely to be damaged and eroded than a place on a gentle or level slope due to topographic position (Riva et al. 2017). The degree of land degradation is determined by the infiltration and percolation capacity of the soil, which varies depending on the type of soil. Sand soil is more prone to degradation than clay soil because it has less porosity and water-holding capacity (Mosaffaei et al. 2020). In relation to the soil and geographic features of the land, heavier rain falls because of more land degradation than in those of lower intensity. The intervention of human activity may generate and accelerate the natural factors of land degradation, aggravating the overall processes of land degradation (Abatenh et al. 2017).

11.5.2 Anthropogenic Factors

Factors related to land use and land-use change, economics (such as market accessibility and investment potential), and social factors (such as infrastructure availability and farmers' access to land that enables them to produce at maximum capacity) are the main human-induced factors influencing land degradation processes. Vegetation reduces the impact of rainfall and protects the soil's integrity from soil erosion through its root networks. The indiscriminate removal of trees and vegetation has an undesirable impact on the environment, leading to ecological disruption, decreased soil fertility, and soil erosion. Expanding agricultural production, logging, and quick urbanization are some of the main reasons for deforestation. Topsoil is swept away or washed away during soil erosion. The nutrients, minerals, and organic materials in topsoil are abundant. Soil erosion can be hastened by elements including poor farming methods, logging, overgrazing, and construction work. Unsustainable agriculture methods reduce land productivity because they cause a number of problems (Bhattacharyya et al. 2015).

The unwarranted use of chemical fertilizers, pesticides, and herbicides not only contaminates soil and water resources but also poses significant risks (Tudi et al. 2021). Overgrazing, improper irrigation techniques, faulty tillage practices, and insufficient soil conservation measures further exacerbate the problem, resulting in extensive soil erosion. These collectively faulty agricultural practices significantly contribute to severe land degradation. The rapid rate of urbanization and industrialization leads to the conversion of productive agricultural, pasture, and forestland into built-up areas. This is responsible for soil erosion; soil sealing reduces soil fertility (Bhattacharyya et al. 2015). Industrial pollution and waste disposal can further contribute to land degradation. Mining activities can affect the biological, chemical, and physical features of soil, which ultimately reduces soil fertility. Mineral mining and extraction from the earth can result in the removal of topsoil, habitat destruction, and contamination of soil and water bodies, causing land degradation (Dehkhordi et al. 2024). The high rate of population growth puts huge pressure on available land for food, water, and other resources, which leads to unsustainable land use, overexploitation, and degradation of available land resources. A variety of direct causes, like bush invasion, forest fires, floods, landslides, and droughts, may appear natural at first, but they are entirely, partially, or indirectly caused by humans. There are frequently other factors that contribute to human-induced land degradation in addition to the apparent ones. These factors include population pressure, poverty, weak institutional frameworks and bad governance, a lack of markets and infrastructure, and inadequate education.

11.6 Extent of Land Degradation

Land degradation is a complex and multifaceted problem that adversely impacts land productivity, ecosystem services, and the overall quality of life. According to FAO (2024) estimates, more than 2 billion hectares of agricultural land worldwide, or roughly 52%, are moderately or severely degraded, negatively affecting 1.5 billion people, particularly women, children, and the poor in rural areas. The UN Convention to Combat Desertification (UNCCD 2023) states that between 2015 and 2019, at least 100 million hectares of productive, healthy land were lost annually worldwide. Over 250 million people were directly impacted by land degradation, and another billion people living in more than 100 countries are in danger. According to UNCCD, land deterioration is expected to cost 3.4 trillion euros a year (ELD Initiative 2013; AbdelRahman 2023). Global scale estimates of degradation (million ha) in Asia, Africa, Europe, North America, South America, and Australia revealed that Asia is the most degraded continent, followed by Africa, North America, South America, and Australia, while Europe is least degraded (Hossain et al. 2020). In India, a country with diverse landscapes and ecosystems, land degradation poses a significant challenge that extends across diverse landscapes and ecosystems, reflecting a complex interplay of natural and anthropogenic factors. The extent of area under land degradation in Indian context is substantial, with various forms of degradation, including soil erosion, salinity, waterlogging, deforestation, desertification, and urbanization-induced land cover changes. These issues manifest differently in different regions, with the arid and semiarid zones of Rajasthan and Gujarat, the fragile ecosystems of the Western and Eastern Ghats, and the agriculturally intensive Indo-Gangetic Plains being among the most affected areas. It is difficult to assess land degradation; hence, many different techniques are employed, such as modeling, remote sensing, and expert judgment. There is also a significant variation in the statistics that are readily available due to variations in definitions and terminologies. Table 11.2 enlists the extent and severity of problems of Land/soil degradation as determined by various agencies.

According to the most recent data from the National Remote Sensing Centre (NRSC)/(ISRO), out of India's 329 million hectare total geographical area (TGA), 91.20 million hectares (Mha) of land are now experiencing land degradation, making up 27.7% of the TGA. According to NRSC's 2019 estimate, this percentage represented 29.32% of the nation's TGA between 2011 and 2013. The percent share of various agents or processes that contribute to land degradation in India is shown in Fig. 11.5 (NRSC 2019). Water erosion was the major process causing land degradation in India, accounting for 55.91% of total degraded land or 15.53% of total

Table 11.2 Estimates of land degradation in India

Agency	Year of Assessment	Area under land degradation (Mha)	Broad methodology adopted	References
National Remote Sensing Centre (NRSC)/(ISRO)	2015–16	91.2	Visual interpretation of multi-season LISS-III data	NRSC (2019)
Space Applications Centre (SAC)/ISRO	2011–13	96.40	Visual interpretation of multi-season AWiFS data NRSC/ISRO	SAC (2016)
NRSC/ISRO	2005–06	91.29	Visual interpretation of multi-season LISS-III data	NRSC (2007)
SAC/ISRO	2003–05	94.53	Visual interpretation of multi-season AWiFS data	SAC (2016)
Department of Land Resources	2005	55.27	Visual interpretation of satellite data at a 1:50,000 scale	–
National Bureau of Soil Survey (NBSS) & Land Use Planning (LUP) (revised)	2005	146.82	Based on a 1:1 million scale soil map	NBSS & LUP (2005)
Department of Land Resources	2000	63.8	Visual interpretation of satellite data at a 1:50,000 scale	–
National Remote Sensing Agency	2000	53.3	Mapping on a 1:1 million scale based on remote sensing techniques	NRSA (2000)
NBSS and LUP	1994	187.7	Mapping on a 1:4 million scale based on the Global Assessment of Soil Degradation (GLASOD) guidelines	NBSS and LUP (1994)
Ministry of Agriculture	1994	107.43	Elimination of duplication of area. Area reclaimed counted	MoA (1994)
National Wasteland Development Board	1985	123	–	NWDB (1985)
Society for Promotion of Wastelands Development	1984	129.58	Based on secondary estimates	Bhumbla and Khare (1985)
Department of Environment	1980	95.0	–	Snedecor and Cochran (1967)

(continued)

Table 11.2 (continued)

Agency	Year of Assessment	Area under land degradation (Mha)	Broad methodology adopted	References
MoA-Soil and Water Conservation Division	1978	175	Based on the National Commission on Agriculture's estimates. No systematic survey was undertaken	MoA (1978)
National Commission on Agriculture	1976	148	Based on the secondary data	NCA (1976)

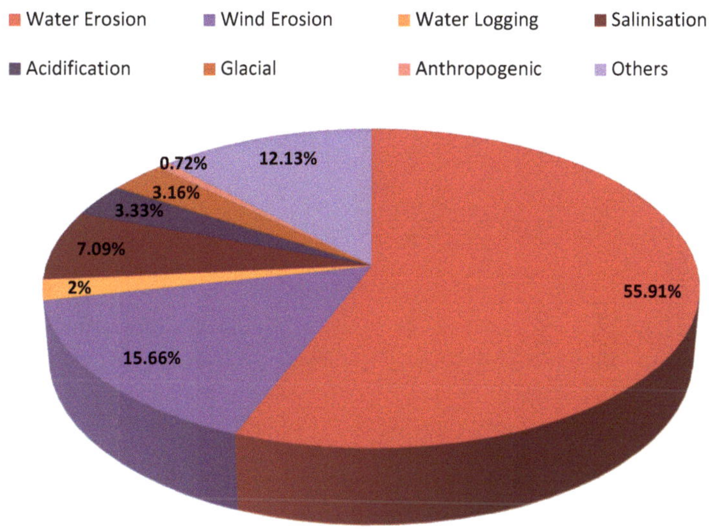

Fig. 11.5 Percent share of different processes of land/soil degradation. (Data has been adapted from Sreenivas et al. (2021))

geographical area (TGA) of the country. Next most important cause was wind erosion spread over nearly 14,279,746 ha and accounting for 15.66% of total degraded land or 4.35% of TGA. The salinization/alkalization (comprising various categories like saline, sodic, and saline–sodic and Rann) was spread over 6.5 Mha, accounting for 7.09% of total LD. Acidification covered over 3.04 Mha (3.33% of total degraded land) (Sreenivas et al. 2021). The main factors contributing to soil degradation are anthropogenic activities like deforestation, mining, intensive agriculture, and erosion by winds, water, and glaciers. The overall state wise contribution of these processes in land degradation is listed in Table 11.3 (NRSC 2019).

Rajasthan has the greatest degraded area among all states, covering an area of more than 18.03 million hectares and making up 19.77% of the nation's total

Table 11.3 Category wise distribution of land degradation (2015–2016) in different States of India (area in ha) (NRSC 2019)

States	Water Erosion	Wind Erosion	Water Logging	Salinisation/ Alkalisation	Acidification	Glacial	Anthropogenic	Others	Total	Land Degradation (%)
Andhra Pradesh	4,773,702	15,556	4399	365,639	0	0	50,619	293,183	5,503,098	34.46
Arunachal Pradesh	442,920	0	35,033	0	336,881	585	0	165,800	981,219	11.23
Assam	359,168	0	268,671	0	73,885	0	6516	25,906	734,146	8.89
Bihar	345,027	0	711,112	113,918	0	0	4731	15,454	1,190,242	12.25
Chhattisgarh	3,236,127	0	0	0	94	0	31,454	356,643	3,624,318	26.86
Goa	24,732	0	769	1347	37	0	8120	7418	42,423	11.49
Gujarat	1,618,709	1328	85,339	3,793,377	0	0	25,127	469,712	5,993,592	30.89
Haryana	107,110	131,052	35,260	95,852	0	0	8402	10,893	388,569	10.25
Himachal Pradesh	315,119	0	41	0	0	73,787	1958	225,087	615,992	22.99
Jammu & Kashmir	747,713	236,600	1187	45,599	0	1,887,379	369	7,353,452	10,272,299	46.22
Jharkhand	1,196,920	0	354	0	0	0	24,825	83,418	1,305,517	16.46
Karnataka	4,876,020	0	2004	116,510	33,797	0	60,136	180,544	5,269,011	27.48
Kerala	74,695	0	69,037	0	54,092	0	5442	94,153	297,419	8.46
Madhya Pradesh	5,630,509	0	92	25,601	0	0	21,109	211,779	5,889,090	19.29
Maharashtra	10,613,283	0	0	183,433	1351	0	30,754	150,554	10,979,375	35.81
Manipur	221,931	0	1343	0	631,463	0	0	22	854,759	37.49
Meghalaya	86,767	0	304	0	458,765	0	6910	83,264	636,010	27.78
Mizoram	266,577	0	0	0	469,594	0	0	0	736,171	30.88
Nagaland	35,843	0	0	0	743,388	0	409	0	779,640	46.24
Odisha	5,832,981	335	16,091	3057	18,835	0	23,231	61,406	5,955,936	38.36

Punjab	62,568	44,311	11,447	4252	0	0	11,364	233,327	367,269	3.21
Rajasthan	2,752,487	13,805,182	8769	536,207	0	0	57,430	850,248	18,010,323	53.04
Sikkim	5362	0	0	0	0	11,271	0	59,667	76,300	10.75
Tamil Nadu	1,766,278	45,382	10,540	346,468	124,141	0	28,484	119,248	2,440,541	18.78
Telangana	2,879,100	0	2499	230,639	249	0	41,989	72,289	3,226,765	28.83
Tripura	6076	0	298	0	95,319	0	600	475	102,768	9.8
Uttar Pradesh	1,695,505	0	444,900	600,630	0	0	20,354	92,973	2,854,362	53.48
Uttarakhand	621,798	0	188	0	0	246,776	1409	69,741	939,912	3.9
West Bengal	341,584	0	111,657	753	0	0	16,608	9095	479,697	5.52
Andaman & Nicobar	47,388	0	0	0	0	0	161	5349	52,898	6.46
Chandigarh	57	0	0	0	0	0	0	0	57	0.5
Dadra Nagar Haveli	5802	0	28	0	0	0	0	0	5830	11.91
Daman and Diu	329	0	364	0	0	0	133	41	867	8.57
Delhi	6098	0	0	0	0	0	21	1179	7298	4.92
Lakshadweep	0	0	0	0	0	0	0	0	0	1.46
Puducherry	16	0	3	1326	0	0	0	0	1345	3.67

degraded land. In this state, wind erosion accelerated due to anthropogenic interventions was the primary cause of degraded land, followed by water erosion. Maharashtra with 10.99 Mha land under degradation accounted for 12.05% of total degraded land. Here, water erosion, followed by salinization/alkalization was the major cause. Water erosion was found to be the dominant process in the majority of the states, especially in the central plateau region of India. Following the order Maharashtra> Odisha > Madhya Pradesh. High rainfall variability and intensity, undulating terrain conditions, and moderate to high erodibility of soils were the major factors for extensive water erosion in these areas. In the red soil region of the southern plateau, the undulating terrain conditions and cultivation practices along the slopes are the main factors for sheet erosion. Nagaland, Manipur, Mizoram, Meghalaya, Arunachal Pradesh, Tamil Nadu, Tripura, Assam, Kerala, Karnataka, Odisha, and Maharashtra had the highest degree of acidification, in decreasing order. This was predominantly limited to areas with heavy rainfall and was mostly caused by alkaline salts leaching to lower depths. Waterlogging was found to be highest in Bihar, followed by Uttar Pradesh and Assam. The salinization/alkalization was spread over 6.46 Mha, accounting for 7.09% of the total degraded land in India. It was found extensively in Gujarat (4.2%), followed by Uttar Pradesh (0.66%) and Rajasthan (0.59%). This process was significant in Andhra Pradesh, Tamil Nadu, Telangana, Maharashtra, Bihar, Karnataka, Haryana, Ladakh, Madhya Pradesh, Punjab, and Odisha, in decreasing order of area. Glacial process also leads to erosion but is confined to high altitude areas. Besides, anthropogenic factors, such as mining, industrial effluent dumping, brick kilns, etc., also caused land degradation, especially in states of Karnataka, Rajasthan, Andhra Pradesh, and Telangana.

11.7 Land Degradation Mechanism

Degradation of the soil indicates a reduction in its ability to offer ecosystem services that are valuable to humans and support natural processes. The main factors that cause soil degradation are drought, nitrogen and carbon (C) depletion, erosion, salinization, loss of soil structure, and tilth. The ecosystem services supplied by soil include provisional (food, fiber, fuel, and forages), regulating (water purification and flow, C sequestration, temperature variations), ecological/supporting (biomass production, nutrient cycling), and cultural (aesthetical, spiritual, and cultural). The ability of soil to produce ecosystem services and sustain ecosystem services is reduced by erosion-induced deterioration. It is estimated that annually, 5–7 Mha farm land is being degraded globally (AbdelRahman 2023).

11.7.1 Mechanisms of Land Degradation

11.7.1.1 Physical Processes

Decline in soil structure leads to crusting, compaction, erosion, desertification, and environmental pollution. Although soil erosion is a natural process, human activity has greatly increased its severity. The global average soil erosion rate per year is estimated to be between 0.5 and 2 tons per acre (AbdelRahman 2023), depending on the type of soil, the degree of land degradation, and the nature of erosion. Soil erosion rates vary from region to region; in the United States of America, for example (AbdelRahman 2023), 44% of agricultural land is affected by erosion to varying degrees, and in India nearly half of the agricultural land is subject to erosion. In some areas—especially in the Middle East—wind is an important factor of erosion affecting about 35% of the land area in the region (AbdelRahman 2023). The most important problems of soil change and perhaps the most serious of these are irrigation increase. Water saturation, salinity, and sodium rise occur if marginal water is used for irrigation or improper water management methods are used, such as poor drainage.

11.7.1.2 Chemical Processes

Chemical soil degradation after erosion is the second most abundant form of soil degradation and, as such, poses a threat to our finite soil resource, as it tends to render it less usable. It is therefore necessary to understand the means by which soils are degraded chemically. Chemical degradation refers to the accumulation of toxic chemicals and chemical processes that impact chemical properties that regulate life processes in the soil (Logan 1990). Estimates in 1991 suggest that about 240 Mha land is chemically degraded (Osman and Osman 2014). Nutrient depletion has affected 136 Mha, salinization damaged 77 Mha, and acidification degraded 6 Mha. Another 11 Mha is affected by soil pollution. Agricultural mismanagement (58%) and deforestation (28%) are the main causes of chemical degradation of soil. Nutrient depletion is the most prevalent in Africa (65 Mha) and South America (68 Mha), while salinization is the major chemical degradation in Asia (53 Mha). Nutrient depletion has caused serious nutrient imbalances in soils under low-input agriculture in marginal lands. In a study, all the African agricultural soils exhibited negative NPK balances. Nutrient depletion is caused by leaching, residue harvest and burning, erosion, and crop removal. Salinization occurs naturally by pedogenic processes in different climatic conditions, but human-induced salinization has compelled to abandon many soils, which were productive earlier. The principal cause of human-induced salinization is an inappropriate irrigation system in arid and semi-arid regions. Leaching of salts by extra irrigation and growing salt-tolerant crops are the strategies for salty soil management. Soils are acidified by acid rains, base leaching, and the use of acidifying fertilizers. Liming is an ancient method of reclaiming acid soils. Growing crops suitable for the current soil pH may be profitable in low to medium-acid soils (Osman 2014).

11.7.1.3 Biological Processes

Soil microorganisms play key roles in the cycling of nutrients, decomposition of wastes and residues, and detoxification of pollutant compounds in the environment. Biological degradation of soil as described herein refers to the impairment or elimination of one or more "significant" populations of microorganisms in soil, often with a resulting change in biogeochemical processing within the associated ecosystem. "Significant" microorganisms are those for which an ecologically significant role is understood. Biological factors refer to the human and plant activities that tend to reduce the quality of soil. Some bacteria and fungi overgrowth in an area can highly impact the microbial activity of the soil through biochemical reactions, which reduces crop yield and the suitability of soil's productivity capacity. Human activities such as poor farming practices may also deplete soil nutrients, thus diminishing soil fertility. The biological factors mainly affect and lessen the microbial activity of the soil.

11.7.2 Causes for Land Degradation

(i) Intensive irrigation leads to waterlogging and salinization, on which crops cannot grow.
(ii) The use of more and more chemical fertilizers poisons the soil so that eventually the land becomes unproductive.
(iii) The roots of trees and grasses bind the soil. If forests are depleted or grasslands overgrazed, the land becomes unproductive, and wasteland is formed.
(iv) Land is also converted into a nonrenewable resource when highly toxic industrial and nuclear wastes are dumped on it.
(v) As urban centers grow and industrial expansion occurs, the agricultural land and forests shrink. This is a serious loss and has long-term ill effects on human civilization.
(vi) Land degradation/soil erosion due to deforestation is more evident on steep hill slopes in the Himalayas and in the Western Ghats. These areas are called "ecologically sensitive areas" or ESAs. To prevent the loss of millions of tons of valuable soil every year, it is essential to preserve what remains of our natural forest cover. It is equally important to reforest denuded areas. The linkage between the existence of forests and the presence of soil is greater than the forest's physical soil-binding function alone. The soil is enriched by the leaf litter of the forest. It is broken down by soil micro-organisms, fungi, worms, and insects, which help to recycle nutrients in the system. Further losses of our soil wealth will impoverish our country and reduce its capacity to grow enough food in the future.
(vii) The rate of mangrove loss is significantly higher than the loss of any other type of forest. If deforestation of mangroves continues, it can lead to severe losses of biodiversity and livelihoods, in addition to salt intrusion in coastal areas and siltation of coral reefs, ports, and shipping lanes.

11.8 Factors Affecting Land Degradation

Throughout the world, desertification and land degradation are the most serious environmental problems. The issues are challenging on all levels, from local to global (Kosmas et al. 2014). The issues associated with land degradation, drought, and desertification are highly complex (Lal 2012; Wijitkosum 2020). Desertification is the persistent degradation of dryland ecosystems driven by human activities, such as mistaken land use and management, as well as climatic factors (Xie et al. 2015). However, climate-related impacts happen gradually, whereas humans accelerate rapidly and are more severe. Besides, the biotic, abiotic, and soil components of the land, as well as the interactions among them, can all be impacted by land degradation processes. Changes in the hydrological system—which are especially significant as a result of climate change—can additionally act as a trigger for land degradation processes. Salinization, although perceived and reported in soils, is typically triggered by water table-level rises, driving salts to the surface under dry to subhumid climates (Schofield and Kirkby 2003), and changes in rainfall pattern and evapotranspiration also influenced the land degradation process. The major factors responsible for land degradation are as follows:

1. *Deforestation*

Deforestation is one of the major causes of land degradation, and forest-degraded areas consist 28% of the total eroded area in India (Weeraratna 2022). Deforestation causes soil degradation on account of exposing soil minerals by removing trees and crop cover, which support the availability of humus and litter layers on the surface of the soil. Vegetation cover primarily promotes the binding of the soil together and soil formation; hence, when it is removed, it considerably affects the capabilities of the soil, such as aeration, water holding capacity, and biological activity. When trees are removed by logging, infiltration rates become elevated, and the soil remains bare and exposed to erosion and the buildup of toxicities. Some of the contributing activities include logging and slash-and-burn techniques used by individuals who invade forest areas for farming, rendering the soils unproductive and less fertile in the end.

2. *Overgrazing*

The main causes of this degradation were overgrazing, responsible for degradation of about 34% of the degraded area, deforestation 29%, agricultural activities 28%, overuse of land 7%, and other irrational activities, 1.5%. Land degradation in dry areas (arid, semiarid and dry subhumid) is characterized by desertification (AbdelRahman 2023), which is defined as "loss of productivity and land biodiversity in arid, semiarid, and dry subhumid areas due to natural factors or processes resulting from human activities." It was estimated that about 30% of irrigated areas in drylands, 47% of cropland, and 73% of rangeland were affected by desertification at least moderately. Africa is at the forefront of areas where land has been affected by desertification (Tesfa and Mekuriaw 2014). Overgrazing is the main cause of land degradation in states like Madhya Pradesh and Maharashtra.

3. *Urbanization*

Urbanization has major implications on the soil degradation process. Foremost of all, it denudates the soil's vegetation cover, compacts soil during construction, and alters the drainage pattern. Secondly, it covers the soil in an impermeable layer of concrete that amplifies the amount of surface runoff, which results in more erosion of the topsoil. Again, most of the runoff and sediments from urban areas are extremely polluted with oil, fuel, and other chemicals. Increased runoff from urban areas also causes a huge disturbance to adjacent watersheds by changing the rate and volume of water that flows through them and impoverishing them with chemically polluted sediment deposits.

4. *Over Irrigation*

Over irrigation is also responsible for land degradation as it leads to waterlogging that leads to increase in salinity and alkalinity of soil. In States like Punjab and Haryana, over irrigation is responsible for land degradation.

5. *Mining*

Mining leaves deep scars and traces of overburdening on the land leading to land degradation. In states like Jharkhand and Odisha, mining has played a significant role in land degradation. Soil is chiefly polluted by industrial and mining activities. As an example, mining destroys crop cover and releases a myriad of toxic chemicals, such as mercury into the soil, thereby poisoning it and rendering it unproductive for any other purpose. Industrial activities, on the other hand, release toxic effluents and material wastes into the atmosphere, land, rivers, and groundwater that eventually pollute the soil, and as such, it impacts on soil quality. Altogether, industrial and mining activities degrade the soil's physical, chemical, and biological properties.

6. *Industrial Effluents*

Industrial effluents are also responsible for the degradation of land and water in many parts of our country.

7. *Improper Cultivation Practices*

There are certain agricultural practices that are environmentally unsustainable, and at the same time, they are the single biggest contributor to the worldwide increase in soil quality decline. The tillage on agricultural lands is one of the main factors since it breaks up soil into finer particles, which increases erosion rates. The soil quality decline is exacerbated more and more as a result of the mechanization of agriculture, which gives room for deep plowing, reduction of plant cover, and the formation of hardpan. Other improper cultivation activities, such as farming on steep slopes and mono-cropping, row-cropping, and surface irrigation, wear away the natural composition of the soil and its fertility and prevent the soil from regenerating.

8. *Misuse or Excess Use of Fertilizers*

The excessive use and misuse of pesticides and chemical fertilizers kill organisms that assist in binding the soil together. Most agricultural practices involving the use of fertilizers and pesticides often entail misuse or excessive application, thereby contributing to the killing of the soil's beneficial bacteria and other microorganisms that help in soil formation. The complex forms of the fertilizer's chemicals are also responsible for denaturing essential soil minerals, giving rise to nutrient losses from the soil. Therefore, the misuse or excessive use of fertilizers increases the rate of soil degradation by destroying the soil's biological activity and building up toxicities through incorrect fertilizer use.

11.9 Agroforestry Approaches to Reclamation of Degraded Land

In India, land degradation affected around 29.3% of the geographical area, with highest contribution coming from water erosion, acidification, wind erosion, desertification, salinization, and mining (ISRO 2018). The process of land degradation further aggravated by major factors such as tree cover loss or deforestation (2.07 Mha) from 2001 to 2021 (Global Forest Watch 2022), intensive rainfall (>7.5 mm ha^{-1}), uncontrolled grazing in 5.65 Mha (Down to Earth 2019), indiscriminate use (32 MT $year^{-1}$) of fertilizers (Fertiliser Association of India 2021), and shifting cultivation in 7.6 Mha (Bhat et al. 2022). As per the total geographical area of India, the states of Rajasthan, Gujarat, Maharashtra, Jammu and Kashmir, and Karnataka constitute the maximum area undergoing degradation/desertification, approx. 18.4% (out of India's total 29.3% degraded land), while less than 2% of degraded lands, shared by each of all the other states. Without conservation measures, both marginal and submarginal lands are brought under the plough. All these led to natural resources degradation, like land, water, and vegetation, causing poor fertility status, soil erosion, low available soil moisture, acidic nonirrigated soil, etc. (Kumar et al. 2019).

The agroforestry system (AFS) proved as an established method for integrated land management, not only comprising renewable resource production but also for land reclamation and mitigation of climate change. The improved biomass production, i.e., partly returned to the soil through litter from both aboveground and belowground parts of the plants, may lead to better storage of carbon in an agroforestry system (Newaj et al. 2017). It is said that in order to have around 53 Mha area under agroforestry in India by 2050, it will be achieved through reclamation of cultivable fallows, fallows, groves, pastures and problematic soils (Dhyani and Handa 2013). Different agroforestry species adapted for erosion prone areas in India have been presented in Table 11.4.

Table 11.4 Different agroforestry systems for wind-eroded areas and their impact on controlling erosion

Agroforestry system	Region	Species	Impact	References
Sand dune stabilization	Karnataka	Canavalia cathartica, *Canavalia maritima*	Act as green manure, mulch, cover crop, pasture, fodder, oil, and medicinal value	Sridhar and Bhagya (2007)
	Thar desert (Rajasthan)	*Calligonum polygonoides, Lasiurus sindicus, Calotropis procera*	Significant reduction in soil erosion through wind because grasses have a massive network of underground roots, which works as an effective "sand binders"	Chauhan (2003)
Shelterbelts	Karnataka	Bamboo	Sorghum gave a higher yield of 4.15 Mg ha^{-1} with a shelterbelt compared with 3.72 Mg ha^{-1} without a shelterbelt	Reddy and Kulkarni (1978)
	Thar desert (Rajasthan)	*Prosopis juliflora, Acacia tortilis, Cassia siamea*	Brought about 50% reductions in wind erosion. Most effective in conserving soil due to thicker branching and more leafiness and, therefore, lesser air permeability provided by this shelterbelt	Chauhan (2003)
	Jaisalmer and Bikaner (Rajasthan)	*Acacia tortilis, Tecomella undulata*	Reduced sand deposition by 0.513 m^3 and 1.023 m^3 per running meter length of the canal, respectively, saved de-silting cost by Rs. 6156–Rs 12,276 per kilometer in 1 year	Upadhyaya (1991)

(continued)

Table 11.4 (continued)

Agroforestry system	Region	Species	Impact	References
Micro-windbreaks	North Western part of Indian desert	*Acacia tortilis, Prosopis juliflora, Calligonum polygonoides, Cassia angustifolia, Cenchrus ciliaris*	Improvement of soil conditions and controls the sand drift when the seedlings attain greater height	Singh and Rathod (2002)
Shelterbelts	Western Rajasthan	*Acacia tortilis, Eucalyptus camaldulensis, Dalbergia sissoo, Tecomella undulata*	Enhanced soil organic carbon (0.11–0.46%) and reduced daily air temperature (3.5 °C–8 °C) in the sheltered area	Prasad et al. (2013)
Live fencing	Bikaner (Rajasthan)	*Calligonum polygonoides, Acacia jacquimontii, Acacia senegal*	Live fences act as windbreaks, reducing wind stress on livestock or crop plants, drying of soils and wind erosion. Add nitrogen and organic matter, thus building up humus in the soil through the breakdown of leaf litter and root hairs	Soni et al. (2016)
Silvopastoral system	Jodhpur (Rajasthan)	Strip cropping of *Cenchrus ciliaris* or *Lasiurus sindicus* with the *Lablab purpureus* in association with *Colophospermum mopane* or *Hardwickia binate*	Improved fodder yield, soil-binding capacity of soil, and reduced wind erosion	Patidar et al. (2008)
Horti-pasture	Jodhpur (Rajasthan)	*Ziziphus rotundifolia* and *C. ciliaris*	Sustain 554 Tharparkar cattle days ha^{-1} with 60% pasture utilization, along with soil conservation	Narain and Bhati (2004)
Silvopastoral system	Kachchh (Gujarat)	*Acacia* + *C. ciliaris, Acacia* + *C. setigerus*	Sequestered carbon from 60.1 to 6.82 Mg C ha^{-1} as well as a reduction in wind erosion due to soil cover of grasses	Shamsudheen et al. (2014)

(continued)

Table 11.4 (continued)

Agroforestry system	Region	Species	Impact	References
Multipurpose tree plantation	Barmer (Rajasthan)	*Acacia senegal*	Improvement in soil fertility by nitrogen fixation and 12,000 trees were treated through CAZRI gum inducer, which resulted in exudation of 5.4 Mg of gum Arabic and farmers earned INR 2.7 million from the sale of gum	Tewari et al. (2014)
Silvopastoral system	Bhuj (Gujarat)	*Acacia tortilis, Azadirachta indica, C. ciliaris, C. setigerus*	18% area under culturable wasteland improved due to improvement in soil physical and chemical properties by silvopastoral system	Shamsudheen et al. (2009)
Silvopastoral system	Jhansi (Uttar Pradesh)	*Acacia tortilis, C. ciliaris*	*A. tortilis* + *C. ciliaris* maintained higher soil moisture, organic carbon content and available N, P, and K for the longer period in the soil	Mishra et al. (2010)
Silvopastoral system	Jhansi (Uttar Pradesh)	Multipurpose tree with *D. annulatum*	Average soil loss from a silvopasture system recorded only 0.9 Mg ha^{-1} but it was recorded 12–43 Mg ha^{-1} in deforested black soil, 4–10 Mg ha^{-1} in red soil and 3.2 Mg ha^{-1} from natural grassland	Kalloo (2003)
Grass-based strip cropping	Western Rajasthan	*C. ciliaris* and *L. sindicus*	Grasses helped in reducing wind speed due to vegetation cover and soil-binding, especially in the top 15 cm soil profile due to their fine root systems. and thus control the soil erosion against the wind	Soni et al. (2006, 2013)
Silvopastoral system	Kurukshetra (Haryana)	*L. leucocephala, C. ciliaris S. hamata*	Increase in organic carbon of 1.7–2.3 times compared to a control	Kaur et al. (2002a)

(continued)

Table 11.4 (continued)

Agroforestry system	Region	Species	Impact	References
Silvopastoral system	Kurukshetra (Haryana)	*A. nilotica, D. sissoo, P. juliflora, D. bipinnata*	About 77–89% of the net annual N uptake by the vegetation was returned to the soil through litter fall and turnover of fine roots. Thus, improved soil fertility and reduced soil erosion	Kaur et al. (2002b)
Agri-silviculture	North Western Gujarat	*Azadirachta indica, Ailanthus excelsa,* cowpea, green gram, cluster bean, and sesame	Fetched 25.7–59.3% more income than sole cropping and also improved soil organic carbon, along with catering diverse needs of farmers, viz., fodder, fuel wood, and timber	Patel et al. (2008)
Bio-fencing	Ladakh (Cold desert)	*Hippophae* spp. (Sea buckthorn)	Bio-fencing of this plant, being thorny in nature, protects crops from stray animals. Its multipurpose utility as a nitrogen fixer, check against soil erosion, conservation of moisture, source of fuel wood, and indigenous drug (rich source of vitamin C) makes it a promising plant for the eco-economic rehabilitation of the region	Khosla et al. (2018)

11.10 Soil Erosion Control

The process of soil erosion primarily affects the topsoil layer through different agents mainly water and air. It is expected that climate change is going to affect rainfall patterns, thus aggravating the situation of erosion. Thus, the practice of agroforestry could act as a barrier for intense storm events, erosion through water by holding the soil through its roots tightly. The windbreaks, alley cropping, buffers riparian were designed for wind erosion reduction (Garrett 2009). The AFS will deliver more flexible opportunities both in terms of socioeconomic aspects as well as ecosystem services. Further, the leaf litter, root decay exudates, etc., from the rhizosphere increase the soil organic matter and thus improve the structure of soil,

Table 11.5 Impact of barrier hedges, trees, as well as grasses on runoff and soil loss

Treatment	Runoff (%)	Soil loss (t ha^{-1} yr^{-1})
Corn on contour	40	21
Leucaena hedges	21.3	12.1
Panicum (0.75 m wide)	36.7	7.0
Eulaliopsis (0.75 m wide)	42.7	10.0
Veteveria (0.75 m wide)	39.6	8.1
Leucaena trees (6–8 years)	20.4	8.4
Eucalyptus trees (6–8 years)	16.3	5.8

Source: Tomar et al. (2023)

thus reducing erosion. A number of studies concluded that developing countries have well established AFS for control of soil erosion from steep slopes (Lal 1989; Pacardo and Montecillo 1983; Raintree and Torres 1986; Sukmana et al. 1985; Wiersum 1981, 1984). In a study conducted at North-Western Himalaya at Dehradun (1740 mm rainfall), different barrier hedges, trees, and grasses on runoff and soil loss at 4% slope were presented in Table 11.5. The grasses were found to be very effective in reducing soil loss in spite of high runoff (Table 11.5). The tree alleys are also found to be effective in soil loss and runoff control.

11.11 Salt-Affected Soils

The guidelines regarding restoration of degraded ecosystems by IUCN define restoration as a process bringing a degraded ecosystem or landscape back to the same prescribed, productive condition in a short process of revitalization (Lamb 1988). The vegetation helps in amelioration of microclimatic conditions and marking a catalytic effect on succession on severely degraded sites (Parrotta 1992). Most of the research results showed that plantation of trees improves the physicochemical environment of soil by decreasing soil pH and EC, improves soil fertility and organic carbon, decreases bulk density, and improves cation exchange capacity of salt-affected soils. Given below is the list if suitable tree species for degraded land management (Table 11.6).

11.12 Mine-Spoilt Rehabilitation

Mainly in the mined soils, due to excessive use of blasting materials, high temperature, and low moisture, the quality of soil highly deteriorates and exposes subsurface soils or bed rocks. The magnitude of impact intensifies when the mine lease is close to the river and drainage flow is towards the river (Chaterjee 1988; Verma 1988; Dadhwal et al. 1992). The rehabilitation of the mine soils is highly challenging. Though the choice of species and site preparation before planting, after care, legal provisions, etc. are important aspects.

Table 11.6 Suitable tree species for degraded lands in India

Degraded lands	Suitable tree species
Saline soils	*Casuarina glauca, Casuarina equisetifolia, Prosopis juliflora, Salvadora persica, Salvadora oleiodes, Tamarix articulata. Tamarix troupii,Parkinsonia aculeata*
Alkali soils	*Prosopis juliflora, P. cineraria, Acacia nilotica, Casuarina equisetifolia, Tamarix articulata, Pongamia pinnata, Butea monosperma, Feronia limonia, Eucalyptus tereticornis, Moringa oleifera, Mangifera indica, Pyrus communis*
Acid soils	*Acacia mangium, Acacia auriculiformis, Cassia siamea, Calliandra calothyrsus, Pinus roxburghii, P. wallichiana, Celtis australis, Acacia mearnsii, A. decurrens, A. dealbata, A. pycnantha, Manilkara hexandra, Phoenix* spp., *Pterocarpus marsupium, Shorea robusta, Hardwickia binata, Terminalia* spp., *Tectona grandis, Gmelina arborea, Xylia xylocarpa, Bamboo* spp., *Alnus* spp. *among agri-horti system-mango, peach, orange, grapes, banana, kiwifruit, pineapple, guava.* etc.
Ravine land	*Eucalyptus species, Dendrocalamus strictus, Morus alba, Broussonetia papyrifera, Dendrocalamus strictus, Gmelina arborea, Dalbergiasissoo, Tectona grandis, Azadirachta indica, Leucaena leucocephala, and Prosopis juliflora*
Mine spoils	*Acacia auriculiformis, A. nilotica, Dalbergia sissoo, E. camaldulensis, Agave americana, Bauhinia retusa, Buddleja asiatica, Grevillea pteridifolia, Salix tetrasperma, Pinus* spp., *Grewiatenax, P. juliflora, Albizia lebbeck, bamboos, and grasses such as Arundo donax, Saccharum spontaneum, Cenchrus setigerus, Cymbopogon* spp., *Cynodon dactylon, Dichanthium annulatum,* etc.
Waterlogged soils	*Species of Casuarina, Pongamia, Eucalyptus, Acacia, Terminalia, Syzygium, Dalbergia*
Sand dunes stabilization	*Prosopis cineraria, Acacia cupressiformis, Pongamia pinnata, Acacia tortilis, Acacia nilotica, Prosopis juliflora, Azadirachta indica, Tecomella undulata, Azadirachta indica, Melia azedarach, Albizia lebbeck, and Ailanthus excelsa*
Coastal soils	*Avicennia officinalis, Ailanthus malabarica, Excoecaria agallocha, Bruguiera cylindrica, B. parviflora, B. gymnorrhiza, B. sexangula; species of Rhizophora, Ceriops, Sonneratia, Xylocarpus; Salicorniabigelovii, Salvador apersica, S. oleoides, Terminalia catappa species of Pandanus, Borassus flabellifer, Anacardium occidentale,* etc.

Source: Adapted from Dagar (2014), Uthappa et al. (2015)

11.13 Acid Soil Reclamation

Amelioration of acid soil by use of lime most favored practice, as it raises the soil pH and thus makes all other essential nutrients available to plant roots and reduces toxicity of Fe and Al (Sharma and Sarkar 2005a, b; Fageria and Baligar 2008; Bhat et al. 2007). The soils are having very low pH and red to yellow in color due to excess of Al and Fe. The acid-tolerant tree and fruit crop species may be advocated for acid soil region. Besides mangroves, tree species such as *Thespesia populnea, Terminalia catappa,* Borassus palm, coconut (*Cocos nucifera*), cashew (*Anacardium occidentale*), arecanut (*Areca catechu*), *Acacia auriculiformis, Pandanus* spp., *Casuarina* spp., and oil-yielding Salicornia are ideal species for these areas (Dagar et al. 2014).

11.14 Waterlogged Soils

The waterlogged soils developed mainly due to the unprecedented use of irrigation water without proper drainage, mostly in the canal areas of western India. This problem further leads to the problem of soil salinization. The amelioration of waterlogged soils through a subsurface drainage system is not always viable. Meanwhile, the bio-drainage plants or trees, through their deep-rooted systems, can take up excess water in their system. Heuperman et al. (2002) concluded that the plantations behave as groundwater pumps, pumping water at the rate of 34,460 m^3 yr^{-1} or 3.93 m^3 hr^{-1} ha^{-1} of plantation, and the water used by plantations in the IGNP command was 3446 mm yr^{-1}, which was about 1.4 Class A pan. Tomar et al. (1998) in long-term studies found that *Acacia tortilis, A. farnesiana, Casuarina glauca, Eucalyptus camaldulensis, Salvadora oleoides, Parkinsonia aculeata*, and Tamarix spp. are the ideal trees for waterlogged saline soils.

11.15 Wastewater Use by Agroforestry

A large amount of wastewater is being used as irrigation water. These are considered as reliable sources of nutrients like N, P, K and organic matter, which are beneficial to soil fertility, plant growth, and production (Libutti et al. 2018). In dry land environment, the wastewater could be used successfully for an afforestation program, provided the proper species are selected. A lot of studies were conducted on treated wastewater use for greening urban areas and reduction of land degradation as well as use of fresh water in arid ecologies. The species like *Eucalyptus camaldulensis*, *S. oleoides*, *S. persica*, and *T. undulata* give better results in survival, adaptation, and growth for wastewater use in Indian deserts (Singh et al. 2021).

11.16 Ravine Land Reclamation

The ravine land topography is typical owing to the fact that there is sparse vegetation or almost no vegetation and severe erosion. The surface soil erodes first and thus making deep cracks that in due course of time form gullies of varying degrees and sizes. It is prevalent in semiarid to arid regions receiving meagre rainfall. Out of the 6 classes of gully reclaimability ratings proposed by Verma and Bhushan (1986), the gullies ascertained to reclaimability Classes 1, 2, and 3 were suited for common crops like pearl millet (*Pennisetum typhoides*), sorghum (*Sorghum durra*), wheat (*Triticum aestivum*), maize (*Zea mays*), green gram (*Vigna radiata*), pigeon pea (*Cajanus cajan*), black gram (*Vigna mungo*), tobacco (*Nicotiana tabacum*), mustard (*Brassica juncea*), and gram (*Cicer arietinum*) that could be grown in the gully beds and on bench terraces made on gully sides. Class 4 may be utilized for growing arable crops along with fruit trees such as mango (*Mangifera indica*), lemon (*Citrus limon*), aonla (*Emblica officinalis*), and ber (*Ziziphus mauritiana*). Classes 5 and 6 may be planted for perennial vegetations like fuel and fodder trees and grasses.

Further, the woody species, like *Acacia eburnea, A. nilotica, A. catechu, A. leucophloea, Albizia lebbeck, Azadirachta indica, Butea monosperma, Balanites roxburghii, Dalbergia sissoo, Dichrostachys cinerea, Dendrocalamus strictus, Eucalyptus* spp., *Feronia limonia, Prosopis juliflora, Pongamia pinnata*, and *Ziziphus mauritiana*, found to adapt easily in the ravines. The grasses, like *Cenchrus ciliaris, Dichanthium annulatum, Chrysopogon fulvus, Heteropogon contortus, Bothriochloa pertusa, Cynodon dactylon, Themeda triandra, Tragus biflorus, Sehima nervosum, Saccharum munja*, and *Iseilema laxum*, flourish well in ravine lands (Dagar 2018a, b). The medicinal species, such as *Ocimum americanum, O. sanctum, Aloevera, Withania somnifera, Adhatoda vasica, Solanum xanthocarpum, Barleria prionitis*, etc., and biofuels, such as *Jatropha curcas, Euphorbia antisyphilitica*, and *Pongamia pinnata* (at the bottom of ravines) can easily be blended in these habitats.

11.17 The Challenges for Agroforestry in Landscape Restoration

There is no doubt that all AFS have tremendous potential in all the agroecological zones of India to sequester carbon, which in turn helps in mitigating adverse effects of climate change. In spite of several advantages, there is still a paucity of quality planting material, problems in selling, harvesting, transportation, legal hurdles on cultivation, nonpayment of ecosystem services, long gestation period of trees, lack of awareness, and technical know-how act as major hurdles toward the promotion and use of AF for land rehabilitation. Moreover, the poor training and extension, non-implementation of policies, lack of organized markets for AF produce, dearth of suitable silviculture and cultivation methodology, low price, problems in tree loan, and nonavailability of insurance schemes are further hindering the AF adoption (Cooper and Denning 2000). Thus, there is an urgent need to focus on developing strategies and devising plans for the successful adoption of AFS and to utilize it fully for rehabilitating degraded lands.

11.18 Future Thrust and Strategies for Restoration of Degraded Land Through Agroforestry Approaches

India's degraded areas have shown enormous potential for reclamation through agroforestry. In order to promote agroforestry in the nation, the Indian government has also developed a number of policies and launched a number of initiatives. As a result, the nation should take into account the following plans of action and methods for the large-scale agroforestry-based regeneration of degraded lands:

- Creation of need-based agroforestry technologies through multilocational trials in diverse agroecological zones with the goal of scaling up the most effective agroforestry models and practices across the nation

- Mass multiplication and cloning of superior tree species' germplasm to create high-quality planting material appropriate for degraded soils
- Establishing high-tech nurseries to produce high-quality planting materials for potential species in the nation's various agroclimatic zones
- Establishing a center for each species' excellence in its respective region through collaboration with state governments and nursery growers
- Creating high-value tree species cultivation and management techniques for degraded soil
- Locating and creating agroforestry systems that are climate resilient in order to mitigate the effects of climate change and adapt to it on degraded lands
- Enhancing the abiotic stress tolerance of tree species through breeding, improvement, and selection techniques
- Developing suitable metrics and indicators to quantify the intangible benefits and environmental externalities of ecosystem payment services offered by agroforestry
- Simplifying the rules surrounding the harvesting and sale of tree produce to fully explore the potential of agroforestry
- Developing agroforestry-based business models by collaborating with local farmers, research institutions, and private industries
- Using mechanization in agroforestry from planting to harvesting to minimize reliance on manual labor
- Postharvest and value addition in trees to maximize benefits
- Easy and subsidized access to quality plating material
- Initiatives to promote agroforestry at all levels, including those of farmers, extension agents, village-level organizations, and cooperative societies for tree producers, are being launched, along with low-interest financing mechanisms, insurance, credit, and subsidy schemes, and low-risk financing mechanisms for farmers adopting agroforestry practices
- Lastly, institutional mechanisms are being created to coordinate and unify the ministries' and schemes' efforts to pursue agroforestry thematically.

Further, Parthiban et al. (2021) have proposed a number of research, technological, and policy interventions that could aid in the spread and promotion of agroforestry models in India, including an organized supply chain, a buyback agreement with MSP, analysis and sensitization of price, the creation of supportive price mechanisms, insurance and credit facilities, tax exemption, processing technology facilities, promotion of mechanization, effective postharvest management, value addition, development of high-yielding varieties and clonal planting material, reducing the rotation, adoption of precision silviculture, and profitable tree crop model. Furthermore, there is an urgent need to concentrate on creating high-value tree species-based agroforestry practices with an emphasis on productivity, profitability, and sustainability, given the limitations and successes of previous and current agroforestry practices. Farmers value and favor certain tree species over others, including *Tectona grandis*, *Santalum album*, *Swietenia Mahagoni*, *Pterocarpus santalinus*, *Aquilaria*, and *Melia dubia*. These species have the potential to significantly boost farm income, restore degraded lands, and offer resilience against climate change. They are easily integrated with a variety of agroforestry practices, including block plantations, agri-silviculture, silvi-pastoral, and agri-horti-silviculture.

11.19 Conclusion

The degradation of land undermines the three fundamental pillars of sustainable development, namely environmental, social, and economic sustainability, posing a serious threat to social stability, economic prosperity, and ecological health. While restoring degraded land to support livelihoods, enhancing food security, restoring ecosystem services, and reducing pressure on forests sustainably are all possible goals of agroforestry. While starting an agroforestry program to minimize land degradation, consideration should be given to the selection of appropriate species, technology, inputs, and organizational structure, among other things. Nowadays, agroforestry research has produced beneficial technological and policy advancements. Agroforestry is now regarded as a science that solves problems as well as a sustainable production system. Agroforestry needs a systems perspective that is easily incorporated into landscape approaches in order to succeed. Agroforestry interventions at the field scale are designed to minimize adverse interactions between crops and trees while optimizing a number of positive outcomes, including enhancing soil health, preserving biodiversity, and producing minor forest products. It is commonly known that restoration can stop the processes of degradation and increase the benefits that ecosystems and landscapes provide to people's quality of life, land productivity, environmental services, and the resilience of both natural and human systems. Numerous agroforestry techniques, including agri-silviculture, agri-horticulture, and silvi-pastroral, have demonstrated a high potential for yielding both financial and environmental gains from these lands. Furthermore, there can be a significant impact on farm income growth from the widespread cultivation and adoption of high-value trees by farmers. More research is needed to create agroforestry techniques that stakeholders and farmers can embrace broadly. Therefore, it is essential to create suitable agroforestry practices, policies, and action plans for the increased promotion and adoption of agroforestry in order to meet the land degradation neutrality target.

References

Abatenh E, Gizaw B, Tsegaye Z, Wassie M (2017) The role of microorganisms in bioremediation-a review. Open J Environ Biol 2(1):038–046

AbdelRahman MAE (2023) An overview of land degradation, desertification and sustainable land management using GIS and remote sensing applications. Rendiconti Lincei Scienze Fisiche e Naturali 34:767–808. https://doi.org/10.1007/s12210-023-01155-3

Abdi OA, Glover EK, Luukkanen O (2013) Causes and impacts of land degradation and desertification: case study of The Sudan. Int J Agric For 3(2):40–51

Abebe S, Taddese TT, Omer AAD (2019) Agroforestry in Ethiopia: a review of agroforestry practices and potentials. Agrofor Syst 93(5):1389–1402. https://doi.org/10.1007/s10457-019-00346-3

Ainsworth N, McCormack BA, McDonald PJ (2012) Agroforestry in Australia: the role of trees in sustainable agricultural landscapes. Aust J Agric Res 63(5):393–405. https://doi.org/10.1071/AR12005

Ajit Dhyani SK, Handa AK, Sridhar KB, Jain A, Uma Sasindran P, Kaza M, Prasad R, Sriram K (2014) Carbon sequestration assessment of block plantations at JSW Steel Limited. In: Compendium of Abstracts, 3rd World Agroforestry Congress, organized by ICAR, WAC and ISAF at Delhi, Feb 10-13, pp 354–355

Amerino J, Apedo CK, Anang BT, Lin C (2024) Factors influencing adoption of cocoa agroforestry in Ghana: analysis based on tree density choice. Sustain Environ 10(1). https://doi.org/10.1080/27658511.2023.2296162

Bai ZG, Dent DL, Olsson L, Schaepman ME (2008) Global assessment of land degradation and improvement. 1. Identification by remote sensing. Report 2008/01. ISRIC – World Soil Information, Wageningen

Bateni C, Ventura M, Tonon G, Pisanelli A (2021) Soil carbon stock in olive groves agroforestry systems under different management and soil characteristics. Agrofor Syst. https://doi.org/10.1007/s10457-019-00367-7

Bhat JA, Mandal B, Hazra GC (2007) Basic slag as a liming material to ameliorate soil acidity in Alfisols of sub-tropical India. Am-Eurasian J Agric Environ Sci 2:321–327

Bhat Y, Nandy S, Padalia H, Nath AJ, Pebam R (2022) Role of geospatial technology in shifting cultivation studies in India: a systematic review. J Indian Soc Remote Sens 50:2359–2379. https://doi.org/10.1007/s12524-022-01607-7

Bhattacharyya R, Ghosh BN, Mishra PK, Mandal B, Rao CS, Sarkar D, Franzluebbers AJ (2015) Soil degradation in India: challenges and potential solutions. Sustainability 7(4):3528–3570

Bhumbla DR, Khare (1985) A Estimates of wastelands in India. Society for Promotion of Wastelands Development (SPWD), New Delhi

Boumenjel A, Papadopoulos A, Ammari Y (2021) Growth response of Moringa oleifera (Lam) to water stress and to arid bioclimatic conditions. Agrofor Syst. https://doi.org/10.1007/s10457-020-00509-2

Cannell MGR (1987) Agroforestr-A decade of development. ICRAF 24:393

Chaterjee PC (1988) Mine spoils and their rehabilitation. In: Mining and environment in India. H.R. Publishers, Nainital, pp 394–409

Chauhan SS (2003) Desertification control and management of land degradation in the Thar Desert of India. Environmentalist 23:219–227

Chavan SB, Dhillon RS, Sirohi C, Uthappa AR, Jinger D, Jatav HS, Chichaghare AR, Kakade V, Paramesh V, Kumari S, Yadav DK, Minkina T, Rajput VD (2023) Carbon sequestration potential of commercial agroforestry systems in Indo-Gangetic Plains of India: poplar and eucalyptus-based agroforestry systems. Forests 14:559. https://doi.org/10.3390/f14030559

Cooper PJM, Denning GL (2000) Scaling up the impact of agroforestry research. World Agroforestry Centre (ICRAF), Nairobi, p 43

Dadhwal KS, Singh B, Narain P (1992) Characteristics of limestone mine-spoil/debris from outer Himalayas of Uttar Pradesh. Indian For 118(9):650–658

Dagar JC (2014) Greening salty and waterlogged lands through agroforestry systems for livelihood security and better environment. In: Dagar JC, Singh AK, Arunachalam A (eds) Agroforestry systems in India: livelihood security & ecosystem services, Advances in Agroforestry, vol 10. Springer, New Delhi, pp 273–332

Dagar JC (2018a) Ravines: formation, extent, classification, evolution and measures of prevention and control. In: Dagar JC, Singh AK (eds) Ravine lands: greening for livelihood and environmental security. Springer Nature, Singapore, pp 19–67

Dagar JC (2018b) Perspectives of vegetation ecology and biodiversity for management of ravine lands. In: Dagar JC, Singh AK (eds) Ravine lands: greening for livelihood and environmental security. Springer Nature, Singapore, pp 69–118

Dagar JC, Pandey CB, Chaturvedi CS (2014) Agroforestry: a way forward for sustaining fragile coastal and Island agro-ecosystems. In: Dagar JC, Singh AK, Arunachalam A (eds) Agroforestry systems in India: livelihood security and ecosystem services, Advances in agroforestry, Springer, vol 10. New Delhi, Dordrecht, pp 185–232

De Leijster V, Santos MJ, Wassen MW, Camargo García JC, Llorca Fernandez I, Verkuil L, Scheper A, Steenhuis M, Verweij PA (2021) Ecosystem services trajectories in coffee agroforestry in Colombia over 40 years. Ecosyst Serv 48:101246. https://doi.org/10.1016/j.ecoser.2021.101246

Dehkordi MM, Nodeh ZP, Dehkordi KS, Khorjestan RR, Ghaffarzadeh M (2024) Soil, air, and water pollution from mining and industrial activities: sources of pollution, environmental impacts, and prevention and control methods. Results Eng 102729

Dhyani SK, Handa AK (2013) India needs agroforestry policy urgently: issues and challenges. Indian J Agrofor 15:1–9

Dixon RK (1995) Agroforestry systems: sources or sinks for greenhouse gases? Agrofor Syst 31:99–116. https://doi.org/10.1007/BF00711719

Dollinger J, Jose S (2018) Agroforestry for soil health. Agrofor Syst 92(2):213–219. https://doi.org/10.1007/s10457-018-0223-9

Down To Earth (2019). Available at: https://www.downtoearth.org.in/news/agriculture/india-lost-31-of-grasslands-in-a-decade-66643

ELD Initiative (2013) Opportunity lost: mitigating risk and making the most of your land assets. An assessment of the exposure of business to land degradation risk and the opportunities inherent in sustainable land management. Available from: www.eld-initiative.org

Fageria NK, Baligar VC (2008) Ameliorating soil acidity of tropical Oxisols by liming for sustainable crop production. Adv Agron 99:345–399

FAO (2020) Food and Agriculture Organization of the United Nations, Rome, Italy

FAO (2022) Healthy soils for a healthy people and planet: FAO calls for reversal of soil degradation. Food and Agriculture Organization, Rome, Italy

FAO (2024). https://www.fao.org/home/en/

FAO & ICRAF (2019) Agroforestry and Tenure. Rome Forestry Working Paper Number 8:40

Fertilizer Association of India (2021). Available at: https://www.faidelhi.org/general/con-npk.pdf

Garrett HE (2009) North American Agroforestry: an integrated science and practice, 2nd edn. American Society of Agronomy, Inc, Madison, WI

Gastauer M, Cavalcante RBL, Caldeira CF, de Nunes SS (2020) Structural hurdles to large-scale forest restoration in the Brazilian Amazon. Front Ecol Evol 8:1–6. https://doi.org/10.3389/fevo.2020.593557

Global Forest Watch (2022). Available at: https://www.globalforestwatch.org/

Gold MA, Garrett HE (2009) Agroforestry nomenclature, concepts, and practices, Chapter 3. In: Garrett HE (ed) North American agroforestry: an integrated science and practice, 2nd edn. Agronomy Society of America, pp 45–46. https://doi.org/10.2134/2009.northamericanagroforestry.2ed.c3

Hamdy A, Aly A (2014) Land degradation, agriculture productivity and food security. In: Proceedings Fifth International Scientific Agricultural Symposium "Agrosym 2014", Jahorina, Bosnia and Herzegovina, pp 708–717

Handa AK, Dhyani SK, Uma (2015) Three decades of agroforestry research in India: retrospection for way forward. Agric Res J PAU 52:1–10

Handayani IP, Prawito P (2011) Agroforestry systems for sustaining rural development and protecting environmental quality. Prosiding Seminar Nasional Budidaya Pertanian, pp 12–23. 7 July 2011, Bengkulu

Harvey CA, Gonzales JG, Somarriba E (2006) Dung beetle and terrestrial mammal diversity in forest, indigenous agroforestry systems and plantain monocultures in Talamanca, Costa Rica. Biodivers Conserv 15:555–585

Heuperman AF, Kapoor AS, Denecke HW (2002) Biodrainage: principles, experiences and applications. International Programme for Technology and Research in Irrigation and Drainage, IPTRID Secretariat, FAO, Rome, p 78

Hossain A, Krupnik TJ, Timsina J, Mahboob MG, Chaki AK, Farooq M, Hasanuzzaman M (2020) Agricultural land degradation: processes and problems undermining future food security. In: Environment, climate, plant and vegetation growth. Springer International Publishing, Cham, pp 17–61

Intergovernmental Science-Policy Platform on Biodiversity and Ecosystem Services (2021) The Global Assessment Report on Biodiversity and Ecosystem Services

International Assessment of Agricultural Science and Technology for Development (2008) Executive summary of the synthesis report. http://www.agassessment.org/docs/SR_Exec_Sum_280508_English.htm. Accessed on March 27, 2009

IPCC (2000) Land-use, land-use change and forestry. Special report of the intergovernmental panel on climate change. Cambridge University press. Retrieved November 28, 2019, from https://archive.ipcc.ch/ipccreports/sres/land_use/index.php?idp=0

IPCC (2022) Climate Change 2022: mitigation of Climate Change – Working Group III contribution to the WGIII Sixth Assessment Report of the Intergovernmental Panel on Climate Change, p 2913. www.ipcc.ch/report/ar6/wg3

ISRO (2018) Desertification and land degradation atlas of selected districts of India (Based on IRS LISS III data of 2011–13 and 2003–05), vol 2. Space Applications Centre, ISRO, Ahmedabad, p 145. www.sac.gov.in

Jara-Rojas R, Russy S, Roco L, Fleming-Muñoz D, Engler A (2020) Factors affecting the adoption of agroforestry practices: insights from silvopastoral systems of Colombia. Forests 11:648. https://doi.org/10.3390/f11060648

Jera R, Ajayi OC (2008) Logistic modelling of smallholder livestock farmers' adoption of tree-based fodder technology in Zimbabwe. Agrekon 47:379–392

Jose S (2009) Agroforestry for ecosystem services and environmental benefits: an overview. Agrofor Syst 76(1):1–10. https://doi.org/10.1007/s10457-009-9229-7

Julian A (2016) Cocoa farming system in Indonesia and its sustainability under climate change. Agric For Fish 5(5):170–180. https://doi.org/10.11648/j.aff.20160505.15

Kalloo G (2003) Forage research-new dimensions. 42nd foundation Day lecture. Indian Grassland and Fodder Research Institute, Jhansi, UP, India

Kapur R (2016) Natural resources and environment issues. J Ecosyst 6:2–5. https://doi.org/10.4172/2157-7625.1000196

Karanja S, Muthuri DKW, Muia LJ (2018) Agroforestry and sustainable land management in Kenya: current practices and future opportunities. Land Use Policy 70:224–233. https://doi.org/10.1016/j.landusepol.2017.10.01

Kaur B, Gupta SR, Singh G (2002a) Carbon storage and nitrogen cycling in silvo-pastoral systems on a sodic soil in North Western India. Agrofor Syst 54:21–29

Kaur B, Gupta SR, Singh G (2002b) Bio-amelioration of a sodic soil by silvo-pastoral systems in North Western India. Agrofor Syst 54:13–20

Khosla PK, Kumar K, Dutt N, Kathuria K, Mehta P, Farhana M (2018) ITKs for land use and water management in cold deserts of Ladakh valley in Jammu and Kashmir. J Pharmacogn Phytochem 7:521–524

Kirby KR, Potvin C (2007) Variation in carbon storage among tree species: implications for the management of a small-scale carbon sink project. For Ecol Manag 246:208–221

Knowler D, Bradshaw B (2007) Farmers' adoption of conservation agriculture: a review and synthesis of recent research. Food Policy 32:25–48

Kosmas C, Kairis O, Karavitis C et al (2014) Evaluation and selection of indicators for land degradation and desertification monitoring: methodological approach. Environ Manag 54:951–970. https://doi.org/10.1007/s00267-013-0109-6

Kumar A (2006a) Agroforestry systems in South Asia. Agrofor Syst 69(2):223–237. https://doi.org/10.1007/s10457-006-9005-6

Kumar BM (2006b) Carbon sequestration potential of tropical homegardens. In: Kumar BM, Nair PKR (eds) Tropical Homegardens: a time-tested example of sustainable agroforestry, Advances in Agroforestry 3. Springer, Dordrecht, pp 185–204

Kumar D, Chavan SB, Ram A, Dev I, Newaj R (2019) Management of hill and mountain soil of North Eastern India through agroforestry interventions. Indian Farming 69(10):47–48
Lal R (1989) Agroforestry systems and soil surface management of a tropical Alfisol. II. Water runoff, soil erosion and nutrient loss. Agrofor Syst 8:97–111
Lal R (2012) Climate change and soil degradation mitigation by sustainable management of soils and other natural resources. Agric Res 1:199–212. https://doi.org/10.1007/s40003-012-0031-9
Lal R (2015) Restoring soil quality to mitigate soil degradation. Sustainability 7(5):5875–5895
Lal R, Smith P, Jungkunst HF, Mitsch WJ, Lehmann J, Nair PKR, McBratney AB, de Moraes Sá JC, Schneider J, Zinn YL (2018) The carbon sequestration potential of terrestrial ecosystems. J Soil Water Conserv 73:145A–152A. https://doi.org/10.2489/jswc.73.6.145A
Lamb D (1988) IUCN guidelines for restoration of degraded ecosystems. Draft-world conservation union (IUCN), Gland
Le H, Thanh DMK, Thao HTL (2017) Agroforestry systems in Vietnam: current status and future prospects. J Trop For Sci 29(3):315–323
Leakey RR, Tchoundjeu Z, Schreckenberg K, Shackleton SE, Shackleton CM (2015) Agroforestry Tree Products (AFTPs): targeting poverty reduction and enhanced livelihoods. Int J Agric Sustain 3:37–41. https://doi.org/10.1080/14735903.2005.9684741
Libutti A, Gatta G, Gagliardi A, Vergine P, Pollice A, Beneduce L et al (2018) Agro-industrial wastewater reuse for irrigation of a vegetable crop succession under Mediterranean conditions. Agric Water Manag 196:1–14. https://doi.org/10.1016/j.agwat.2017.10.015
Logan TJ (1990) Chemical degradation of soil. Adv Soils Sci Springer 11:187–221
Marsden C, Martin-Chave A, Cortet J, Hedde M, Capowiez Y (2019) How agroforestry systems influence soil fauna and their functions – a review. Plant Soil 453:29–44. https://doi.org/10.1007/s11104-019-04322-4
McNeely JA (2004) Nature vs nurture: managing relationships between forests, agroforestry and wild biodiversity. Agrofor Syst 61:155–165
Miah MG, Hussain MJ (2010) Homestead agroforestry: a potential resource in Bangladesh. In: Lichtfouse E (ed) Sociology, organic farming, climate change and soil science. Sustainable Agriculture Reviews, vol 3. Springer, Dordrecht. https://doi.org/10.1007/978-90-481-3333-8_16
Millennium Ecosystem Assessment (2005) Ecosystems and human well-being: biodiversity synthesis. World Resources Institute, Washington, DC
Mishra AK, Tiwari HS, Bhatt RK (2010) Growth, biomass production and photosynthesis of *Cenchrus ciliaris* L. under *Acacia tortilis* (Forssk.) Hayne based silvo-pastoral systems in semiarid tropics. J Environ Biol 31:987–993
MoA (1978) Indian Agriculture in Brief, 20th edn. Directorate of Economics and Statistics, Ministry of Agriculture, Department of Agriculture and Cooperation, Ministry of Agriculture, Government of India, New Delhi
MoA (1994) Indian Agriculture in Brief, 25th edn. Directorate of Economics and Statistics, Ministry of Agriculture, Department of Agriculture and Cooperation, Government of India, New Delhi
Mosaffaei Z, Jahani A, Chahouki MAZ, Goshtasb H, Etemad V, Saffariha M (2020) Soil texture and plant degradation predictive model (STPDPM) in national parks using artificial neural network (ANN). Modeling Earth Syst Environ 6:715–729
Mosquera-Losada MR, Santiago-Freijanes JJ, Rigueiro-Rodrıguez A, Rodriguez-Rigueiro FJ, Arias Martinez D, Pantera A, Ferreiro-Dominguez N (2020) The importance of agroforestry systems in supporting biodiversity conservation and agricultural production: a European perspective, Reconciling agricultural production with biodiversity conservation. Burleigh Dodds Science Publishing Limited, Cambridge, p 282
Mziray W, Ng'walali LH, Ngowi RTM (2020) Agroforestry practices in Tanzania: potential for sustainable land management. J Land Use Sci 15(2):132–145. https://doi.org/10.1080/1747423X.2020.1713441
Nachtergaele F, Petri M, Biancalani R, van Lynden G, van Velthuizen H (2010) Global land degradation information system (GLADIS), Beta version, an information database for land degrada-

tion assessment at the global level, land degradation assessment in drylands technical report 17. Food and Agriculture Organization of the United Nations, Rome

Nair PKR, Asare RA, Suyam TA (2009a) Agroforestry systems in the tropics: a review. Agrofor Syst 76(1):49–67. https://doi.org/10.1007/s10457-009-9204-4

Nair PKR, Kumar BM, Nair VD (2009b) Agroforestry as a strategy for carbon sequestration. J Plant Nutr Soil Sci 172:10–23. https://doi.org/10.1002/jpln.200800030

Nair PKR, Kumar BM, Nair VD (2021) Global distribution of agroforestry systems. In: An introduction to agroforestry. Springer, Cham. https://doi.org/10.1007/978-3-030-75358-0_4

Narain P, Bhati TK (2004) Alternative farming system: issues and opportunities in arid ecosystem. In: Singh AK, Gangwar B, Sharma SK (eds) Alternative farming systems. Farming System Research and Development Association, Modipuram, pp 57–64

Nath TK, Inoue M, Myant H (2005) Small-scale agroforestry for upland community development: a case study from Chittagong Hill Tracts, Bangladesh. J For Res 10(6):443–452. https://doi.org/10.1007/s10310-005-0171-x

NBSS&LUP (1994) Global Assessment of Soil Degradation (GLASOD) Guidelines. National Bureau of Soil Survey and Land Use Planning, Nagpur

NBSS&LUP (2005) National Bureau of Soil Survey and Land Use Planning. Annual Report. NBSS &LUP, Nagpur

NCA (1976) Report of the National Commission on Agriculture. National Commission of Agriculture, Government of India, New Delhi, India, pp 427–472

Newaj R, Chaturvedi OP, Kumar D, Rajendra P, Rizvi RH, Badre A, Handa AK, Chavan SB, Singh AK, Mayank C, Karmakar PS, Abhishek M, Abhishek S, Gargi G, Kedari S (2017) Soil organic carbon stock in agroforestry systems in western and southern plateau and hill regions of India. Curr Sci 112(11):2191–2193

Nkamleu GB, Manyong VM (2005) Factors affecting the adoption of agroforestry practices by farmers in Cameroon. Small-Scale For Econ Manage 4:135–148

NRSA (2000) Waste Land Atlas of India. National Remote Sensing Centre, Government of India, Balanagar, Hyderabad

NRSC (2007) Nationwide mapping of Land degradation using multitemporal satellite data. In: Project Manual. Soil and Land Resources Assessment Division, Earth Resources Group, RS & GIS application area, Department of Space, Govt of India, Balanagar, Hyderabad

NRSC (2019) Status of Land degradation in India: 2015–16 (ATLAS). National Remote Sensing Centre, ISRO, Govt of India, Hyderabad

NWDB (1985) Guidelines for Action. National Wasteland Development Board. Ministry of Environment and Forests, Government of India, New Delhi

Oostendorp RH, Zaal F (2012) Land acquisition and the adoption of soil and water conservation techniques: a Duration Analysis for Kenya and The Philippines. World Dev 40:1240–1254

Osman KT (2014) Chemical Soil Degradation. In: Soil Degradation, Conservation and Remediation. Springer, Dordrecht. https://doi.org/10.1007/978-94-007-7590-9_5

Osman KT, Osman KT (2014) Chemical soil degradation. In: Soil degradation, conservation and remediation, pp 125–148

Othman H, Darus FM, Hashim Z (2012) Best management practices for oil palm cultivation on peat: Mucuna bracteate as ground cover crop. MPOB Information Series

Pacardo EP, Montecillo L (1983) Effects of corn/ipil-ipil cropping systems on productivity and stability of upland agro-ecosystems, Annual Report, UPLB PCARRD Research Project, Philippines

Parodi A, Villamonte-Cuneo G, Loboguerrero AM, Martínez-Barón D, Vázquez-Rowe I (2022) Embedding circularity into the transition towards sustainable agroforestry systems in Peru. Sci Total Environ 838(3):156376. https://doi.org/10.1016/j.scitotenv.2022.156376

Parrotta JA (1992) The role of plantation forests in rehabilitating degraded tropical ecosystems. Agric Ecosys Environ 41:115–133

Parthiban KT, Fernandaz CC, Sudhagar RJ, Sekar I, Kanna SU, Rajendran P et al (2021) Industrial agroforestry—A sustainable value chain innovation through a consortium approach. Sustainability 13:7126. https://doi.org/10.3390/su13137126

Patel JM, Jamini SN, Patel SB (2008) Evaluation of neem and ardu based Agri-silvi system. In: Narain P, Singh MP, Kar A, Kathju S, Kumar P (eds) Diversification of arid farming systems. Arid Zone Research Association of India and Scientific Publisher, Jodhpur, pp 114–118

Patidar M, Rajora MP, Singh R (2008) Forage production potential of different silvipastoral systems under arid conditions of Rajasthan. Indian J Agron 53:234–238

Perfecto I, Vandermeer JA, Wright AR (2005) Silvopastoral systems in the tropics: a review of the status and prospects. Agrofor Syst 63(1):37–51. https://doi.org/10.1007/s10457-005-0142-5

Prasad R, Mertia RS, Gajja BL, Singh D, Chandra P (2013) Morphological characteristics of tree shelterbelts and their impact on wind regimes, air temperature and soil quality. Indian J Agrofor 15:74–81

Raintree JB, Torres F (1986) The agroforestry research in farming systems perspectives: the ICRAF approach, IARC's workshop on FSR, 17–21 Feb 1986, ICRISAT, Hyderabad, India

Reddy MG, Kulkarni GN (1978) Effect of shelterbelt on water use and yield of CSH-1 sorghum. Ann Arid Zone 17:343–347

Reisner Y, de Filippi R, Herzog F, Palma J (2007) Target regions for silvoarable agroforestry in Europe. Ecol Eng 29:401–418. https://doi.org/10.1016/j.ecoleng.2006.09.020

Riva MJ, Daliakopoulos IN, Eckert S, Hodel E, Liniger H (2017) Assessment of land degradation in Mediterranean forests and grazing lands using a landscape unit approach and the normalized difference vegetation index. Appl Geogr 86:8–21

SAC I (2016) Desertification and land degradation atlas of India (Based on IRS AWiFS data of 2011–13 and 2003–05). Space Applications Centre, ISRO, Ahmedabad, India, Ahmedabad, p 219

Santini NS, Lovelock CE, Hua Q, Zawadzki A, Mazumder D, Mercer TR (2019) Natural and regenerated saltmarshes exhibit similar soil and belowground organic carbon stocks, root production and soil respiration. Ecosystems 22:1803–1822. https://doi.org/10.1007/s10021-019-00373-x

Sathaye JA, Ravindranath NH (1998) Climate change mitigation in the energy and forestry sectors of developing countries. Annu Rev Energy Environ 23:387–437

Scherr SJ (1999) Soil degradation: a threat to developing-country food security by 2020? Int Food Policy Res Inst 27:20–27

Schofield RV, Kirkby MJ (2003) Application of salinization indicators and initial development of potential global soil salinization scenario under climatic change. Glob Biogeochem Cycles 17. https://doi.org/10.1029/2002GB001935

Sena A, Ebi L (2021) When land is under pressure health is under stress. Int J Environ Res Public Health 18(1):136. https://doi.org/10.3390/ijerph18010136

Shamsudheen M, Devi D, Meena SL, Ram B (2009) Improvement of soil properties under silvipastoral systems in the Kachchh region of arid Gujarat. In: Abstract retrieved from 4th world congress on conservation agriculture. National Academy of Agricultural Sciences, New Delhi

Shamsudheen M, Devi D, Meena SL, Ram B (2014) Carbon sequestration in agroforestry and pasture systems in arid northwestern India. Curr Sci 107:1290–1293

Sharma PD, Sarkar AK (2005a) Managing acid soils for enhancing productivity. In: Proceedings IPIOUAT-IPNI International Symposium. Natural Resource Management Division ICAR, New Delhi, pp 01–23

Sharma PD, Sarkar AK (2005b) Managing acid soils for enhancing productivity, Technical Bulletin, vol 2005. New Delhi, NRM Division, KAB-II, Pusa Campus, p 23

Sharrow SH, Ismail S (2004) Carbon and nitrogen storage in agroforests, tree plantations, and pastures in western Oregon, USA. Agrofor Syst 60:123–130

Singh G, Rathod TR (2002) Plant growth, biomass production and soil water dynamics in a shifting dune of Indian desert. For Ecol Manag 171:309–320

Singh G, Nagora PR, Haksar P, Chauhan AR (2021) Utilizing treated wastewater in tree plantation in Indian desert: part I–species suitability, plant growth and biomass production. Int J Phytoremediation 24:1–11. https://doi.org/10.1080/15226514.2021.1993784

Sklenicka P (2016) Classification of farmland ownership fragmentation as a cause of land degradation: a review on typology, consequences, and remedies. Land Use Policy 57:694–701

Snedecor GW, Cochran WG (1967) Statistical methods, 6th edn. The Iowa State University

Soni ML, Beniwal RK, Talwar HS, Yadava ND, Singh JP, Kumar S (2006) Root distribution pattern of two predominant grasses viz., *Lasiurus sindicus* and *Cenchrus ciliaris* of arid ecosystem of western Rajasthan in relation to their soil binding capacity. Indian J Agri Sci 76:716–720
Soni ML, Yadava ND, Beniwal RK, Singh P, Kumar S, Birbal (2013) Grass based strip cropping systems for controlling soil erosion and enhancing system productivity under drought situations of hot arid western Rajasthan. Int J Agric Stat Sci 9:685–692
Soni ML, Subbulakshmi V, Yadava ND, Tewari JC, Dagar JC (2016) Silvopastoral agroforestry systems: lifeline for dry regions. In: Dagar JC, Tewari JC (eds) Agroforestry research developments. Nova Science Publishers, New York, pp 245–305
Sreenivas K, Sujatha G, Mitran T, Suresh KJR, Ravisankar T, Rao PVN (2021) Decadal changes in land degradation status of India. Curr Sci 121(4):25
Sridhar KR, Bhagya B (2007) Coastal sand dune vegetation: a potential source of food, fodder and pharmaceuticals. Livestock Res Rural Deve 19:1–2
Stock E, Standish RJ, Munoz-Rojas M, Bell RW, Erickson TE (2020) Field-deployed extruded seed pellets show promise for perennial grass establishment in arid zone mine rehabilitation. Front Ecol Evol 8:1–15. https://doi.org/10.3389/fevo.2020.576125
Sukmana S, Suwardjo H, Abdurachman A, Dai J (1985) Prospects of Flemengia congesta for reclamation and conservation of volcanic skeletal soils, Pembr. Pen, Tanah, Dan. Pipuk 4:50–54
Tabora PC (1991) Analysis and evaluation of agroforestry as an alternative environmental design in The Philippines. Agrofor Syst 14:39–63. https://doi.org/10.1007/BF00141596
Tesfa A, Mekuriaw S (2014) The effect of land degradation on farm size dynamics and crop-livestock farming system in Ethiopia: a review. Open J Soil Sci 4(1):1–5. https://doi.org/10.4236/ojss.2014.41001
Tewari JC, Ram M, Roy MM, Dagar JC (2014) Livelihood improvements and climate change adaptations through agroforestry in hot arid environments. Adv Agrofor 10:155–183
Thevathasan NV, Gordon AM (2004) Ecology of tree intercropping systems in the North temperate region: experiences from southern Ontario, Canada. Agrofor Syst 61:257–268
Tomar OS, Gupta RK, Dagar JC (1998) Afforestation techniques and evaluation of different tree species for waterlogged saline soils in semiarid tropics. Arid Soil Res Rehabil 12(4):301–316
Tomar JMS, Ahmed A, Bhat JA, Kaushal R, Shukla G (2023) Potential and opportunities of agroforestry practices in combating land degradation. In: Agroforestry - small landholder's tool for climate change resiliency and mitigation. Intech Open. https://doi.org/10.5772/intechopen.97843
Tudi M, Daniel Ruan H, Wang L, Lyu J, Sadler R, Connell D, Chu C, Phung DT (2021) Agriculture development, pesticide application and its impact on the environment. Int J Environ Res Public Health 18(3):1112
Turner KG, Anderson S, Gonzales-Chang CM, Courville R (2016) A review of methods, data, and models to assess changes in the value of ecosystem services from land degradation and restoration. Ecol Model 319:19–207. https://doi.org/10.1016/j.ecolmodel.2015.07.017
UNCCD (2023). https://www.unccd.int/sites/default/files/2023 10/ Data% 20 Dashboard % 20press% 20release%20EN.pdf
UNEP (2022) Making peace with nature: a scientific blueprint to tackle the climate, biodiversity and pollution emergencies
Upadhyaya AK (1991) Shelterbelt plantations effectively check sand deposition in Indira Gandhi canal. Indian Forest 117:511–514
Uthappa AR, Chavan SB, Dhyani SK, Handa AK, Newaj R (2015) Trees for soil health and sustainable agriculture. Indian Farming 65(3):2–5
Verchot LV, Van Noordwijk M, Kandji S, Tomich T, Ong C, Albrecht A, Mackensen J, Bantilan C, Anupama KV, Palm C (2007) Climate change: linking adaptation and mitigation through agroforestry. Mitig Adapt Strateg Glob Chang 12(5):901–918. https://doi.org/10.1007/s11027-007-9105-6
Verma SK (1988) Environmental aspects of mining activities. In: Mining and environment in India. H.R. Publishers, Nainital, pp 388–393
Verma B, Bhushan LS (1986) Management of gullied watersheds. Indian J Soil Conserv 14(3):105–110

Weeraratna S (2022) Factors causing land degradation. In: Understanding land degradation: an overview. Springer International Publishing, Cham, pp 5–22

Wiersum KF (1981) Observations on agroforestry in Java, Indonesia. Agricultural University, Wageningen, p 131

Wiersum KF (1984) Surface erosion under various tropical agroforestry systems. In: O'Loughlin CL, Pearce AJ (eds) Effects of forest land use on erosion and slope stability. Univ. of Hawaii, Honululu

Wijitkosum S (2020) Reducing vulnerability to desertification by using the spatial measures in a degraded area in Thailand. Land 9:49. https://doi.org/10.3390/land9020049

Williams-Guilleen K, Perfecto I, Vandermeer J (2008) Bats limit insects in a tropical agroforestry system. Science 320:70

World Bank (2004) Sustaining forests: a development strategy. The World Bank, Wash., D.C.. ISBN 0-8213-5755-7

Xie S et al (2015) Key evidence of the role of desertification in protecting the underlying permafrost in the Qinghai–Tibet Plateau. Sci Rep 5:15152. https://doi.org/10.1038/srep15152

Zoomer R, Neufeldt H, Xu J (2016) Global Tree Cover and Biomass Carbon on Agricultural Land: the contribution of agroforestry to global and national carbon budgets. Sci Rep 6:29987. https://doi.org/10.1038/srep29987

Invasive Species-Based Phytomanagement of Contaminated Land

12

Nayan Sahu and Poornima Garg

Abstract

Phytoremediation, the use of plants to remediate contaminated lands, has gained considerable traction globally, particularly through the application of invasive plant species. This sustainable and cost-effective approach leverages the natural capabilities of invasive and other suitable plant species, which are resilient and adaptable, possess unique traits that enable them to thrive in polluted environments and efficiently uptake heavy metals and organic pollutants, and other hazardous substances, effectively reducing environmental risks and restoring ecosystem health. There have been published reports of certain invasive species capable of reducing soil heavy metal concentrations by up to 90% within a year (Zhao et al. 2015). Various phytoremediation strategies—including phytoextraction, phytostabilization, rhizodegradation, and phytovolatilization—have been successfully implemented in diverse contaminated environments, from industrial sites to agricultural lands. While the problem of contaminated lands is a global challenge that requires concerted action by governments, industries, and individuals by harnessing the ecological strengths of invasive plants, phytoremediation can offer a cost-effective and sustainable solution for restoring degraded landscapes, particularly in areas where traditional remediation methods are financially or environmentally prohibitive. While the use of invasive species presents a promising opportunity for enhancing remediation efforts, it is crucial to manage their deployment carefully to mitigate potential ecological risks, such as biodiversity loss.

While advancements in environmental management practices continue to enhance the efficacy of phytoremediation, challenges such as site-specific condi-

N. Sahu (✉) · P. Garg
Department of Botany, Indira Gandhi National Tribal University, Amarkantak, Madhya Pradesh, India
e-mail: nayansahu@igntu.ac.in

V. C. Pandey (ed.), *Plantation-Based Land Restoration*,
https://doi.org/10.1007/978-981-95-2860-8_12

tions, invasive plant selection, survival rates, and public acceptance remain pertinent. Overall, phytoremediation presents a novel approach promising, eco-friendly solution for rehabilitating contaminated lands by integrating invasive plant species, contributing to food security, and balancing environmental recovery with economic feasibility on a global scale.

Keywords

Phytoremediation · Invasive plants · Allelopathy · Fly ash · Soil contamination

12.1 Introduction

Contaminated lands, a legacy of rapid industrialization, urbanization, mining, technological advancements, inadequate regulations, and unsustainable practices, pose a significant threat to human health and the environment worldwide. Contaminated lands can harm ecosystems, pollute water sources, and disrupt natural processes. These lands, polluted with hazardous substances, range from abandoned industrial sites and landfills to areas affected by pesticide runoff and improper waste disposal. The extent of contaminated lands globally is difficult to quantify precisely, but estimates suggest that millions of sites exist worldwide. The European Union is home to 3.42 million contaminated sites and an additional 2.5 million that are potentially contaminated (Panagos et al. 2013). Among European nations, the Netherlands stands out for its proactive approach in creating legislation focused on soil quality assessment and protection, but is still struggling to manage its 2.65 million potentially identified contaminated sites; of these, 11,000 require immediate remedial action (Horta et al. 2014). An estimated 1.6 million sites contaminated with hazardous substances might be lurking across the vast landscapes of Australia (State of the Environment Committee, 2011). Among these, locations tainted by total petroleum hydrocarbons (TPH) represent a significant share of the nation's polluted lands (Clements and Rohr, 2009). In the United States, the urgency of this issue is underscored by the presence of around 1200 sites on the National Priority List (NPL), awaiting the critical intervention of soil remediation (Mulligan et al. 2001; Horta et al., 2014). Alarmingly, about 63% of these NPL sites are burdened by contamination from toxic heavy metals, raising serious concerns about public health and environmental safety. These sites contain a wide array of contaminants, including heavy metals, persistent organic pollutants, radioactive materials, and petroleum hydrocarbons. The health and environmental risks associated with contaminated lands are substantial. As the dual pressures of food security and rapid urbanization mount, the quest to identify and remediate these contaminated sites has never been more crucial (Liu et al. 2013; Chen et al. 2007; Luo et al. 2012; Cai et al. 2008). Many of these polluted areas are being eyed as prime opportunities for redevelopment into high-value commercial and residential properties, yet their hidden dangers pose significant health and environmental risks. The pace of remediation, however, remains dishearteningly slow; in a sobering statistic from 33 European countries, only

58,000 out of 1.27 million recognized contaminated sites—merely 5%—have experienced any form of cleanup (Panagos et al. 2013). In Australia, the Cooperative Research Centre for Contamination Assessment and Remediation of the Environment (crcCARE) reveals that a mere 1000 sites are cleaned up each year, representing only about 0.5% of the total—a figure that reflects a nearly stagnant status quo, where the number of newly contaminated sites equals those being remediated. The staggering estimated costs for clean-up soar to around $2 billion each year, but the true expenses for thorough remediation are likely much more substantial (State of the Environment Committee, 2011). The staggering costs of cleaning up the contaminated lands have shifted the attention of the responsible authorities in recent years towards the utilization of invasive plant species, an innovative and cost-effective strategy for the remediation of contaminated lands. As urbanization and industrial activities increasingly threaten soil health and ecosystem integrity, traditional remediation methods, such as excavation and chemical treatments, can prove to be prohibitively expensive and environmentally disruptive (Rosenfeld et al. 2010). In contrast, certain invasive plant species exhibit unique physiological and ecological characteristics that enable them to thrive in contaminated environments, often facilitating the uptake and accumulation of pollutants (Zhao et al. 2015). These plants can fulfill multiple roles within the remediation process, particularly through phytoremediation, where they assist in the detoxification of contaminated soil by absorbing heavy metals and other toxic substances (Vangronsveld et al. 2009). Additionally, the growth of selected invasive species can enhance soil structure and health, subsequently promoting beneficial microbial activity that is crucial for natural bioremediation processes (Stefanowicz et al. 2017). The strategic selection of appropriate invasive species tailored to site-specific conditions may lead to improved remediation efficacy while simultaneously reducing costs associated with conventional methods. Moreover, the employment of invasive plants presents opportunities for landscape restoration and habitat creation following the completion of the cleanup process. However, it is imperative to consider the ecological risks associated with invasive species, as their unchecked proliferation may result in biodiversity loss and other ecological disruptions (Richardson et al. 2011). Nevertheless, when managed judiciously, invasive plant species can represent a sustainable and economically viable approach to addressing the critical issue of land contamination, yielding both environmental and financial benefits. A large portion of land and soil contamination is due to heavy metal contaminants that are poisonous and persistent and have the capacity to bioaccumulate in ecosystems; they pose a serious environmental risk. Zn, Cd, Pb, Hg, Cr, and Ar are among the toxic metals that are frequently detected in contaminated soil. The main sources of these pollutants include mining, agriculture, industrial processes, and inappropriate waste disposal. Heavy metals can persist in the soil for extended periods after their introduction, posing significant risks to the health of humans, animals, and plants. Research has shown that some heavy metals, when present in high quantities, can inhibit vegetative growth and decrease crop yield (Sipos et al. 2023; Ngayila et al. 2009; Rascio and Navari-Izzo 2011). Soil disturbances brought on by human activity frequently foster the growth of invasive plants. These disturbances, which include mining, deforestation, agriculture, and construction, upset natural ecosystems and lessen the capacity of nearby species to compete (IUCN 2012; Hess et al. 2019; Richardson et al. 2000; Daehler 2003). The application of plants or soil for lowering volume of toxic metals in the air, water, and soil is known as phytoremediation. Compared to typical remediation approaches, which can be costly and disruptive, this green technology is both economically practical and ecologically sustainable. Remarkably, because of their strong growth and high tolerance to contaminated environments, invasive alien species (IAS) have attracted attention for use in phytoremediation (Dominguez-Rosado and Pichtel 2004; Sipos et al. 2023). One such technological solution is phytoremediation, which involves decontamination (Leguizamo et al. 2017). Phytoremediation uses plants to absorb, collect, and detoxify pollutants, including heavy metals, providing a potential remedy for contaminated soil. Non-native species, sometimes known as invasive alien species (IAS), can enter new habitats in a variety of ways. Upon establishment, the invasive species have varied detrimental effects on the economy, ecosystem, and human health (Singh et al. 2021; Rai and Singh 2020). These species, which are frequently seen as an annoyance, can be effectively used to clean contaminated soils, converting an issue into a benefit. To ensure that the application of IAS in phytoremediation yields benefits to the economy and environment without aggravating ecological issues, however, cautious control is necessary to stop the unintentional spread of IAS (Rachmadiarti et al. 2022; Nouri et al. 2011). Phytoremediation, a technique that IASs may do, lessens the damage pollution does to ecosystems. IASs are able to store more heavy metals (HMs) in their vacuoles because of their special qualities, which lessen the HMs' harmful effects on nearby plants. Due to their rapid growth and longer root systems that enhance metal absorption, they are beneficial in removing dangerous heavy metals (HMs) from the environment. Furthermore, a range of biotechnological strategies are reportedly employed by invasive or native weeds to lessen the harmful effects of heavy metals on the environment (Khan et al. 2023). Due to their remarkable adaptability and resilience, invasive plants can swiftly take advantage of these disrupted settings and proliferate (Catford 2012; Lowe 2000; Hellmann 2008; Rejmánek 2015; Pyšek 2012; Wan and Wang 2018). Through outcompeting and displacing native plants, changing habitat architecture, and upsetting ecological processes, their rapid growth and spread can have a large negative impact on biodiversity. As a result, there may be a cascade impact on the ecosystem as a whole, harming animals, soil health, and water quality in addition to decreasing native biodiversity. Implementing effective management strategies is essential to mitigate the impacts of invasive species and safeguard biodiversity (Middleton 2019; Ehrenfeld 2010; Wei et al. 2018a, b).

12.2 Contaminated Sites

Contaminated places like soil, water, or air that has been polluted by hazardous substances lead to the degradation of ecosystems, loss of plant diversity, and contamination of water supplies. Production areas, dumps, landfills, waste treatment

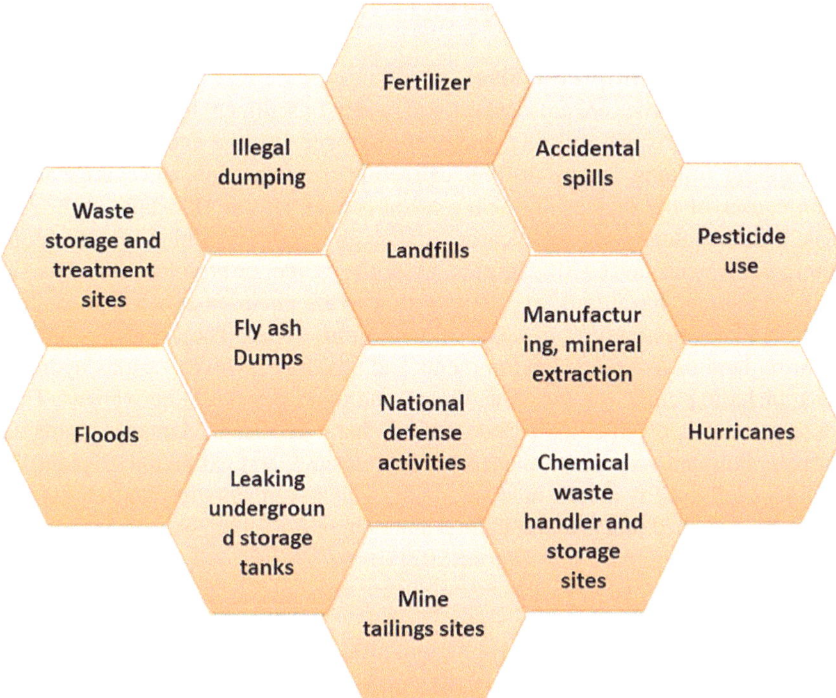

Fig. 12.1 Contaminated sites

and storage facilities, leaks and spills of chemicals or garbage tailings sites, spill sites, chemical waste, agriculture, mining, industry, commerce, handling and storage facilities situated in a variety of land uses and production regions such as these are examples of contaminated sites (EPA, 2020, 2023) (Fig. 12.1). Polluted locations can provide complex environmental and health issues. Hazardous waste and chemical spills have the potential to negatively influence the near environment, especially the soil, surface water, and water of ground. Contaminated sites often suffer from various forms of waste, including chemical, biological, and physical waste, arises from anthropogenic activities such as manufacturing operations, scoop out of metals from soil, and improper disposal practices (CPCB, 2022). Chemical waste, including heavy metals and toxic compounds, can persist in the environment for extended periods, posing ongoing risks to human health and ecosystems (Akhtar et al. 2020; Masud et al. 2023; Sarker et al. 2023; Sakshi et al. 2019; Jaishankar et al. 2014). Biological waste, including organic matter and pathogens, can also contribute to contamination, while physical waste, like plastics and debris, further degrades the site. Mining activities frequently lead to severe contamination, releasing hazardous substances into the soil and water (Srikanth 2019; Bali et al. 2021; Bayabil et al. 2022).

12.2.1 Types of Contaminated Sites

12.2.1.1 Oil Contaminated Sites

Oil-contaminated sites are areas where soil, water, or sediment has been polluted by petroleum products due to spills, leaks, or improper disposal and industrial activities. Mining, refining, and oil drilling, extraction, transportation, and storage are the main causes of the increase in heavy metal concentrations (Mustafa et al. 2015; EPA, 2023). Halogenated hydrocarbons, such as carbon tetrachloride, which contain inorganic substances like fluoride, chloride, iodide, or bromide, are often said to be dangerous, hazardous, and long-lasting in the environment (Klamerus-Iwan et al. 2015; Kuppusamy et al. 2017; Daâssi et al. 2022). The contamination can occur on land or in aquatic environments and typically involves various hydrocarbons, including crude oil, gasoline, diesel, and other petroleum derivatives. There are various sorts of oil fields based on their characteristics. It damages plants and animals. Polluted soil causes plant mutation, disease, and extinction of plant life; declining soil fertility results in lower yields, which eventually lead to a shortage of food, and poisoned plants cause animal and human starvation, as well as increased morbidity and mortality (Abu-Khasan 2021).

12.2.1.2 Mining Sites

Heavy metals, like lead, nickel, tin, titanium, zinc, arsenic, cadmium, chromium, copper, mercury, and nickel, are frequent contaminants from mining, especially during metallurgical extraction stages. These elements, even at low concentrations, can cause extensive pollution. Mining operations, processing and clustering, and open tailings discharge that is carried by wind and flood all contribute to environmental contamination. The areas surrounding mining operations usually have higher concentrations of heavy metals due to the release and disposal of waste materials in various forms, including elemental, organic, and inorganic substances (Karn et al. 2021). Mining waste is a major concern associated with the extractive industry. The chemical composition of mining deposits, often containing very high concentrations of metals, can inhibit plant growth (Nriagu and Pacyna 1988; Salomons 1995; Boularbah et al. 2006; Wong et al. 1998; Probst et al. 2009; El Khalil et al. 2008). The primary harmful substances produced during mining operations include lead, nickel, tin, titanium, zinc, arsenic, cadmium, copper, mercury, chromium, and nickel (Aguilar-Carrillo et al. 2018; Dragun et al. 2019; Terrones-Saeta et al. 2021). Heavy metals migration facilitates the pollution from minerals with acidic reaction activity (Yi and Huang 2019).

12.2.1.3 Underground Storage Tank

When there is a problem with contamination from an underground storage tank, local environmental agencies get involved. The most common source of pollution is underground leaking at petrol stations, which are a useful technique to store hazardous materials (Geo Forward, 2023). Subterranean storage tanks are used to store toxic items like spent fuel and oil. These materials include dangerous substances (Henke 1994). One serious environmental problem that is of international concern leaking underground fuel storage tanks (UFSTs) can have a negative impact on flora, groundwater, and drinking water, among other environmental concerns (Rosales et al. 2012). Fuel leakage from UFSTs, including gasoline and diesel oil, can contaminate drinking water and groundwater with a variety of hydrocarbon pollutants, potentially having a negative effect on the surroundings and human health (Alharbi et al. 2018).

12.2.1.4 Fly Ash Site

Globally, the production of fly ash is projected to exceed 600 million tons annually by 2015, which will require the use of up to 3235 square kilometers of land for its disposal (Pandey and Singh 2012). Fly ash is the fine residue produced after coal is burned in thermal power plants. It consists of finely ground particles that result from the combustion of coal in these facilities (Paswan et al. 2017; Dwivedi et al. 2014). Fly ash is generally considered harmful to living organisms. When coal is burned, heavy metals present in the coal can be released into the environment through flue gases (Yuan 2009). Research indicates that the concentration of heavy metals in fly ash is higher than in the feed coal. Fly ash consists of fine particles that can contain hazardous levels of metals, such as Ni, Cr, Cd, Pb, Hg, As, and Se (Verma et al. 2016). FA landfill is a hazardous waste disposal site (Pandey 2015). It is anticipated that over time, the environmental conditions of structures containing ash will improve as the ash ages (Brännvall and Kumpiene 2016; Gajic et al. 2018).

12.2.1.5 Agriculture Site

Heavy metal contamination of agricultural land can result from either a direct metal entry point or from the metals' slow migration and buildup in the soil. Heavy metal pollution of agricultural land can result from either a direct metal entry point or from the metals' slow migration and buildup in the soil. Many substances can contaminate agricultural soils, arising from point sources, like fertilizer and pesticide application, as well as diffuse sources of pollution, such as air deposition and flooding (FAO, 2021; EC, 2021). Anthropogenic and natural sources account for the majority of heavy metal contamination of agricultural land worldwide. The main human-caused sources of environmental release of arsenic, cadmium, copper, lead, and zinc are agricultural practices (Ross 1994; Official Gazette 2010; Radočaj et al. 2020).

12.2.1.6 Industrial Sites

Industrial waste refers to the byproducts generated from manufacturing or other industrial processes. This waste includes items such as cafeteria refuse, dirt and gravel, masonry and concrete, scrap metals, general junk, oil, solvents, chemicals, vegetation, timber, and scrap lumber. Industrial solid waste is classified into two types: hazardous and non-hazardous, and it can exist in solid, liquid, or gaseous form. Hazardous waste, produced during manufacturing and industrial activities, includes commercial materials like paints, insecticides, and cleaning solutions. Soil pollution occurs when harmful compounds—known as pollutants or contaminants—are present in the soil at levels high enough to endanger the environment or human health. Even at lower contaminant levels, soil pollution remains a significant concern (SDWF, 2017). As industries expand, they release an increasing number of contaminants into the environment. Industrial processes can generate hazardous waste, such as chemicals and heavy metals, which can significantly threaten human health and soil fertility. Industrial pollution involves the discharge of various contaminants that can harm the quality of soil, water, and air (EPE, 2023; Washington, 1981).

12.3 Sources of Contamination in Soil

Concern over the new class of pollutants in soil and water bodies has grown in recent years (Bolong et al. 2009; Houtman 2010; Bai et al. 2018, Bayabil et al. 2022). These are extremely hazardous and are found in trace amounts (Fig. 12.2). They are often released into soil and surface and groundwater bodies as a result of treated wastewater disposal (Loos et al. 2013; Petrie et al. 2015). Solids, dissolved or suspended particles, nutrients, and heavy metals are the most prevalent pollutants (Wuana and Okieimen, 2011; Gasser et al. 2014). Factors such as rapid industrialization, unplanned urbanization, and prolonged use of heavy fertilizers and pesticides contribute to this issue. Common hazardous contaminants include petroleum hydrocarbons, polynuclear aromatic hydrocarbons (like naphthalene and benzo(a) pyrene), solvents, pesticides, lead, and other heavy metals (Ghazi and Tawfiq 2018). Soil is one of the ultimate ecological destinations of polycyclic aromatic hydrocarbons (PAHs) (Agarwal 2009) known for their high rate of bioconcentration and quick inflow in the food chain due to their lipophilic and hydrophobic properties (Li et al. 2019; Yang et al. 2022). 16 PAHs are identified by the US Environmental Protection Agency (US EPA) as top contaminants (Cao et al. 2017; Li et al. 2019) that can stay in the soil for long periods (Agarwal 2009) not burning or breaking down very easily the water (Li et al. 2019). The PAH are mostly found in two denominations: one with ring numbers 5 or more (low solubility and low volatility, higher environmental presence) and PAH with low ring numbers (high bioavailability and degradation due to high water solubility) (Mackay and Callcott, 1998; Johnsen et al. 2005). In addition to soil contamination, PAH also shows carcinogenic properties (cyclopenta [cd]pyrene, dibenz [a,h] anthracene, and dibenzo [a,l] pyrene) (Van Poucke et al. 2020; Jameson, 2019).

12.3.1 Lead Paint

Paint and automobile exhaust are major sources of lead. Because of this, lead is now widely dispersed and persistent in urban environments (Body et al. 1991). Lead is probably present in certain paints made before 1978. Lead can leak into the ground (near homes or buildings) after color expiry or during removal by techniques, like stripping, scraping, or sandblasting. Lead concentrations in soil typically peak immediately next to a building and then gradually decline as one moves farther away from the source of contamination (Shayler et al. 2009).

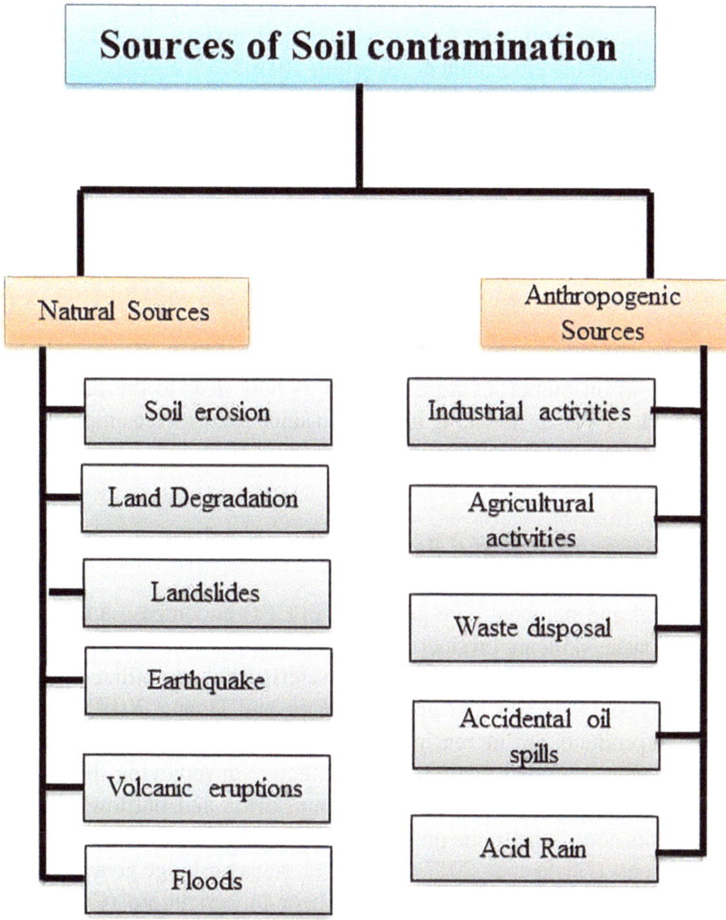

Fig. 12.2 Natural and anthropogenic sources of soil contamination

12.3.2 Pesticides

A pesticide is a substance that is used against insects, plant pathogens, and microorganisms. Besides their envisioned destinations >95% of pesticides wind up in the air, water, and soils. Pesticides are among the highly persistent chemicals in soil. The excessive use of these chemicals can lead to the formation of soil contaminants (Pullagurala et al. 2018). Nine pesticides (desthyl-atrazine (DEA), fipronil, chlorpyrifos, metolachlor, aminomethylphosphonic acid (AMPA), dieldrin, atrazine, desulfinyl, and glyphosate) and two pharmaceutical medications (caffeine, metformin) had detection rates ranging from 66% to 84%, according to a US Geological Survey (USGS) study on the surface water of 38 streams in the United States (Bradley et al. 2017). When these compounds are used excessively, pollutants may accumulate in the soil. Despite being prohibited, it has been shown that the usage of pesticides is rising in some places (Pan et al. 2019). Organochlorine pesticides (OCPs) are among the persistent organic pollutants known to have high toxicity, bioaccumulation, and biomagnification in the environment (Sparling 2016).

12.3.3 Industrial/Commercial Site Use

The precise compounds that may be involved owing to commercial or business activity will rely on the kind of industry and the certain techniques performed on site. If industrial or commercial operations are currently operating on or near the land, or might've taken place in the past, it could be helpful to explore what substances might have been utilized for a specific activity. The amount of contamination will depend on numerous aspects, such as how near to the land the specific event occurred, as well as how long it has been since toxins were employed (Shayler et al. 2009).

12.3.4 Pharmaceutical and Personal Care Products

Pharmaceutical and personal care products (PPCPs) encompass a range of items, such as toothpaste, skincare products, fragrances, antibiotics, and medications used for health and cosmetic purposes, as well as veterinary drugs utilized in agriculture to promote livestock health and growth (Wang and Huang 2019; Bishnoi et al. 2022). These products are increasingly prevalent in environments where conventional wastewater treatment facilities are ineffective at removing them (Wang and Wang 2016). Additionally, runoff from animal farms and outflows from sewage treatment plants contribute to the presence of emerging contaminants in aquatic and soil environments (Ebele et al. 2017). Moreover, sewage sludge generated by wastewater treatment is often repurposed as fertilizer in agriculture (Corradini 2014). While it contains beneficial nutrients such as potassium, nitrogen, manganese, and iron (Mtshali et al. 2014), treated sewage sludge may also harbor emerging contaminants, including antibiotics and engineered nanomaterials, which end up in soil (Koumaki et al. 2021). PPCPs are significant emerging contaminants in sewage sludge, posing potential ecological threats and health risks to humans when introduced into the soil ecosystem (Van et al. 2021). Due to the limited understanding of PPCPs' environmental effects, they are classified as contaminants of emerging concern (CECs) (Vasilachi et al. 2021). Another source of PPCPs in soils arises from treated wastewater or contaminated river water utilized for irrigation (Gallego et al. 2021). Soils often accumulate CECs with low hydrophobicity through interactions with organic material (Beltrán et al. 2020). A study conducted in Spain analyzed 166 emerging contaminants and heavy metals (e.g., Cd, Ni, Pb, and Hg) and detected 38 pharmaceuticals in low concentrations within tertiary treated wastewater (Cabeza et al. 2012). This aligns with findings from other research (Ziylan and Ince 2011), indicating that current wastewater treatment methods are insufficient for eliminating certain emerging contaminants. The inadequate removal of these substances, particularly pharmaceuticals, during treatment processes largely accounts for their presence in water bodies receiving treated wastewater (Kasprzyk-Hordern et al. 2008a, b).

12.3.5 Phthalate Esters

Phthalate esters (PAEs) are commonly utilized as additives to enhance the flexibility of certain polyvinyl chloride (PVC) resins, and they also find applications in other types of resins, like cellulose, vinyl acetate, and polyurethanes. These compounds are synthesized from phthalic anhydride and are integrated with plastics to enhance various properties, including resilience, plasticity, and clarity (Thomas and Brogat 2017). The end-user applications of these substances encompass resin manufacturing, agricultural adjuvants, cosmetics, soap and laundry detergents, toys, and several other uses. PAEs are characterized by their poor solubility in water, which significantly impacts their biodegradability, aquatic toxicity, and distribution in environmental settings. Despite their low aqueous solubility, PAEs can be readily absorbed by organic matter and solid surfaces within environmental systems. This slow and steady accumulation and subsequent release of these substances may influence the ecological health of aquatic environments. These processes particularly affect areas such as sludge-amended soils and wastewater treatment facilities (Staples et al. 1997). Due to their widespread use, PAEs have become ubiquitous, resulting in their accumulation across various ecological compartments. This buildup in agricultural soils poses risks of contaminating food chains and crops, which could lead to direct or indirect human exposure (Zeng et al. 2008).

12.3.6 Hormones

The advancement of industrial activities worldwide has positioned steroidal estrogen as a critical environmental concern. Research has demonstrated that these compounds can have severe repercussions for aquatic life and soil fertility (Singh et al. 2021). Steroidal hormones may be classified as either synthetic or naturally occurring estrogens, produced by the adrenal cortex and various other organs in both animals and humans (Biga et al. 2019). Many of these emerging contaminants, encompassing both synthetic and natural hormones, are identified as hormone disruptors (Preisendanz et al. 2021). Annually, human activities result in the discharge of approximately 30,000 kg of natural steroidal estrogens and 700 kg of synthetic estrogens, predominantly attributable to the usage of birth control pills. The excretion of estrogens from livestock contributes significantly to this issue, with estimates indicating an annual release of about 83,000 kg in the United States and the European Union (Adeel et al. 2017). The presence of natural estrogens in waste from both animals and humans poses a substantial environmental threat (Arnon et al. 2008), particularly in light of the increasing utilization of biosolids, such as animal manure, in organic farming practices (Xuan et al. 2008). A systematic study conducted by the US Geological Survey (USGS) and the Environmental Protection Agency (EPA) in 1999 and 2000 assessed 139 streams, revealing that 82 different chemicals were detected in 80 percent of these water bodies, with steroid hormones, antibiotics, and insect repellents identified as the most commonly occurring contaminants (USGS, 2002).

12.3.7 Petroleum Spills and Automobile Repair/Storage

Gasoline spills or leaks from fuel pumps, containers, or other sources can contain high levels of pollutants like toluene, xylene, and benzene. Many of these volatile compounds are unlikely to remain on the soil's surface as they are absorbed by vegetation or come into direct contact with humans, unless the spill is substantial. However, this does not apply to all toxins or spills, especially those from underground leaks, which can generate vapors that reach the surface soil. If the leak originates from a subterranean oil heating tank, surface soil contamination with these pollutants is less likely. Vehicle maintenance activities can result in unintentional spills or deliberate disposal of toxins into household or public soils. Many potential pollutants could be linked to such operations, such as fuels, PAHs, solvents like the chemical (TCE), recycled tires and rubber goods, metals as molybdenum, chromium, lead, or nickel, or used battery packs that release lead or mercury. Junk vehicles may potentially be a source of these substances or other toxins, depending on their state and where and how they are kept (Shayler et al. 2009).

12.3.8 Landfills

A municipal solid waste landfill (MSWLF) is defined as a designated area of land or excavation specifically designated for the deposition of household waste. Currently, millions of hectares worldwide are utilized as disposal sites for municipal solid waste. The Ghazipur Landfill in Delhi, India, ranks as one of Asia's largest landfills, with a height exceeding 73 m and covering an area of more than 100 acres. Other significant landfills include the Bantar Gebang Landfill in Indonesia (near Jakarta) and the Payatas Landfill in the Philippines (Quezon City, Manila). As per the published report for the year 2020–2021 by CPCB New Delhi, India, Delhi produced a total of 160,038.9 metric tons per day (TPD) of solid waste, with 29,427.2 TPD (18.4%) allocated to landfill disposal (Singh et al. 2024). The process of landfilling encompasses a complex array of biological, chemical, and physical interactions leading to waste degradation (Ma et al. 2022), which may involve toxic substances such as petroleum-based chemicals, herbicides, lead, and other heavy metals. These degradation processes result in the release of biogas and the leaching of materials from the landfill. Effective management of gas and leachate is essential to safeguard the ecosystem (Ren et al. 2022; Srivastava and Chakma, 2022). The presence of chemicals in soils adjacent to disposal sites, whether current or historical, is contingent upon the specific characteristics of each site and the types of materials disposed of therein (Shayler et al. 2009; Yaashika et al. 2022).

12.3.9 Fertilizers

Fertilizers are essential for modern agriculture, providing vital nutrients that support robust crop growth and help ensure bountiful harvests. However, it's important to recognize the challenges that accompany their use. The indiscriminate and long-term application of fertilizers has led to significant concerns about soil and water pollution (Hudak 2000; Hanson 2002; Almasri and Kaluarachchi 2004). Many farmers strive for healthy crops but may not be aware that excessive fertilizer use can lower soil pH, making harmful heavy metals (HMs) more available in the soil. Additionally, fertilizers can introduce various heavy metals from their raw materials into our ecosystems, contributing to the presence of copper, zinc, cadmium, and lead, as well as other persistent contaminants (PBTs) (Huang and Jin 2008). It's troubling that products derived from cement kiln dust may also carry heavy metals and dioxins, complicating the issue further. While many rely on animal manure and chemical fertilizers to improve soil health, it's unfortunate that these practices can sometimes elevate levels of certain contaminants. For instance, phosphate fertilizers often contain cadmium sourced from rock phosphate, and some manures can be unexpectedly high in copper or zinc (Shayler et al. 2009). Rainfall can sadly mobilize the nutrients found in fertilizers, leading to problems, like eutrophication in our precious water bodies. This condition fosters overgrowth of aquatic plants and harmful algal blooms, which can significantly deplete dissolved oxygen levels and threaten aquatic ecosystems. The soluble nutrients can wash away into groundwater through irrigation runoff, making it increasingly difficult to ensure safe drinking water for both people and livestock (Khan et al. 2018). Recent applications of nano-engineered fertilizers like nano urea might actually help mitigate heavy metal toxicity by binding to metals in the soil and reducing their bioavailability to plants; however, further research is needed to fully understand the potential risks and benefits of nano-fertilizers in contaminated environments. It is critical for us to find a balance that supports agricultural productivity while also protecting our environment and water resources for future generations.

12.4 Phytoremediation

Phytoremediation is a process of removing, breaking down, or immobilizing different chemicals from contaminated soils, sediments, groundwater, or surface water by using plants and the rhizospheric microbes that are linked with them. Despite its limitations, phytoremediation is a possible cleanup strategy for contaminated areas. Numerous drawbacks and benefits of phytoremediation can be traced back to the biologic component of this treatment (Marques et al. 2009).

12.4.1 Techniques of Phytoremediation for Contamination

Many phytoremediation technologies (Fig. 12.3) are available for removing heavy metals from the environment. Some plants can accumulate not just important elements like magnesium, Molybdenum, copper, iron, manganese, selenium, and zinc, which are required for plant growth, but also heavy metals like Pb, Hg, Cd, and Ag, which are acknowledged to have no biological role. When these HMs accumulate in large quantities, most flora may become toxic (Nedjimi 2021; Yan et al. 2020).

12.4.1.1 Phytostabilization

Microorganisms that dwell in the rhizosphere, like bacteria and mycorrhiza, can help in phytostabilization. Phytostabilization can assist contaminated lands in regaining their vegetative cover by utilizing plants that are resistant to the metal contamination (Pang et al. 2023; Yan et al. 2020). The plants' primary purpose is to prevent the toxic metal from moving to other areas. A robust root system prevents erosion and stabilizes the soil. The three parts of the phytostabilization mechanism are physical soil stabilization, metal precipitation in the rhizosphere, and metal absorption and accumulation by roots. To change them into nontoxic molecules, HMs and some organic pollutants can conjugate with results of sugar, protein, and amino acids or form compounds in the rhizosphere (Jian et al. 2011; Ghosh and Singh 2005; Hammond et al. 2018).

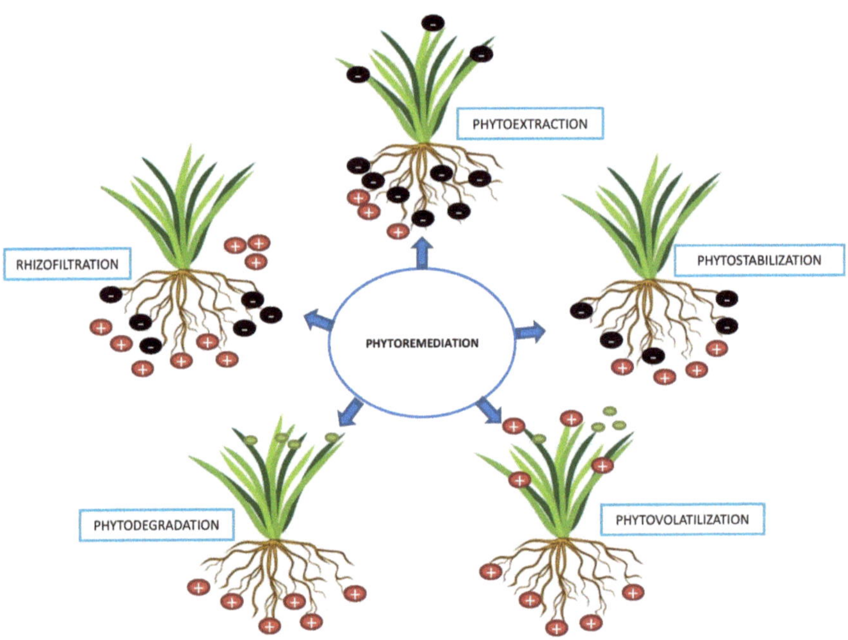

Fig. 12.3 Types of phytoremediation techniques

12.4.1.2 Phytoextraction

HMs are taken up through crop roots from the soil and transmitted to the aerial part of the plant during the phytoextraction process. It is a great and effective site remediation approach that reduces the mass of heavily polluted soil. The only aspect under control in plants is the number of times they grow; adding artificial chelates to improve the uptake of metals and their intrinsic ability for repair determines how much metal is eliminated (Pang et al. 2023; Ghosh and Singh 2005; Yan et al. 2020). Plants with fast development rates (high biomass) and deeper root systems are more adept at phytoextraction. Phytoextraction is a useful remediation technique for contaminated areas close to the surface because plants can only grow their root systems in the top layer of soil (Jian et al. 2011; Nedjimi 2021).

12.4.1.3 Phytovolatilization

Phytovolatilization is mostly focused on organic acid remediation process by which pollutants are absorbed and transformed into less dangerous volatile forms; it can also be applied to selenium (Se) and mercury (Hg) remediation. The main objective of phytovolatilization has been to eliminate mercury by turning the mercuric ion into elemental mercury, which is less harmful. Phytovolatilization is a process where plants take up pollutants from the soil, convert them into volatile compounds, and release them into the atmosphere through transpiration (Ghosh and Singh 2005; Pang et al. 2023; Nedjimi 2021). This technique requires very little maintenance (Jian et al. 2011). The volatilization approach is most effective; a plant absorbs and transpires a pollutant, and it releases the pollutant into the atmosphere (US EPA & TTE, 2001; Kafle et al. 2022).

12.4.1.4 Rhizofiltration

The process where pollutants are taken up by plant roots is known as phytoremediation. In this technique, plant roots either adsorb or absorb contaminants from the soil or surface water, aiding in the cleaning and remediation of these resources (US EPA & TTE, 2001; Jadia and Fulckar 2009). Rhizofiltration and phytoextraction operate through similar mechanisms (Pang et al. 2023; Yan et al. 2020). Nickel, zinc, lead, cadmium, copper, and chromium are among the elements that can be utilized, as they are mostly found in roots (Ghosh and Singh 2005). Rhizofiltration technique for removing contamination from water and liquid waste of large root surface areas and a fibrous root structure are important components of this approach (Nedjimi 2021).

12.4.1.5 Phytotransformation/Phytodegradation

Using their matching microbes, plants engage in enzymatic metabolism to break down pollutants into simpler molecules that are not harmful a process called phytodegradation or phytotransformation (Pang et al. 2023; Doty 2008; Nedjimi 2021; US EPA & TTE, 2001). These organic pollutants are taken up via passive absorption because plants lack active transporters. Rhizodegradation is used to describe the breakdown of pollutants in the rhizosphere. Enzymes found in plants are capable of breaking down and converting various herbicides, munitions wastes, and chlorinated solvents, like trichloroethylene (Ghosh and Singh 2005; Kafle et al. 2022).

12.5 Invasive Species for Phytomanagement

A higher rate of pollution uptake is typically observed in invasive plant species because of their competitive growth and adaptation. These species can act like bioremediation agents and frequently exhibit powerful heavy metal accumulation and tolerance capacities, enormous biomass, extensive adaptation, and fast growth, a great capacity to acquire metals, and tolerance and ability to survive in a variety of environmental circumstances (Fig. 12.4). These characteristics, which facilitate invasion, are also in charge of invasive plants' ability to effectively perform phytoremediation (Xue et al. 2022; Miletic et al. 2024a, b; Pilon 2005; Pandey and Singh 2011; Rachmadiarti et al. 2022, Pandey 2012; Singh et al. 2021). India hosts 1599 alien species across 842 genera and 161 families. Developed countries report 30 times more invasive plant species compared to low-income or developing nations (Seebens et al. 2018; Khuroo et al. 2011; Singh et al. 2021). Research has shown

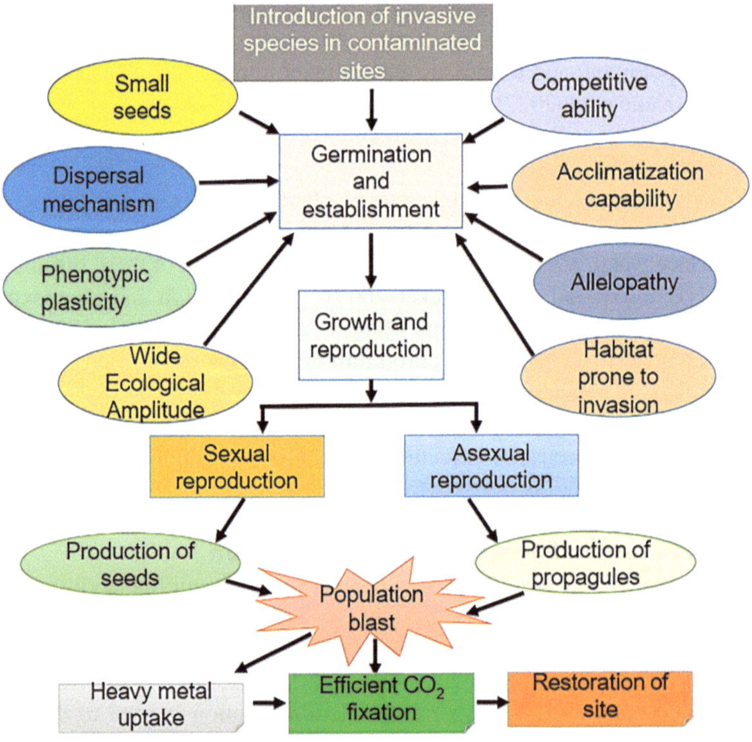

Fig. 12.4 Attributes of invasive plant species useful in the restoration of contaminated sites

that weeds are an excellent indicator of soil pollution and provide a variety of ecosystem services (Le Maitre et al. 2011, Ramírez-Satoyo et al. 2021, Sipos et al. 2023; Potgieter et al. 2019). Invasive plants are those that have been brought to an environment where they are not natural and are toxic to the environment, economy, or public health (Daehler 2003; IUCN 2012; Richardson et al. 2000; Prabakaran et al. 2019).

12.5.1 Characteristics of Invasive Species

Invasive species are exceptionally capable of infiltrating and establishing themselves in new ecosystems, primarily due to their remarkable ability to endure a variety of environmental conditions. They typically exhibit rapid growth and reproduction, gain distinct reproductive advantages, aggressively compete for available resources, and thrive without the constraints of natural predators or pests in their new surroundings. The success of an invasive species hinges on its ability to prosper and displace native species, encompassing both flora and fauna, within a specific locale. A significant indicator of a species' invasiveness is its potential to disrupt the existing ecological equilibrium (IUCN, 2013). Several key characteristics define invasive alien species (IAS):

12.5.1.1 Allelopathy

It is a phenomenon in which certain invasive plant species (IPS) release allelochemicals into their environment, which can inhibit or restrict the growth of surrounding native species. This process is particularly relevant to plants, given their stationary nature and dependence on soil as their growth medium. Consequently, allelopathy may contribute to the establishment of monocultures among IPS, thereby illustrating their potential for resource monopolization. The competitive interactions arising from allelopathic effects are known to enhance the invasiveness of invasive alien species (IAS) (Sheng and Bao 2006). For example, *P. hysterophorus* has been observed to release allelochemicals such as lactones and phenols into the soil through leaching, root exudation, and the decomposition of organic materials, which cause growth inhibition. These compounds are capable of significantly reducing the growth and yield of native plant species (Sankaran 2008). Moreover, the incorporation of excessive amounts of *Lantana camara* leaves has demonstrated detrimental effects on both the vegetative and reproductive development of rice plants (Wickramasinghe and Ranwala 2008).

12.5.1.2 Rapid Growth

One of the defining traits of invasive plant species is their ability to grow at an astonishing rate. This rapid growth is supported by specific adaptations that confer a competitive edge over indigenous plant species. Invasive plants often demonstrate features such as exceptional seed production volumes, quick germination rates, and the ability to thrive in disturbed environments, which collectively enhance their dispersal capabilities. For instance, species like *Solidago gigantea* (giant goldenrod) and *Impatiens* are particularly adept at flourishing in nutrient-rich, disturbed sites, allowing them to effectively outcompete native flora. Furthermore, invasive plants are known to alter soil nutrient dynamics, which can lead to ecological shifts that favor their own growth while hindering native species (Afreen et al. 2023; IUCN, 2013).

12.5.1.3 Extended Phenophases

A significant advantage for many invasive species is their extended periods of flowering and fruiting, which greatly enhances their pollination success. By producing flowers over long durations, these plants can increase their interactions with various pollinators, thereby boosting seed set rates. Additionally, longer fruiting periods can help sustain future generations during unfavorable environmental conditions, thus securing their survival. However, this adaptive strategy does carry associated reproductive costs that are critical to understanding its ecological impact (Ratnayake 2014).

Prolific Seed and Offspring Production Invasive species are known for their astonishing reproductive capacity, with prolific flowering leading to an abundance of seeds. Such high seed output allows them to establish numerous progenies across extensive areas. Many invasive plants are characterized by their ability to produce seeds that are easily dispersed by wind, animals, or other natural mechanisms. An exemplary case is *Parthenium hysterophorus*, commonly known as Congress grass, which can produce an astonishing 100,000 seeds in a single lifecycle. Furthermore, the surface soil in infested regions can contain over 340 million seeds per hectare, significantly enhancing the likelihood of successful colonization (Ratnayake 2014; Sankaran 2008; IUCN, 2013).

12.5.1.4 Short Life Cycle

Invasive species often possess shorter life cycles, enabling them to outpace native species in disturbed habitats. For example, *Parthenium hysterophorus* tends to decline rapidly, often within a month after establishing itself in a given area (Sankaran 2008). This life cycle pattern is typical among several invasive plants, including *Alternanthera philoxeroides* (alligator weed), *Eichhornia crassipes* (water hyacinth), and *Parthenium hysterophorus* (Ratnayake 2014).

12.5.1.5 Early Maturity

The ability to achieve early maturity is a significant advantage for invasive alien species, allowing them to produce seeds and establish dominance before native plants can do the same. Some annual IAS can reach reproductive maturity within just a few weeks after germination. This rapid progression is often linked to a shorter juvenile phase and a higher rate of photosynthesis, which promotes quick growth. For instance, *Chromalena odorata* reaches reproductive maturity swiftly, while *Leucaena leucocephala* starts flowering within a year of germination (Witkoeski and Wilson, 2001; Binggeli 1997).

12.5.1.6 Efficient Seed Dispersal Mechanisms

Many invasive plants utilize highly efficient seed dispersal mechanisms, allowing them to spread rapidly and occupy large areas. Among these methods, abiotic seed dispersion is notable for its energy efficiency, with many IAS showing adaptability to wind dispersal. Depending on the seed size and structure, these invasive species can achieve dispersal over varying distances. For instance, the seeds of *Prosopis juliflora* (mesquite) are known to be spread not only by livestock, such as cattle, but also by elephants that consume the pods. Additionally, seeds of *Panicum maxima* are disseminated by birds. Soil movement can also inadvertently facilitate the transfer of seeds for various invasive species. The seeds of *Parthenium hysterophorus* are primarily spread by animals, vehicles, and agricultural machinery, with wind playing a lesser role in their dispersal (Wijesundara 2010; Sankaran 2008). Consequently, the ability to utilize multiple modes of seed dispersal offers a distinct advantage for these IAS in their quest for effective proliferation (Ratnayake 2014).

12.5.1.7 High Plasticity and Adaptability to Contaminated Ecosystems

Generally, IAS exhibit a greater degree of phenotypic plasticity compared to their noninvasive counterparts. For instance, *Clidemia hirta* (Koster's curse) has shown remarkable adaptability in response to harsh soil nutrient conditions, demonstrating a more flexible phenotypic expression compared to co-occurring noninvasives (Weerasinge et al. 2008). *Parthenium hysterophorus* seeds can germinate over a temperature range from 8 degrees Celsius to 30 degrees Celsius (Sankaran 2008), enabling them to thrive in diverse environments. This species is notorious for its vigorous growth in various settings, including wastelands, neglected lands, orchards, forest edges, floodplains, agricultural fields, shrublands, urban areas, overgrazed pastures, and along roadways and railways. Furthermore, conditions such as drought—marked by reduced pasture cover—create particularly favorable environments for *Parthenium* to thrive (Sankaran 2008). *Mimosa pigra* showcases similar resilience, enduring both extended dry seasons of up to 7 months and periods of flooding (Marko 1999). Similarly, *Sphagneticola trilobata* displays a broad ecological tolerance, underscoring the adaptability and invasion potential of many invasive species. These characteristics allow them to effectively colonize and dominate a wide range of ecosystems. Table 12.1 lists invasive species currently utilized for phytoremediation of different heavy metals across different contaminated sites globally.

12.6 Case Studies and Applications

1. Pandey et al. (2012) screened four invasive plant species (*Typha latifolia, Saccharum spontaneum, Prosopis juliflora, and Ipomea carnea*) to assess their physiological tolerances to the harsh microclimatic and edaphic conditions found in fly ash deposits. Among these, *I. carnea* and *T. latifolia* were found as

Table 12.1 Invasive species for phytoremediation of different heavy metals

S. No.	Species name	Heavy metals	Contaminated site	Reference
1	*Ageratina adenophora* (Spreng.) R.M. King & H. Rob.	Cd, Cr	–	Dai et al. (2020)
2	*Ageratum conyzoides* L.	Zn, Pb, Cd	–	Zhu et al. (2017)
3	*Agrostis capillaris* L.	Multi-metal contaminants	–	Langella et al. (2014)
4	*Alternanthera philoxeroides* (Mart.) Griseb.	Pb, Cd	–	Yang et al. (2014)
5	*Amaranthus retroflexus* L.	Cu, Pb	–	Wang et al. (2020)
6	*Amaranthus spinosus* L.	Cd, Pb	–	Zhou et al. (2024)
7	*Amaranthus spinosus* L.	Pb, Cd	–	Chinmayee et al. (2012)
8	*Ambrosia artemisiifolia* L.	Zn, Cu, Cr, Pb, As	–	Laptiev et al. (2024)
9	*Bidens pilosa* L.	Cd	–	Sun et al. (2008)
10	*Bidens pilosa* L.	Cd, Zn	–	Zhu et al. (2018)
11	*Calotropis procera* (Aiton) Dryand.	Al, Cr, B, Ba, Cu, Mn, Fe, Pb, Zn, Cd, Ni, Pb	Fly ash site, cement factory	Abdullatif et al. (2016) and Pathak et al. (2019)
12	*Cassia tora* L.	Cr	–	Jena et al. (2016)
13	*Catharanthus roseus* (L.) G. Don	Cr	–	Ahmad and Misra (2014)
14	*Ceratophyllum demersum* L.	Ba, Cd, Cr, Cu, Fe, Mn, Ni	Oil spillage soil contamination	Atagana (2011) and Polechońska et al. (2018)
15	*Chromolaena odorata* (L.) R.M. King & H. Rob.	Cd	–	PlazaCazon et al. (2013)
16	*Chromolaena odorata* (L.) R.M. King & H. Rob.	Zn, Cd, Co, Fe, Mn, Cr, Cu, Ni, Pb	–	Wei et al. (2018)
17	*Cichorium intybus* L.	Pb	–	Pirzadah et al. (2020)
18	*Cyperus brevifolius* (Rottb.) Hassk.	–	Oil spillage soil contamination	Angulu et al. (2018)
19	*Cyperus rotundus* L.	–	Oil spillage soil contamination	Angulu et al. (2018)
20	*Diplachne fusca* (L.) P. Beauv. ex Roem. & Schult.	–	Saline soil	Jesus et al. (2015)

(continued)

Table 12.1 (continued)

S. No.	Species name	Heavy metals	Contaminated site	Reference
21	*Epipremnum aureum* (Linden & André) G.S. Bunting	–	Crude oil site	Tang and angela (2020)
22	*Egeria densa* (Planch.) Casp.	Se	–	Ohlbaum et al. (2018)
23	*Elodea Canadensis* Fraser ex Spach	Co	–	Mosoarca et al. (2018)
24	*Eucalyptus camaldulensis* Dehnh.	Cu	–	Hadiani et al. (2018)
25	*Fatsia japonica* (Thunb.) Decne. & Planch.	Cd, Zn, Cr, Pb	–	Sołtysiak (2020)
26	*Gmelina arborea* Roxb.	–	Oil spillage soil contamination	Agbogidi et al. (2007)
27	*Heliotropium curassavicum* L.	–	Saline soil	Ravindran et al. (2007)
28	*Impatiens balsamina* L.	–	Oil spillage soil contamination	Angulu et al. (2018)
29	*Impatiens glandulifera* Royle	B, Cd, Cu and Zn	–	Miletic et al. (2024a, b)
30	*Ipomoea aquatica* Forssk.	Pb	–	Bedabati (2016)
31	*Ipomoea carnea* Jacq.	Cd, Cr, Cu, Mn, Pb, Ni	Fly ash site	Pathak et al. (2019) and Singh et al. (2021)
32	*Jatropha gossypiifolia* L.	Fly ash site	–	Pathak et al. (2019)
33	*Lantana camara* L.	Cd	–	Singh et al. (2021)
34	*Lolium multiflorum* Lam.	–	Oil spillage soil contamination	Angulu et al. (2018)
35	*Lolium perenne* L.	–	Oil spillage soil contamination	Angulu et al. (2018)
36	*Mirabilis jalapa* L.	–	Oil spillage soil contamination	Peng et al. (2009)
37	*Neltuma juliflora* (Sw.) Raf.	Cd, Fe, Zn, Cu, Ni, Pb, and Cr	Fly ash site	Pandey et al. (2014)
38	*Nerium oleander* L.	Cd, Fe, Zn, Cu, Ni, Pb, and Cr	Fly ash site	Pathak et al. (2019)
39	*Myriophyllum aquaticum* (Vell.) Verdc.	Nutrients and organic matter	–	Souza et al. (2013)
40	*Nicotiana tabacum* L.	Cd, Zn	–	Nahar et al. (2017)
41	*Parthenium hysterophorus* Adans.	Pb, Ni, Cd	Fly ash site	Ahmad and Al-Othman, (2013) and Pathak et al. (2019)

(continued)

Table 12.1 (continued)

S. No.	Species name	Heavy metals	Contaminated site	Reference
42	*Paspalum distichum* L.	Hg	Petroleum hydrocarbons	Angulu et al. (2018) and Ding et al. (2019)
43	*Phragmites australis* (Cav.) Trin. ex Steud.	Pb, As	Oil-polluted water	Rahman and Hasegawa (2011) and Liu et al. (2017)
44	*Populus alba* L.	–	Terrestrial land site	Simakov (1957)
45	*Phragmites australis* (Cav.) Trin. ex Steud.	B	–	Podlipná et al. (2013)
46	*Pistia stratiotes* L.	Ag nanoparticles	–	Hanks et al. (2015)
47	*Pluchea indica* Less.	40K and 262Ra, and Pb	–	Prabakaran et al. (2019)
48	*Polygonum perfoliatum* L.	Mn	Petroleum hydrocarbons	Angulu et al. (2018) and Liu et al. (2020)
49	*Pontederia crassipes* Mart.	Cd	–	Zahari et al. (2021)
50	*Pontederia crassipes* Mart.	Fe, Pb, Cd, Co, Mn, Ni, Cr, Cu	–	Singh et al. (2021)
51	*Praxelis clematidea* R.M.King & H.Rob.	Cd	–	Wei et al. (2018)
52	*Pteris vittata* L.	Cd, Cr	–	Ma et al. (2001) and Tang and Angela (2019)
53	*Ranunculus sceleratus* L.	Cd, Cu, Co, Ni, Pb, Zn, Fe	–	Singh et al. (2021)
54	*Reynoutria japonica* Houtt.	Pb, Zn, and especially B,	–	Miletic et al. (2024a, b)
55	*Ricinus communis* L.	Ni, Cu, Mn, Fe	–	Kumar et al. (2021)
56	*Robinia pseudoacacia* L.	Cd, Fe, Zn, Cu, Ni, Pb, and Cr	Fly ash site	Kozuharova et al. (2017)
57	*Saccharum spontaneum* L.	Cd, Fe, Zn, Cu, Ni, Pb, and Cr	Fly ash site	Pandey et al. (2014)
58	*Scolochloa festucacea* (Willd.) Link	–	Oil spillage soil contamination	Angulu et al. (2018)
59	*Senna tora* (L.) Roxb.	Cd, Fe, Zn, Cu, Ni, Pb, and Cr	Fly ash site	Pathak et al. (2019)
60	*Sesuvium portulacastrum* (L.) L.	–	Saline soil	Rabhi et al. (2009)
61	*Sorghum bicolor* (L.) Moench	–	Saline soil	Jesus et al. (2015)
62	*Sagittaria montevidensis* Cham. & Schltdl.	Zn, Cu, Cr, Ni, Pb	–	Gomez et al. (2020)

(continued)

Table 12.1 (continued)

S. No.	Species name	Heavy metals	Contaminated site	Reference
63	*Solidago ser. Canadenses* Semple & J.B.Beck	B and Zn	–	Miletic et al. (2024a, b)
64	*Spartina alterniflora* (Loisel.) P.M.Peterson & Saarela	Pb, Fe, Cu, Cr, Zn	–	Prabakaran et al. (2019)
65	*Sporobolus montevidensis* (Arechav.) P.M. Peterson & Saarela	Co	–	Prabakaran et al. (2019)
66	*Tamarix smyrnensis* Bunge	Pb, Cd	Fly ash site	Kadukova et al. (2008) and Kosti'c et al. (2021)
67	*Tithonia diversifolia* (Hemsl.) A. Gray	Zn, Cd, Pb	–	Adesodum et al. (2010)
68	*Typha latifolia* L.	Cd, Zn, Cu	–	Pandey et al. (2014) and Bonanno and Cirelli (2017)
69	*Urochloa brizantha* (A.Rich.) R.D.Webster	–	Oil spillage soil contamination	Angulu et al. (2018)
70	*Xanthium strumarium* Lour.	Cd, Cr	–	Xue et al. (2021)

efficient phytoremediators of the pond region of the fly ash dyke, while *S. spontaneum* displayed the highest efficiency in sequestering CO_2, as evidenced by its maximum carboxylation rates and increased below-ground biomass. The species also demonstrated the best water-use efficiency attributed to its high carboxylation, along with the lowest transpiration and stomatal conductance values. These factors contributed to its dense growth throughout the fly ash dumps, unlike the other three species. This study offered valuable insights into leveraging invasive plant species for effective management and phytoremediation of fly ash deposits, transforming them into net carbon sinks, an area otherwise devoid of native vegetation.

2. Sipos et al. (2023) investigated the European weed species *Amaranthus retroflexus* and its capacity to accumulate metals in areas with mild and severe metal contamination. The study focused on metal deposition in the roots, stems, and leaves of the plant. The findings revealed that concentrations of copper (Cu), aluminum (Al), iron (Fe), chromium (Cr), and lead (Pb) were highest in the roots, while barium (Ba), manganese (Mn), strontium (Sr), and zinc (Zn) concentrations were highest in the leaves. *Amaranthus retroflexus*, a rapidly spreading weed throughout Europe, is particularly effective for phytoremediation and phytoextraction of toxic metals.

3. Rachmadiarti et al. (2022) examined the ability of *Salvinia molesta* to thrive in a lead-contaminated environment and absorb lead from its surroundings, indicating its potential as a phytoremediator for Pb-contaminated water. The study found that the plant's roots contain more lead than its leaves. Additionally, Pb-exposed plants exhibited a reduction in total proteins and various amino acids, suggesting a plant response mechanism to environmental changes. To better understand the phytoremediation potential of *S. molesta*, it is essential to enhance the concentration of heavy metals in the medium more rapidly and investigate the remediation mechanisms occurring within the plant.
4. Miletic et al. (2024a,b) investigated the ability of various plant species, including invasive species, like *Reynoutria japonica*, *Amorpha fruticosa*, *Impatiens glandulifera*, and *Solidago canadensis*. Along with invasive species, native species such as *Populus alba*, *Salix alba*, *Juglans regia*, *Ulmus glabra,* and *Populus nigra* can absorb and accumulate lithium (Li) and strontium (Sr), along the Sava River. The study focused on the phytoremediation and bio-indication potential of these plants. Samples were collected from riparian zones in Slovenia, Croatia, and Serbia, which were subject to various pollution sources. The researchers analyzed the physical and chemical properties of the soil and measured the lithium (Li) and strontium (Sr) content in the soil, as well as in the roots and leaves of selected plants. Invasive species, particularly *Amorpha fruticosa* and another species, demonstrated an exceptional ability to translocate Sr and Li, respectively, to their leaves. These findings provide valuable insights into the effectiveness of these plants for biomonitoring soil pollution and their potential use in phytoremediation efforts.
5. Trueman and Erber (2013) found that the non-native curly leaf pondweed (*Potamogeton illinoensis* Morong) assembled, on average, 66% more estrogenic compounds and 94% more Diphenylpropane compared to the native Illinois pondweed (*Potamogeton crispus* L.). Specifically, the non-native species collected 76%, 55%, and 31% more estrone, estradiol, and ethynylestradiol, respectively, compared to the indigenous species. Additionally, the non-indigenous plants accumulated 72% more of these compounds overall compared to the native Illinois pondweed. These findings suggest that invasive species can be utilized by ecosystem managers to remove contaminants and help restore the biogeochemistry of affected ecosystems.
6. Rai (2021) addressed the issue of metallic pollutants by examining the heavy metal content and phytoremediation capacity of invasive alien plants (IAP) such as *Phragmites karka* (Retz.), *Arundo donax* L., and *Trin. ex Steud.* through microcosm experiments. The study highlighted the antioxidative defense potential and variable effectiveness of these IAPs in remediating heavy metals. *Phragmites karka* exhibited the highest translocation factor (TF = 0.89) for zinc (Zn). The primary phytoremediation method observed was rhizofiltration, characterized by a TF of less than 1. The results indicated that *Phragmites karka* was a more effective phytoremediation tool compared to *Arundo donax* in the experimental setup.

12.7 Challenges in Application Challenges of Invasive Alien Species

The usage of non-native plant species to bioremediate toxic soils has a number of drawbacks, including growth periods, improper disposal, loss of native floral diversity, and more. Although using non-native species for human wellness may be good, native or endemic species' survival is affected by certain functional characteristics related to function of physiology, accumulation rate of biomass, fast growth rate, brief life span, and strong homeostatic capacity to tolerate stress circumstances (Colautti et al. 2017). When a plant is employed in phytoremediation, it should have a few such as high biomass growth at a rapid rate, an unpleasant nature, an abundant root system, and a hyperaccumulation of the target toxins paired with stress-tolerant traits that can endure a variety of environmental circumstances. Unfortunately, the toxic effects of the chemicals present in contaminated locations prevent many plants from growing, which makes invading species the better candidates for phytoremediation. When it comes to restoration, biological incursions constitute an ecosystem-level issue. Exotic invasive species hinder the successful restoration of many natural habitats by causing ecological changes that could be permanent (Doren et al. 2009). To properly manage invasive plants' use within a sustainable development framework, it is essential to focus on their benefits rather than solely on their drawbacks. Interfering species, however, are tough to eradicate once they have settled into a new ecosystem. Even so, they either resist them from getting stable (Lu and Zhang 2013; Chen et al. 2016; Prabakaran et al. 2019). Transport of products between nations during export and import is one of the key drivers of the spread of interfering species. Material screening is a crucial first step in solving this issue. It is evident from the foregoing that invasive alien species degrade the native species in the habitat. The aforementioned traits of interfering species enable them to adapt to environmental, physiological, and climatic conditions. These traits include the ability to withstand various diseases, extreme heat or cold, drought or floods, modifications to their silvicultural practices, and various human-oriented chemicals. Evaluate the toxicity of using interfering species, make decisions about their use, and comprehend potential hazards to the ecosystem, potential management risks, and other resulting issues; a number of measures can be followed (Singh et al. 2021).

12.8 Conclusion

Over many years, the issue of invasive species has been hotly debated. While there have been discussions about how plant invasions affect ecosystems and homogenize the world's flora, there are many cases where plant invasions have been proven to be advantageous because many invasive species are deliberately planted to serve horticultural and agricultural purposes. Furthermore, once invasive plants establish themselves in a naturalized area, they are opportunists that cannot be completely eliminated. It is therefore crucial to assess its ability to offer ecologically sustainable services for effective overall ecosystem management under current conditions.

Given that invasive plants may thrive in environments damaged by a range of pollutants and are typically tolerant of a wide range of stresses, phytoremediation of toxins by invasive plants is proven to be a workable approach. The literature search makes it clear that the majority of invasive plant-based pollution remediation trials have been determined to be effective; as a result, invading plants may be taken into consideration as viable options for the botanical rehabilitation of contaminated land or wasteland. It is well acknowledged that using invasive plants shouldn't be approached passively and that appropriate management techniques must be used to prevent invasive plants from spreading too far into untargeted regions (Prabakaran et al. 2019). However, invasive plants can be utilized extensively in the making of biofuels, animal nutrition, charcoal, medicines, pesticides, and other products. Several years ago, advocates for plant preservation attempted to maintain native plant species through a variety of means, including the retention, restoration, and development of multiple plans for quick recovery in areas vulnerable to fire, flooding, and drought. But because invading plants can better adapt to biotic or abiotic difficulties, scientists these days concentrate on using them for sustainable care and exploitation. Given the opposite impacts of non-indigenous species on agricultural landscapes, woodlands, and wastelands, it's crucial to develop effective methods for commercially utilizing these plants in a way that does not harm biodiversity. Furthermore, the application of scientific methodology, societal connections, and the involvement of decision support frame workers should be used in the control of IAS. Even though IAS are dangerous for biosphere, they can also be used for a variety of social purposes and may even be able to help clean up areas that have been poisoned by heavy metals (Singh et al. 2021).

References

Abu-Khasan M (2021) Analysis of soil contamination with oil and petroleum products. Springer International Publishing. International Scientific Publications Siberian Transport Forum, Cham, pp 1514–1522. https://doi.org/10.1088/1755-1315/937/2/022046

Abdullatif M (2016) Phytoremediation ability of Calotropis procera in reducing air pollution in Jeddah City, Kingdom of Saudi Arabia. Int J Curr Microbiol Appl Sci 5:289–297. (check pagination, issue varies)

Adeel M, Song X, Wang Y, Francis D, Yang Y (2017) Environmental impact of estrogens on human, animal and plant life: a critical review. Environ Int 99:107–119

Adesodun JK, Atayese MO, Agbaje TA, Osadiaye BA, Mafe OF, Soretire AA (2010) Phytoremediation potentials of sunflowers (Tithonia diversifolia and Helianthus annuus) for metals in soils contaminated with zinc and lead nitrates. Water Air Soil Pollut 207:195–201. https://doi.org/10.1007/s11270-009-0128-0

Afreen T, Kumari S, Bhadouria R, Devi RS, Singh S, Tripathi S (2023) Plant invasion and soil processes: a mechanistic understanding. In: Tripathi S, Singh A (eds) Plant invasions and global climate change. Springer, Singapore, pp 227–246. https://doi.org/10.1007/978-981-99-0367-6_12

Agarwal T (2009) Concentration level, pattern and toxic potential of PAHs in traffic soil of Delhi, India. J Hazard Mater 171:894–900. https://doi.org/10.1016/j.jhazmat.2009.06.073

Agbogidi MO, Dolor DE, Okechukwu ME (2007) Evaluation of Tectona grandis (Linn.) and Gmelina arborea (Roxb.) for phytoremediation in crude oil contaminated soils. Agric Conspec Sci 72:149–152

Aguilar-Carrillo J, Herrera L, Gutiérrez EJ, Reyes-Domínguez IA (2018) Solid-phase distribution and mobility of thallium in mining-metallurgical residues: environmental hazard implications. Environ Pollut 243:1833–1845. https://doi.org/10.1016/j.envpol.2018.10.014
Ahmad R, Misra N (2014) Evaluation of phytoremediation potential of Catharanthusroseus with respect to chromium contamination. Am J Plant Sci 2014
Akhtar FZ, Archana KM, Krishnaswamy VG, Rajagopal R (2020) Remediation of heavy metals (Cr, Zn) using physical, chemical and biological methods: a novel approach. SN Appl Sci 2:1–14
Alharbi BH, Pasha MJ, Alhudhodi AH, Alduwais AK (2018) Assessment of soil contamination caused by underground fuel leakage from selected gas stations in Riyadh, Saudi Arabia. Soil Sediment Contam Int J 27(8):674–691
Almasri MN, Kaluarachchi JJ (2004) Assessment and management of long-term nitrate pollution of groundwater in agriculture-dominated watersheds. J Hydrol 295:225–245. https://doi.org/10.1016/j.jhydrol.2004.03.013
Al Masud MA, Shin WS, Sarker A, Septian A, Das K, Deepo DM et al (2023) A critical review of sustainable application of biochar for green remediation: research uncertainty and future directions. Sci Total Environ 16:166813
Angulu C, Atuluku S, Oladoja E, Balogun A, Attah-Olottah R, Adeniyi K (2018) Application of phytoremediation in the management of oil spillage: a review. Global J Earth Environ Sci 3(3):16–22
Arnon S, Dahan O, Elhanany S, Cohen K, Pankratov I, Gross A et al (2008) Transport of testosterone and estrogen from dairy-farm waste lagoons to groundwater. Environ Sci Technol 42(15):5521–5526
Atagana HI (2011) Bioremediation of co-contamination of crude oil and heavy metals in soil by phytoremediation using Chromolaenaodorata (L) King & HE Robinson. Water Air Soil Pollut 215:261–271
Bai L, Wang J, Ma X, Lu H (2018) Air pollution forecasts: an overview. Int J Environ Res Public Health 15:780. https://doi.org/10.3390/ijerph15040780
Bali AS, Sidhu GPS, Kumar V (2021) Plant enzymes in metabolism of organic pollutants. In: Hasanuzzaman M, Prasad MNV (eds) Handbook of bioremediation. Academic Press, pp 465–474. https://doi.org/10.1016/B978-0-12-819382-2.00029-6
Bayabil HK, Teshome FT, Li YC (2022) Emerging contaminants in soil and water. Front Environ Sci 10:873499
Bedabati GA (2016) Phytoremediation of lead using Ipomoea aquatica Forsk. in hydroponic solution. Chemosphere 156:407–411
Beltrán Rodríguez EM, Pablos MV, Fernández-Torija C, Porcel MA, González Doncel M (2020) Uptake of atenolol, carbamazepine and triclosan by crops irrigated with reclaimed water in a Mediterranean scenario, vol 191. Departamento de Medio Ambiente y Agronomía
Biga LM, Bronson S, Dawson S, Harwell A, Hopkins R, Kaufmann J et al (2019) 17.2 hormones. Anatomy & Physiology
Binggeli P (1997) Leucaena leucocephala (Lam.) de Wit (Mimosaceae). University of Wales, Bangor. (online monograph, source details to be verified)
Bishnoi MM, Verma A, Kushwaha A, Goswami S (2022) Social factors influencing household waste management. In: Emerging trends to approaching zero waste. Elsevier, pp 197–213
Body PE, Inglis G, Dolan PR, Mulcahy DE (1991) Environmental lead: a review. Crit Rev Environ Sci Technol 20(5–6):299–310
Bolong N, Ismail AF, Salim MR, Matsuura T (2009) A review of the effects of emerging contaminants in wastewater and options for their removal. Desalination 239(1–3):229–246
Bonanno G, Cirelli GL (2017) Comparative analysis of element concentrations and translocation in three wetland congener plants: Typha domingensis, Typha latifolia and Typha angustifolia. Ecotoxicol Environ Saf 143:92–101. https://doi.org/10.1016/j.ecoenv.2017.05.015
Boularbah A, Schwartz C, Bitton G, Aboudrar W, Ouhammou A, Morel JL (2006) Heavy metal contamination from mining sites in South Morocco: 2. Assessment of metal accu-

mulation and toxicity in plants. Chemosphere 63(5):811–817. https://doi.org/10.1016/j.chemosphere.2005.07.076

Bradley PM, Journey CA, Romanok KM, Barber LB, Buxton HT, Foreman WT et al (2017) Expanded target-chemical analysis reveals extensive mixed-organic-contaminant exposure in US streams. Environ Sci Technol 51(9):4792–4802

Brännvall E, Kumpiene J (2016) Fly ash in landfill top covers—a review. Environ Sci Process Impact 18(1):11–21. https://doi.org/10.1039/c5em00419e

Cabeza Y, Candela L, Ronen D, Teijon G (2012) Monitoring the occurrence of emerging contaminants in treated wastewater and groundwater between 2008 and 2010. The Baix Llobregat (Barcelona, Spain). J Hazard Mater 239:32–39

Cai QY, Mo CH, Zeng QY, Wu QT, Férard JF, Antizar-Ladislao B (2008) Potential of Ipomoea aquatica cultivars in phytoremediation of soils contaminated with di-n-butyl phthalate. Environ Exp Bot 62:205–211. https://doi.org/10.1016/j.envexpbot.2007.08.002

Cao H, Chao S, Qiao L, Jiang Y, Zeng X, Fan X (2017) Urbanization-related changes in soil PAHs and potential health risks of emission sources in a township in Southern Jiangsu, China. Sci Total Environ 575:692–700. https://doi.org/10.1016/j.scitotenv.2016.09.139

Catford JA, Vesk PA, Richardson DM, Pyšek P (2012) Quantifying levels of biological invasion: towards the objective classification of invaded and invasible ecosystems. Glob Chang Biol 18(1):44–62

Central Pollution control board waste management division-I report, 2022. https://cpcb.nic.in/uploads/hwmd/Brief_Contmainated_sites_in_india.pdf

Chen C, Huang D, Wang Q, Wu J, Wang K (2016) Invasions by alien plant species of the agro-pastoral ecotone in northern China: species-specific and environmental determinants. J Nat Conserv 34:133–144

Chen A, Lin C, Lu W, Wu Y, Ma Y, Li J, Zhu L (2007) Well water contaminated by acidic mine water from the Dabaoshan Mine, South China: chemistry and toxicity. Chemosphere 70:248–255. https://doi.org/10.1016/j.chemosphere.2007.06.047

Chinmayee MD, Mahesh B, Pradesh S, Mini I, Swapna TS (2012) The assessment of phytoremediation potential of invasive weed Amaranthusspinosus L. Appl Biochem Biotechnol 167:1550–1559

Clements WH, Rohr JR (2009) Community responses to contaminants: using basic ecological principles to predict ecotoxicological effects. Environ Toxicol Chem 28:1789–1800. https://doi.org/10.1897/09-140.1

Colautti RI, Alexander JM, Dlugosch KM, Keller SR, Sultan SE (2017) Invasions and extinctions through the looking glass of evolutionary ecology. Philos Trans R Soc B Biol Sci 372(1712):20160031. https://doi.org/10.1098/rstb.2016.0031

Corradini C (2014) Soil moisture in the development of hydrological processes and its determination at different spatial scales. J Hydrol 516:1–5

Council on Environmental Quality, & Executive Office of the President (July 1981). Environment trends p. 84. US Government Printing Office

Daâssi D, Qabil Almaghribi F (2022) Petroleum-contaminated soil: environmental occurrence and remediation strategies. 3. Biotech 12(6):139. https://doi.org/10.1007/s13205-022-03198-z

Daehler CC (2003) Performance comparisons of co-occurring native and alien invasive plants: implications for conservation and restoration. Annu Rev Ecol Evol Syst 34(1):183–211

Dai ZC, Cai HH, Qi SS, Li J, Zhai DL, Wan JSH, Du DL (2020) Cadmium hyperaccumulation as an inexpensive metal armor against disease in Crofton weed. Environ Pollut 267:115649

Ding W, Zhang J, Wu SC, Zhang S, Christie P, Liang P (2019) Responses of the grass Paspalumdistichum L. to Hg stress: a proteomic study. Ecotoxicol Environ Saf 183:109549

Dominguez-Rosado E, Pichtel J (2004) Phytoremediation of soil contaminated with used motor oil: II. Greenhouse studies. Environ Eng Sci 21:169–180. https://doi.org/10.1089/109287504773087345

Doren RF, Richards JH, Volin JC (2009) A conceptual ecological model to facilitate understanding the role of invasive species in large-scale ecosystem restoration. Ecol Indic 9(6):S150–S160

Doty SL (2008) Enhancing phytoremediation through the use of transgenics and endophytes. New Phytol 179:318–333. https://doi.org/10.1111/j.1469-8137.2008.02446.x
Dragun Z, Tepić N, Ramani S, Krasnići N, Filipović Marijić V, Valić D, Kapetanović D, Erk M, Rebok K, Kostov V et al (2019) Mining waste as a cause of increased bioaccumulation of highly toxic metals in liver and gills of Vardar chub (Squalius vardarensis Karaman, 1928). Environ Pollut 247:564–576. https://doi.org/10.1016/j.envpol.2019.01.068
Dwivedi A, Jain MK (2014) Fly ash–waste management and overview: a review. Recent Res Sci Technol 6(1)
Ebele AJ, Abdallah MAE, Harrad S (2017) Pharmaceuticals and personal care products (PPCPs) in the freshwater aquatic environment. Emerg Contaminant 3(1):1–16
Ehrenfeld JG (2010) Ecosystem consequences of biological invasions. Annu Rev Ecol Evol Syst 41(1):59–80
El Khalil H, El Hamiani O, Bitton G, Ouazzani N, Boularbah A (2008) Heavy metal contamination from mining sites in South Morocco: monitoring metal content and toxicity of soil runoff and groundwater. Environ Monit Assess 136(1–3):147–160. https://doi.org/10.1007/s10661-007-9671-9
Environment pollution education, 2023. https://pollutioneducation.com
Environmental Protection Agency 2020. https://www.epa.vic.gov.au/about-epa/laws/laws-and-your-business/manage-contaminated-land/about-contamination/what-causes-contamination
Environmental Protection Agency 2023. https://www.epa.vic.gov.au/about-epa/annual-report
European Commission (EC) (2021) Roadmap to a new soil strategy—healthy soil for a healthy life. Author
FAO and UNEP (2021) Global assessment of soil pollution: report, Rome
Gajić G, Djurdjević L, Kostić O, Jarić S, Mitrović M, Pavlović P (2018) Ecological potential of plants for phytoremediation and ecorestoration of fly ash deposits and mine wastes. Front Environ Sci 6:124
Gallego S, Montemurro N, Béguet J, Rouard N, Philippot L, Pérez S, Martin-Laurent F (2021) Ecotoxicological risk assessment of wastewater irrigation on soil microorganisms: fate and impact of wastewater-borne micropollutants in lettuce-soil system. Ecotoxicol Environ Saf 223:112595
Gasser CA, Ammann EM, Shahgaldian P, Corvini PFX (2014) Laccases to take on the challenge of emerging organic contaminants in wastewater. Appl Microbiol Biotechnol 98:9931–9952. https://doi.org/10.1007/s00253-014-6176-1
Ghazi FF, Tawfiq LNM (2018) Mathematical models for contamination soil. Lap Lambert Acadmic Publishing, Mauritius
Ghosh M, Singh SP (2005) A review on phytoremediation of heavy metals and utilization of it's by products. Asian J Energy Environ 6(4). https://doi.org/10.15666/aeer/0301_001018
Gomez BM, Reale M, El Kassisse Y, Mujica C, Gómez C, de Cabo L, Rodríguez Salemi V (2020) Metals uptake by Sagittariamontevidensis in contaminated riparian area of Matanza-Riachuelo River (Argentina). SN Appl Sci 2:1–10
Hadiani MR, Darani KK, Rahimifard N, Younesi H (2018) Biosorption of low concentration levels of lead (II) and cadmium (II) from aqueous solution by Saccharomyces cerevisiae: response surface methodology. Biocatal Agric Biotechnol 15:25–34. https://doi.org/10.1016/j.bcab.2018.05.002
Hammond CM, Root RA, Maier RM, Chorover J (2018) Mechanisms of arsenic sequestration by Prosopis juliflora during the phytostabilization of metalliferous mine tailings. Environ Sci Technol 52(3):1156–1164
Hanks NA, Caruso JA, Zhang P (2015) Assessing Pistiastratiotes for phytoremediation of silver nanoparticles and Ag (I) contaminated waters. J Environ Manag 164:41–45
Hanson CR (2002) Nitrate concentrations in Canterbury groundwater. Canterbury Reg Counc Rep 2:1–20. (grey literature report)
Hellmann JJ, Byers JE, Bierwagen BG, Dukes JS (2008) Five potential consequences of climate change for invasive species. Conserv Biol 22(3):534–543

Henke S (1994) Underground storage tanks: the environmental health role. J Environ Health:13–15
Hess MCM, Mesléard F, Buisson E (2019) Priority effects: emerging principles for invasive plant species management. Ecol Eng 127:48–57. https://doi.org/10.4060/cb4894en
Horta NC, Moreira DA, Aguiar KT, Morais JL (2014) Potenciais de desgaste biopsíquico nos profissionais de Enfermagem que atuam na Estratégia de Saúde da Família. Percurso Acadêmico 4:105–119
Houtman CJ (2010) Emerging contaminants in surface waters and their relevance for the production of drinking water in Europe. J Integr Environ Sci 7:271–295. https://doi.org/10.1080/1943815X.2010.511648
Huang SW, Jin JY (2008) Status of heavy metals in agricultural soils as affected by different patterns of land use. Environ Monit Assess 139:317–327. https://doi.org/10.1007/s10661-007-9838-4
Hudak PF (2000) Regional trends in nitrate content of Texas groundwater. J Hydrol 228:37–47. https://doi.org/10.1016/S0022-1694(99)00211-5
IUCN (2012) International Union for Conservation of Nature and Natural Resources
Jadia CD, Fulekar MH (2009) Phytoremediation of heavy metals: recent techniques. Afr J Biotechnol 8(6)
Jaishankar M, Tseten T, Anbalagan N, Mathew BB, Beeregowda KN (2014) Toxicity, mechanism and health effects of some heavy metals. Interdiscip Toxicol 7:60–72. https://doi.org/10.2478/intox-2014-0009
Jameson CW (2019) Polycyclic aromatic hydrocarbons and associated occupational exposures. In: Baan RA, Stewart BW, Straif K (eds) Tumour site concordance and mechanisms of carcinogenesis. IARC Sci Publ, vol 165. IARC, Lyon, pp 59–64
Jena P, Pradhan C, Patra HK (2016) Cr+ 6 induced growth, biochemical alterations and chromium bioaccumulation in Cassia tora (L.) Roxb. Ann. Plant Sci 5(7):1368–1373
Jesus JM, Danko AS, Fiúza A, Borges MT (2015) Phytoremediation of salt-affected soils: a review of processes, applicability, and the impact of climate change. Environ Sci Pollut Res 22:6511–6525. https://doi.org/10.1007/s11356-015-4205-4
Jian C, Yang Z, Su Y, Han FX, Monts DL (2011) Phytoremediation of heavy metal/metalloid-contaminated soils. Contaminated soils R. V. Steinberg (Ed.)
Johnsen AR, Wick LY, Harms H (2005) Principles of microbial PAH-degradation in soil. Environ Pollut 133:71–84. https://doi.org/10.1016/j.envpol.2004.04.015
Kadukova J, Manousaki E, Kalogerakis N (2008) Pb and Cd accumulation and phyto-excretion by salt cedar (Tamarixsmyrnensis Bunge). Int J Phytoremediation 10(1):31–46
Kafle A, Timilsina A, Gautam A, Adhikari K, Bhattarai A, Aryal N (2022) Phytoremediation: mechanisms, plant selection and enhancement by natural and synthetic agents. Environ Adv 8:100203. https://doi.org/10.1016/j.envadv.2022.100203
Karn R, Ojha N, Abbas S, Bhugra S (2021) A review on heavy metal contamination at mining sites and remedial techniques. In IOP Conference Series. IOP Conference Series: Earth and Environmental Science (Vol. 796, No. 1, p. 012013). IOP Publishing, 796(1). https://doi.org/10.1088/1755-1315/796/1/012013
Kasprzyk-Hordern B, Dinsdale RM, Guwy AJ (2008a) Multiresidue methods for the analysis of pharmaceuticals, personal care products and illicit drugs in surface water and wastewater by solid-phase extraction and ultra performance liquid chromatography–electrospray tandem mass spectrometry. Anal Bioanal Chem 391:1293–1308
Kasprzyk-Hordern B, Dinsdale RM, Guwy AJ (2008b) The occurrence of pharmaceuticals, personal care products, endocrine disruptors and illicit drugs in surface water in South Wales. Water Res 42(13):3498–3518
Khan MN, Mobin M, Abbas ZK, Alamri SA (2018) Fertilizers and their contaminants in soils, surface and groundwater. Encyclopedia of the Anthropocene 5:225–240
Khan IU, Qi SS, Gul F, Manan S, Rono JK, Naz M et al (2023) A green approach used for heavy metals 'phytoremediation' via invasive plant species to mitigate environmental pollution: a review. Plants 12(4):725
Khuroo AA, Reshi ZA, Weber E, Reshid I (2011) Alien flora of India: taxonomic composition, invasion status and biogeographical affiliations. Biol Invasions 14:99–113. https://doi.org/10.1007/s10530-011-9981-2

Klamerus-Iwan A, Błońska E, Lasota J, Kalandyk A, Waligórski P (2015) Influence of oil contamination on physical and biological properties of forest soil after chainsaw use. Water Air Soil Pollut 226:1–9

Kostić O, Gajić G, Jarić S, Vukov T, Matić M, Mitrović M, Pavlović P (2021) An assessment of the phytoremediation potential of planted and spontaneously colonized woody plant species on chronosequence fly ash disposal sites in Serbia—case study. Plants 11:110. https://doi.org/10.3390/plants11010110

Koumaki E, Noutsopoulos C, Mamais D, Fragkiskatos G, Andreadakis A (2021) Fate of emerging contaminants in high-rate activated sludge systems. Int J Environ Res Public Health 18(2):400

Kozuharova E, Matkowski A, Woźniak D, Simeonova R, Naychov Z, Malainer C et al (2017) Amorphafruticosa—a noxious invasive alien Plant in Europe or a medicinal plant against metabolic disease? Front Pharmacol 8:333

Kumar V, Ferreira LFR, Sonkar M, Singh J (2021) Phytoextraction of heavy metals and ultrastructural changes of Ricinuscommunis L. grown on complex organometallic sludge discharged from alcohol distillery. Environ Technol Innov 22:101382

Kuppusamy S, Thavamani P, Venkateswarlu K, Lee YB, Naidu R, Megharaj M (2017) Remediation approaches for polycyclic aromatic hydrocarbons (PAHs) contaminated soils: technological constraints, emerging trends and future directions. Chemosphere 168:944–968

Langella F, Grawunder A, Stark R, Weist A, Merten D, Haferburg G et al (2014) Microbially assisted phytoremediation approaches for two multi-element contaminated sites. Environ Sci Pollut Res 21:6845–6858

Laptiev V, Apori SO, Giltrap M, Tian F, Ryzhenko N (2024) Bioaccumulation of Cr, Zn, Pb and Cu in Ambrosia artemisiifolia L. and Erigeron canadensis L. Resources 13(3):43

Le Maitre DC, Gaertner M, Marchante E et al (2011) Impacts of invasive Australian acacias: implications for management and restoration. Divers Distrib 17:1015–1029. https://doi.org/10.1111/j.1472-4642.2011.00785.x

Leguizamo MAO, Gómez WDF, Sarmiento MCG (2017) Native herbaceous plant species with potential use in phytoremediation of heavy metals, spotlight on wetlands—a review. Chemosphere 168:1230–1247

Li Z, Zhang Y, Ma J, Meng Q, Fan J (2019) Modeling interactions between liposomes and hydrophobic nanosheets. Small 15:1804992. https://doi.org/10.1002/smll.201804992

Liu YN, Guo ZH, Xiao XY, Wang S, Jiang ZC, Zeng P (2017) Phytostabilisation potential of giant reed for metals contaminated soil modified with complex organic fertiliser and fly ash: a field experiment. Sci Total Environ 576:292–302. https://doi.org/10.1016/j.scitotenv.2016.10.090

Liu J, Zeng Y, Sun G, Yu S, Xu Y, He C et al (2020) Polygonumperfoliatum L., an excellent herbal medicine widely used in China: A review. Front Pharmacol 11:581266

Liu X, Zhang Y, Han W, Tang A, Shen J, Cui Z, Zhang F (2013) Enhanced nitrogen deposition over China. Nature 494:459–462. https://doi.org/10.1038/nature11917

Loos R, Carvalho R, António DC, Comero S, Locoro G, Tavazzi S, Gawlik BM (2013) EU-wide monitoring survey on emerging polar organic contaminants in wastewater treatment plant effluents. Water Res 47:6475–6487. https://doi.org/10.1016/j.watres.2013.08.024

Lowe S, Browne M, Boudjelas S, De Poorter M (2000) 100 of the world's worst invasive alien species: a selection from the global invasive species database, vol 12. Invasive Species Specialist Group, Auckland, p 12

Lu J, Zhang Y (2013) Spatial distribution of an invasive plant Spartina alterniflora and its potential as biofuels in China. Ecol Eng 52:175–181

Luo XS, Ding J, Xu B, Wang YJ, Li HB, Yu S (2012) Incorporating bioaccessibility into human health risk assessments of heavy metals in urban park soils. Sci Total Environ 424:88–96. https://doi.org/10.1016/j.scitotenv.2012.02.053

Ma LQ, Komar KM, Tu C, Zhang W, Cai Y, Kennelley ED (2001) A fern that hyperaccumulates arsenic. Nature 409:579. https://doi.org/10.1038/35054664

Ma S, Zhou C, Pan J, Yang G, Sun C, Liu Y, Zhao Z (2022) Leachate from municipal solid waste landfills in a global perspective: characteristics, influential factors and environmental risks. J Clean Prod 333:130234. https://doi.org/10.1016/j.jclepro.2021.130234

Mackay D, Callcott D (1998) Partitioning and physical chemical properties of PAHs. In: Neilson AH (ed) PAHs and related compounds: chemistry. Springer, Berlin, pp 325–345. https://doi.org/10.1007/978-3-540-49697-7_10

Marko M (1999) Controlling invasion of the exotic shrub Mimosa pigra in tropical Australia. (Report/monograph, check publication source for Springer style completion)

Marques AP, Rangel AO, Castro PM (2009) Remediation of heavy metal contaminated soils: phytoremediation as a potentially promising clean-up technology. Crit Rev Environ Sci Technol 39(8):622–654

Middleton BA (2019) In: encyclopedia of ecology. Elsevier, pp 431–440. https://doi.org/10.1016/B978-0-12-409548-9.11175-3

Miletić Z, Jarić S, Jonjev M, Mitrović M, Pavlović D, Matić M, Pavlović P (2024a) Phytoremediation potential of invasive plant species for potentially toxic elements along the Sava River upstream. Heliyon 10(13):e33798

Miletić Z, Marković M, Jarić S, Radulović N, Sekulić D, Mitrović M, Pavlović P (2024b) Lithium and strontium accumulation in native and invasive plants of the Sava River: implications for bioindication and phytoremediation. Ecotoxicol Environ Saf 270:115875

Mosoarca G, Vancea C, Popa S, Boran S (2018) Adsorption, bioaccumulation and kinetics parameters of the phytoremediation of cobalt from wastewater using Elodea canadensis. Bull Environ Contam Toxicol 100:733–739

Mtshali JS, Tiruneh AT, Fadiran AO (2014) Sewage sludge, nutrient value, organic fertilizer, soil amendment, sludge reuse, nitrogen, phosphorus; sewage sludge, nutrient value, organic fertilizer, soil amendment, sludge reuse, nitrogen, phosphorus. Resour Environ 4:190–199

Mulligan CN, Yong RN, Gibbs BF (2001) Remediation technologies for metal-contaminated soils and groundwater: an evaluation. Eng Geol 60(1–4):193–207. https://doi.org/10.1016/S0013-7952(00)00101-0

Mustafa AD, Juahir H, Yunus K, Amran MA, Hasnam CNC, Azaman F et al (2015) Oil spill related heavy metal: a review. Malaysia J Analytic Sci 19(6):1348–1360

Nahar N, Rahman A, Nawani NN, Ghosh S, Mandal A (2017) Phytoremediation of arsenic from the contaminated soil using transgenic tobacco plants expressing ACR2 gene of Arabidopsis thaliana. J Plant Physiol 218:121–126

Nedjimi B (2021) Phytoremediation: A sustainable environmental technology for heavy metals decontamination. SN Appl Sci 3(3):286. https://doi.org/10.1007/s42452-021-04301-4

Ngayila N, Botineau M, Baudu M, Basly JP (2009) Myriophyllum alterniforum DC. Effect of low concentrations of copper and cadmium on somatic and photosynthetic endpoints: a chemometric approach. Ecol Indic 9:307–312. https://doi.org/10.1016/j.ecolind.2008.05.006

Nouri J, Lorestani B, Yousefi N, Khorasani N, Hasani AH, Seif F, Cheraghi M (2011) Phytoremediation potential of native plants grown in the vicinity of Ahangaran lead-zinc mine (Hamedan, Iran). Environ Earth Sci 62:639–644

Nriagu JO, Pacyna JM (1988) Quantitative assessment of worldwide contamination of air, water and soils by trace metals. Nature 333(6169):134–139

Official Gazette (2010) Ordinance on the protection of agricultural land from contamination (N. N. 32/10). Retrieved from http://www.propisi.hr/print.php?id=6842. Accessed on 22 Sept 2020

Ohlbaum M, Wadgaonkar SL, van Bruggen JJ, Nancharaiah YV, Lens PN (2018) Phytoremediation of seleniferous soil leachate using the aquatic plants Lemna minor and Egeria densa. Ecol Eng 120:321–328. https://doi.org/10.1016/j.ecoleng.2018.06.021

Pan H, Lei H, He X, Xi B, Xu Q (2019) Spatial distribution of organochlorine and organophosphorus pesticides in soil-groundwater systems and their associated risks in the middle reaches of the Yangtze River Basin. Environ Geochem Health 41:1833–1845

Panagos P, Van Liedekerke M, Yigini Y, Montanarella L (2013) Contaminated sites in Europe: review of the current situation based on data collected through a European network. J Environ Public Health 2013:158764. https://doi.org/10.1155/2013/158764

Pandey VC (2012) Invasive species based efficient green technology for phytoremediation of fly ash deposits. J Geochem Explor 123:13–18

Pandey VC (2015) Assisted phytoremediation of fly ash dumps through naturally colonized plants. Ecol Eng 82:1–5. https://doi.org/10.1016/j.ecoleng.2015.04.002
Pandey VC, Singh K (2011) Is Vigna radiata suitable for the revegetation of fly ash landfills? Ecol Eng 37(12):2105–2106. https://doi.org/10.1016/j.ecoleng.2011.08.010
Pandey VC, Singh B (2012) Rehabilitation of coal fly ash basins: current need to use ecological engineering. Ecol Eng 49:190–192
Pandey VC, Singh N, Singh RP, Singh DP (2014) Rhizoremediation potential of spontaneously grown Typha latifolia on fly ash basins: study from the field. Ecol Eng 71:722–727. https://doi.org/10.1016/j.ecoleng.2014.08.009
Pang YL, Quek YY, Lim S, Shuit SH (2023) Review on phytoremediation potential of floating aquatic plants for heavy metals: a promising approach. Sustainability 15(2):1290. https://doi.org/10.3390/su15021290
Paswan D, Shabiimam MA, Shaikh M, Nadar A, Maniyar SD (2017) Heavy metals in fly ash; its impact on human health and environment. In NICMAR 2nd International Conference on Construction, Real Estate, Infrastructure and Project (CRIP) Management, pp 10–11
Pathak B, Rawat K, Fulekar MH (2019) Heavy metal accumulation by plant species at fly-ash dumpsites: thermal power plant, Gandhinagar, Gujarat. Int J Plant Environ 5(02):111–116
Peng S, Zhou Q, Cai Z, Zhang Z (2009) Phytoremediation of petroleum contaminated soils by Mirabilis Jalapa L. in a greenhouse plot experiment. J Hazard Mater 168(2–3):1490–1496
Petrie B, Barden R, Kasprzyk-Hordern B (2015) A review on emerging contaminants in wastewaters and the environment: current knowledge, understudied areas and recommendations for future monitoring. Water Res 72:3–27. https://doi.org/10.1016/j.watres.2014.08.053
Pilon SE (2005) Phytoremediation. Annu Rev Plant Biol 56:15–39
Pirzadah TB, Malik B, Tahir I, Hakeem KR, Alharby HF, Rehman RU (2020) Lead toxicity alters the antioxidant defense machinery and modulates the biomarkers in Tartary buckwheat plants. Int Biodeterior Biodegrad 151:104992. https://doi.org/10.1016/j.ibiod.2020.104992
Plaza Cazón J, Viera M, Donati E (2013) Dynamic Cr (III) uptake by Macrocystis pyrifera and Undaria pinnatifida biomasses. Electron J Biotechnol 16(3):6. https://doi.org/10.2225/vol16-issue3-fulltext-9
Podlipná R, Skálová L, Seidlová H, Szotáková B, Kubíček V, Stuchlíková L, Vokřál I (2013) Biotransformation of benzimidazole anthelmintics in reed (Phragmites australis) as a potential tool for their detoxification in environment. Bioresour Technol 144:216–224. https://doi.org/10.1016/j.biortech.2013.06.092
Polechońska L, Klink A, Dambiec M, Rudecki A (2018) Evaluation of Ceratophyllum demersum as the accumulative bioindicator for trace metals. Ecol Indic 93:274–281. https://doi.org/10.1016/j.ecolind.2018.05.012
Potgieter LJ, Gaertner M, O'Farrell PJ, Richardson DM (2019) Perceptions of impact: invasive alien plants in the urban environment. J Environ Manag 229:76–87
Prabakaran K, Li J, Anandkumar A, Leng Z, Zou CB, Du D (2019) Managing environmental contamination through phytoremediation by invasive plants: A review. Ecol Eng 138:28–37
Preisendanz HE, Barnes RG, Mashtare ML, Lintern A, Mina O, Williams C, Elliott HA (2021) The emergence, trajectory, and impacts of emerging contaminants publications. J Environ Quality 50(6):1339–1346
Probst A, Liu H, Fanjul M, Liao B, Hollande E (2009) Response of Vicia faba L. to metal toxicity on mine tailing substrate: geochemical and morphological changes in leaf and root. Environ Exp Bot 66(2):297–308
Pullagurala VLR, Rawat S, Adisa IO, Hernandez-Viezcas JA, Peralta-Videa JR, Gardea-Torresdey JL (2018) Plant uptake and translocation of contaminants of emerging concern in soil. Sci Total Environ 636:1585–1596
Pyšek P et al (2012) A global assessment of invasive plant impacts on resident species, communities and ecosystems: the interaction of impact measures, invading species' traits and environment. Glob Change Biol 18(5):1725–1737
Rabhi M, Hafsi C, Lakhdar A, Hajji S, Barhoumi Z, Hamrouni MH, Smaoui A (2009) Evaluation of the capacity of three halophytes to desalinize their rhizosphere as grown

on saline soils under nonleaching conditions. Afr J Ecol 47(4):463–468. https://doi.org/10.1111/j.1365-2028.2008.01004.x
Rachmadiarti F, Trimulyono G, Utomo WH (2022) Analyzing the efficacy of Salvinia molesta Mitchell as phytoremediation agent for lead (Pb). Nat Environ Pollut Technol 21(2):505–510. https://doi.org/10.46488/NEPT.2022.v21i02.027
Radočaj D, Velić N, Jurišić M, Merdić E (2020) The remediation of agricultural land contaminated by heavy metals. Poljoprivreda 26(2):30–42
Rahman MA, Hasegawa H (2011) Aquatic arsenic: phytoremediation using floating macrophytes. Chemosphere 83(5):633–646
Rai PK (2021) Heavy metals and arsenic phytoremediation potential of invasive alien wetland plants Phragmites karka and Arundo donax: Water-Energy-Food (WEF) Nexus linked sustainability implications. Bioresource Technol Report 15:100741
Rai PK, Singh JS (2020) Invasive alien plant species: their impact on environment, ecosystem services and human health. Ecol Indic 111:106020. https://doi.org/10.1016/j.ecolind.2019.106020
Ramírez-Satoyo LF, Guzmán-Mendoza R, Leyte-Manrique A, Salas-Arazia MD (2021) Weed plants used as bioindicators of special soil characteristics. Hortic Int J 5:21–23. https://doi.org/10.15406/hij.2021.05.00197
Rascio N, Navari-Izzo F (2011) Heavy metal hyperaccumulating plants: how and why do they do it? And what makes them so interesting? Plant Sci 180:169–181. https://doi.org/10.1016/j.plantsci.2010.08.016
Ratnayake RMCS (2014) Address on "Why plant species become invasive?–Characters related to successful biological invasion"
Ravindran KC, Venkatesan K, Balakrishnan V, Chellappan KP, Balasubramanian T (2007) Restoration of saline land by halophytes for Indian soils. Soil Biol Biochem 39(10):2661–2664
Ren X, Zhang X, Yan C, Gozgor G (2022) Climate policy uncertainty and firm-level total factor productivity: evidence from China. Energy Econ 113:106209. https://doi.org/10.1016/j.eneco.2022.106209
Rejmánek M (2015) Global trends in plant naturalization. Nature 525:39
Richardson DM, Pyšek P, Carlton JT (2011) A compendium of essential concepts and terminology in invasion ecology. In: Richardson DM (ed) Fifty years of invasion ecology: the legacy of Charles Elton. Wiley-Blackwell, Oxford, pp 409–420
Richardson DM, Pyšek P, Rejmanek M, Barbour MG, Panetta FD, West CJ (2000) Naturalization and invasion of alien plants: concepts and definitions. Divers Distrib 6(2):93–107
Rosales RM, Martínez-Pagan P, Faz A, Moreno-Cornejo J (2012) Environmental monitoring using electrical resistivity tomography (ERT) in the subsoil of three former petrol stations in SE of Spain. Water Air Soil Pollut 223:3757–3773. https://doi.org/10.1007/s11270-012-1146-0
Rosenfeld L, Reddi AR, Leung E, Aranda K, Jensen LT, Culotta VC (2010) The effect of phosphate accumulation on metal ion homeostasis in Saccharomyces cerevisiae. JBIC J Biol Inorg Chem 15(7):1051–1062. https://doi.org/10.1007/s00775-010-0661-2
Ross SM (1994) Sources and forms of potentially toxic metals in soil-plant systems. Toxic Metals in Soil Plant Systems:3–25
Safe Drinking water foundation, 2017. https://www.safewater.org/fact-sheets-1/2017/1/23/industrial-waste
Sakshi SK, Singh SK, Haritash AK (2019) Polycyclic aromatic hydrocarbons: soil pollution and remediation. Int J Environ Sci Technol 16:6489–6512. https://doi.org/10.1007/s13762-019-02414-3
Salomons W (1995) Environmental impact of metals derived from mining activities: processes, predictions, prevention. J Geochem Explor 52(1–2):5–23. https://doi.org/10.1016/0375-6742(94)00039-E
Sankaran KV (2008) Invasive pest fact sheet – Parthenium hysterophorus, carrot weed. FAO. http://www.fao.org/forestry/media/13378/1/0/. Accessed 30 May 2014
Sarker A, Al Masud MA, Deepo DM, Das K, Nandi R, Ansary MWR et al (2023) Biological and green remediation of heavy metal contaminated water and soils: a state-of-the-art review. Chemosphere 13:138861

Seebens H, Blackburn TM, Dyer EE, Genovesi P, Hulme PE, Jeschke JM et al (2018) Global rise in emerging alien species results from increased accessibility of new source pools. Proc Natl Acad Sci USA 115(10):E2264–E2273. https://doi.org/10.1073/pnas.1719429115
Shayler H, McBride M, Harrison E (2009) Sources and impacts of contaminants in soils
Sheng YY, Xue Bao T (2006) Biological characteristics of alien plants successful invasion. Chin J Appl Ecol 17(4):727–732. (in Chinese)
Simakov VN (1957) The use of phenylanthranilic acid in the determination of humus by Tyurin's method
Singh G, Gupta D, Shukla R, Mishra VK (2021) Application of constructed wetlands for the safe and sustainable treatment of emerging contaminants. In: Sustainable environmental clean-up. Elsevier, pp 85–104
Singh M, Singh M, Singh SK (2024) Tackling municipal solid waste crisis in India: insights into cutting-edge technologies and risk assessment. Sci Total Environ 170453:170453
Sipos B, Bibi D, Magura T, Tóthmérész B, Simon E (2023) High phytoremediation and translocation potential of an invasive weed species (Amaranthus retroflexus) in Europe in metal-contaminated areas. Environ Monit Assess 195(6):790
Sołtysiak J (2020) Heavy metals tolerance in an invasive weed (Fallopia japonica) under different levels of soil contamination. J Ecol Eng 21(7):194–202. https://doi.org/10.12911/22998993/125581
Souza FA, Dziedzic M, Cubas SA, Maranho LT (2013) Restoration of polluted waters by phytoremediation using Myriophyllum aquaticum (Vell.)Verdc., Haloragaceae. J Environ Manag 120:5–9
Sparling DW (2016) Organochlorine pesticides. Ecotoxicology Essentials:69–107
Srikanth K (2019) Emerging contaminants effect on aquatic ecosystem: human health risks. Art 19. https://doi.org/10.19080/ARTOAJ.2019.19.556104
Srivastava AN, Chakma S (2022) Bioavailability reduction of heavy metals through dual mode anaerobic colandfilling of municipal solid waste and industrial organic sludge. Chem Eng J 439:135725. https://doi.org/10.1016/j.cej.2022.135725
Staples CA, Peterson DR, Parkerton TF, Adams WJ (1997) The environmental fate of phthalate esters: a literature review. Chemosphere 35(4):667–749
State of the Environment Committee (2011) Australia State of the environment 2011—in brief. Independent report to the Australian Government Minister for Sustainability, Environment, Water, Population and Communities, DSEWPaC, Canberra
Stefanowicz AM, Stanek M, Nobis M, Zubek S (2017) Few effects of invasive plants Reynoutria japonica, Rudbeckia laciniata and Solidago gigantea on soil physical and chemical properties. Sci Total Environ 574:938–946. https://doi.org/10.1016/j.scitotenv.2016.09.120
Sun Y, Zhou Q, Diao C (2008) Effects of cadmium and arsenic on growth and metal accumulation of Cdhyperaccumulator Solanum nigrum L. Bioresour Technol 99(5):1103–1110. https://doi.org/10.1016/j.biortech.2007.02.027
Tang KHD, Angela J (2019) Phytoremediation of crude oil-contaminated soil with local plant species. In: IOP Conference Series: Materials Science and Engineering, vol 495. IOP Publishing, p 012054
Terrones-Saeta JM, Suárez-Macías J, Bernardo-Sánchez A, Álvarez de Prado L, Menéndez Fernández M, Corpas-Iglesias FA (2021) Treatment of soil contaminated by mining activities to prevent contamination by encapsulation in ceramic construction materials. Materials 14(22):6740
Thomas O, Burgess C (eds) (2017) UV-visible spectrophotometry of water and wastewater. Elsevier
Trueman RJ, Erber L (2013) Invasive species may offer advanced phytoremediation of endocrine disrupting chemicals in aquatic ecosystems. EJFA 25(9):648
Underground Storage Tank Soil Contamination, by Geo Forward, 2023
United States (2001) Brownfields technology primer: selecting and using phytoremediation for site cleanup. Environmental Protection, (Agency). Office of Solid Waste, Emergency Response, & Tetra Tech Education Minnesota, Inc. United States Environmental Protection Agency, Office of Solid Waste and Emergency Response

Van DA, Ngo TH, Huynh TH, Nakada N, Ballesteros F, Tanaka H (2021) Distribution of pharmaceutical and personal care products (PPCPs) in aquatic environment in Hanoi and Metro Manila. Environ Monit Assess 193:1–15

Van Poucke R, Egene CE, Allaert S et al (2020) Application of biochars and solid fraction of digestate to decrease soil solution Cd, Pb and Zn concentrations in contaminated sandy soils. Environ Geochem Health 42:1589–1600. https://doi.org/10.1007/s10653-019-00475-4

Vangronsveld J, Herzig R, Weyens N, Boulet J, Adriaensen K, Ruttens A et al (2009) Phytoremediation of contaminated soils and groundwater: lessons from the field. Environ Sci Pollut Res 16(7):765–794. https://doi.org/10.1007/s11356-009-0213-6

Vasilachi IC, Asiminicesei DM, Fertu DI, Gavrilescu M (2021) Occurrence and fate of emerging pollutants in water environment and options for their removal. Water 13(2):181

Verma C, Madan S, Hussain A (2016) Heavy metal contamination of groundwater due to fly ash disposal of coal-fired thermal power plant, Parichha, Jhansi, India. Cogent Eng 3(1):1179243

Wan J-Z, Wang C-J (2018) Expansion risk of invasive plants in regions of high plant diversity: a global assessment using 36 species. Eco Inform 46:8–18

Wang Y, Huang H (2019) Carbon nanotube composite membranes for microfiltration of pharmaceuticals and personal care products. In: Advanced nanomaterials for membrane synthesis and its applications. Elsevier, pp 183–202

Wang J, Wang S (2016) Removal of pharmaceuticals and personal care products (PPCPs) from wastewater: a review. J Environ Manag 182:620–640

Wang S, Wei M, Cheng H, Wu B, Du D, Wang C (2020) Indigenous plant species and invasive alien species tend to diverge functionally under heavy metal pollution and drought stress. Ecotoxicol Environ Saf 205:111160. https://doi.org/10.1016/j.ecoenv.2020.111160

Weerasinge SM, Priyasinghe IPK, Wijesinghe DC (2008) Plastic responses and competitive ability of invasive plant Clidemia hirta to varying soil nutrient status. In: Proceedings of the national symposium on Invasive Alien Species (IAS 2008). Sri Lanka Foundation Institute, Colombo, pp 55–68

Wei H, Huang M, Quan G, Zhang J, Liu Z, Ma R (2018) Turn bane into a boon: application of invasive plant species to remedy soil cadmium contamination. Chemosphere 210:1013–1020. https://doi.org/10.1016/j.chemosphere.2018.07.099

Wei H, Huang M, Quan G, Zhang J, Liu Z, Ma R (2018a) Turn bane into a boon: application of invasive plant species to remedy soil cadmium contamination. Chemosphere 210:1013–1020

Wei H, Xu J, Quan G, Zhang J, Qin Z (2018b) Invasion of Praxelisclematidea increases the chemically non-labile rather than labile soil organic carbon in a tropical savanna. Arch Agron Soil Sci 64(3):441–447

Wickramasinghe LS, Ranwala SMW (2008) Effect of Keppetiya (Croton lacciferus) on vegetative and reproductive growth of selected rice varieties (BW 361, BW267/3, BW364). In: Proceedings of the 28th annual sessions of the Institute of Biology, Sri Lanka

Wijesundara S (2010) Invasive alien plants in Sri Lanka. In: Marambe B, Silva P (eds) Invasive alien species in Sri Lanka. Biodiversity Secretariat, Colombo, pp 63–72

Witkowski ETF, Wilson M (2001) Changes in density, biomass, seed production and soil seed banks of the nonnative invasive plant Chromolaena odorata, along a 15-year chronosequence. Plant Ecol 152(1):13–27. https://doi.org/10.1023/A:1011402118955

Wong JWC, Ip CM, Wong MH (1998) Acid-forming capacity of lead–zinc mine tailings and its implications for mine rehabilitation. Environ Geochem Health 20(3):149–155. https://doi.org/10.1023/A:1006589124204

Wuana RA, Okieimen FE (2011) Heavy metals in contaminated soils: a review of sources, chemistry, risks and best available strategies for remediation. ISRN Soil Sci 2011:402647. https://doi.org/10.5402/2011/402647

Xuan R, Blassengale AA, Wang Q (2008) Degradation of estrogenic hormones in a silt loam soil. J Agric Food Chem 56(19):9152–9158

Xue C, Sun L, Qu B, Gao Y, Liu Z, Guo C, Liu W, Chang W, Tai P (2022) Grafting with an invasive Xanthium strumarium improves tolerance and phytoremediation of native congener

X. sibiricum to cadmium/copper/nickel tailings. Chemosphere 308. https://doi.org/10.1016/j.chemosphere.2022.136561

Xue C, Gao Y, Qu B, Tai P, Guo C, Chang W, Zhao G (2021) Hybridization with an invasive plant of Xanthium strumarium improves the tolerance of its native congener X. sibiricum to cadmium. Front Plant Sci 12:696687. https://doi.org/10.3389/fpls.2021.696687

Yaashika PR, Kumar PS, Jeevanantham S, Saravanan R (2022) A review on bioremediation approach for heavy metal detoxification and accumulation in plants. Environ Pollut 301:119035. https://doi.org/10.1016/j.envpol.2022.119035

Yan A, Wang Y, Tan SN, Mohd Yusof ML, Ghosh S, Chen Z (2020) Phytoremediation: a promising approach for revegetation of heavy metal-polluted land. Front Plant Sci 11:359. https://doi.org/10.3389/fpls.2020.00359

Yang Y, Wei Z, Zhang X, Chen X, Yue D, Yin Q et al (2014) Biochar from Alternantheraphiloxeroides could remove Pb (II) efficiently. Bioresour Technol 171:227–232

Yang Q, Zhang L, Wang H, Martín JD (2022) Bioavailability and health risk of toxic heavy metals (as, Hg, Pb and Cd) in urban soils: a Monte Carlo simulation approach. Environ Res 214:113772. https://doi.org/10.1016/j.envres.2022.113772

Yi Q, Huang C (2019) Review of heavy metal pollution by mining. In: E3S Web of Conferences. EDP Sciences, 118. https://doi.org/10.1051/e3sconf/201911804028

Yuan CG (2009) Leaching characteristics of metals in fly ash from coal-fired power plant by sequential extraction procedure. Microchim Acta 165:91–96. https://doi.org/10.1007/s00604-008-0103-5

Zahari NZ, Tuah PM, Rahim SA (2021) Inoculation of Bacillus cereus enhances phytoremediation efficiency of Pistia stratiotes and Eichhornia crassipes in removing heavy metal Pb. IOP Conf Ser Earth Environ Sci 847:012012. https://doi.org/10.1088/1755-1315/847/1/012012

Zhao FJ, Ma Y, Zhu YG, Tang Z, McGrath SP (2015) Soil contamination in China: current status and mitigation strategies. Environ Sci Technol 49:750–759. https://doi.org/10.1021/es5047099

Zheng Y, Xu W, He Z, Ma M (2008) Plant regeneration of the arsenic hyperaccumulator Pterisvittata L. from spores and identification of its tolerance and accumulation of arsenic and copper. Acta Physiol Plant 30:249–255

Zhou Y, Lan W, Yang F, Zhou Q, Liu M, Li J et al (2024) Invasive Amaranthus spp. for heavy metal phytoremediation: investigations of cadmium and lead accumulation and soil microbial community in three zinc mining areas. Ecotoxicol Environ Saf 285:117040

Zhu G, Xiao H, Guo Q, Song B, Zheng G, Zhang Z et al (2018) Heavy metal contents and enrichment characteristics of dominant plants in wasteland of the downstream of a lead-zinc mining area in Guangxi, Southwest China. Ecotoxicol Environ Saf 151:266–271. https://doi.org/10.1016/j.ecoenv.2018.01.021

Zhu GX, Xiao HY, Guo QJ, Zhang ZY, Yang X, Kong J (2017) Subcellular distribution and chemical forms of heavy metals in three types of compositae plants from lead-zinc tailings area. Huan Jing Ke Xue 38(7):3054–3060

Ziylan A, Ince NH (2011) The occurrence and fate of anti-inflammatory and analgesic pharmaceuticals in sewage and fresh water: treatability by conventional and non-conventional processes. J Hazard Mater 187(1–3):24–36

Plant Beneficial Rhizosphere Microorganisms Help in Degraded Land Restoration

13

Tushar Goyal, Gowardhan Kumar Chouhan,
Anand Kumar Gaurav, Arpan Mukherjee,
Bansh Narayan Singh, and Jay Prakash Verma

Abstract

Land degradation, often caused by natural and anthropogenic activities, poses a threat toward global food security, biodiversity, and overall ecosystem health. Plant beneficial rhizosphere microorganisms (PBRM), involving plant growth-promoting bacteria (PGPR) and mycorrhizal fungi, confront these issues through improvement of soil fertility, structure, and endurance. These microbes possess the abilities of nitrogen fixation, mineral solubilization, and produce plant growth hormones such as indole-3 acetic acid, gibberellins, and cytokinin. Additionally, PBRMs have a unique, diversified secondary metabolic profile that indirectly helps mitigate drought, salinity, and soil pollution, further improving soil water dynamics and its structure. Beyond this, PBRMs exhibit biocontrolling traits by releasing antibiotics, enzymes, and siderophores that suppress many pathogens and pests, thereby reducing the need for toxic chemicals in agriculture. Moreover, they aid in the bioremediation of soil pollution through heavy metal immobilization and detoxification, fostering a healthier and more sustainable ecosystem. Bioprospecting the role of plant beneficial rhizosphere microorganisms offers an economical and sustainable approach to degraded land restoration. Their incorporation into land management is not only limited to degraded land restoration, but they also support in achieving sustainable development goals related to food security, biodiversity conservation, and climate change mitigation.

T. Goyal · G. K. Chouhan · A. K. Gaurav · A. Mukherjee · B. N. Singh · J. P. Verma (✉)
Plant-Microbe Interaction Lab, Institute of Environment and Sustainable Development, Banaras Hindu University, Varanasi, Uttar Pradesh, India
e-mail: jpv.iesd@bhu.ac.in

V. C. Pandey (ed.), *Plantation-Based Land Restoration*,
https://doi.org/10.1007/978-981-95-2860-8_13

Keywords

Sustainable development goals · Plant hormone · Soil fertility · Bioremediation · Nitrogen fixation

13.1 Introduction

Soil, water, and biodiversity are the three building blocks of land. The synergistic effect of these three elements results in the formation of products and services that provide a base for sustainable livelihoods and peaceful coexistence between beings. The imbalance in these building blocks, termed "land degradation," can cause a deleterious effect on humans and other forms of life. Land degradation is a result of the increased anthropogenic activities like excessive nutrient mining, soil erosion, water pollution, and aggregation of xenobiotic compounds in soil, such as pesticides, drugs, or other toxic chemicals made by humans. In addition to this, land degradation involves a decline in soil carbon, which in turn causes a reduction in soil biodiversity and soil erosion. In particular, soil is one of three building blocks that are severely affected by land degradation which in turn responsible for degrading water and biodiversity as; soil not only acts as a water storage and filter but it also provides habitat for various organisms. Overall, a substantial area of Earth's arable land is severely affected by land degradation, which leads to detrimental effects on natural processes involving water flow, vegetation, nutrient absorption, storage, and recycling (Quinton and Fiener 2024). These factors are responsible for the decline in overall land productivity, which ultimately results in a decline in wealth and economic development worldwide, with intensifying competition for depleting resources (Mahata and Sharma 2021). Land degradation adversely affects soil water dynamics, involving water infiltration and retaining soil moisture. The indispensable role of soil moisture in terrestrial life is well documented in many studies (Green et al. 2019; Seneviratne et al. 2010). Unsustainable land management practices and climate change have significantly reduced the soil moisture retention capacity, thereby reducing the biodiversity of soil inhabitants (Robinson et al. 2019). Microorganisms, as dominant inhabitants of soil, are the key players in maintaining its biological properties and fertility. The multiple roles of these microbes highlight their importance in soil as they have the ability to solubilize nutrients, nitrogen fixation, and the decomposition of organic matter in soil, which collectively improve soil fertility. Additionally, microbes can release compounds (volatiles or nonvolatiles), which can suppress the activity of various soil pathogens or neutralize the hazardous chemicals, contributing to a healthier and productive soil ecosystem (Goyal et al. 2024).

Land degradation has significantly altered the key functions of soil, such as water filtration, carbon cycling, nutrient transportation, biological activities, plant growth, and various other critical ecosystem services. The abovementioned effects of land degradation are quite an alarming situation for people's livelihood, also highlighting the need for sustainable approaches to restore the degraded land. To cope with this problem, plant beneficial rhizospheric microorganisms (PBRMs) are considered as a potential sustainable approach for the restoration of degraded land. Plant beneficial rhizospheric microorganisms are the microbes residing in the zones of root-associated soil parts within a defined range. As the name suggests, these PBRMs are potential plant growth enhancers by improving the soil quality via regulating nutrient cycling, nitrogen fixation, mineral (potassium, phosphorus, and zinc) solubilization, organic matter decomposition, and biocontrolling various soil-borne pathogens (Khademian et al. 2019). Therefore, PBRMs help to improve vital soil activities which in turn helps to restore the degraded lands in multiple ways. In a recent study, it was revealed that several strains of *Pseudomonas* enhance plant growth in saline soil via modulating soil properties, including electrical conductivity (EC), pH, cation exchange capacity (CEC), sodium absorption ratio (SAR), and increasing the fertility status of soil by increasing in NPK (nitrogen, phosphorous, and potassium) levels (Ullah and Bano 2019). However, the potential role of these PBRMs in the restoration of degraded lands has been neglected for decades, but various case studies suggest to reframe the current approach toward more sustainable approach leveraging these microorganisms for effective land restoration.

In 2015, the Sustainable Development Summit, a total of 17 United Nations Sustainable Development Goals (UN-SDGs) was proposed, aiming to achieve them by 2030. Plant beneficial rhizospheric microorganisms could be a promising approach to achieving these goals, as majorly all SDGs are firmly connected, either directly or indirectly, with sustainable soil management and the utilization of natural resources (Hurni et al. 2015). Therefore, more precise use of these microbes could be a sustainable and economical approach for restructuring the degraded lands by improving soil health (Rahina et al. 2018; Upadhyay et al. 2023). This chapter aims to provide authors with a comprehensive understanding toward the importance of plant beneficial rhizospheric microorganisms in the restoration of degraded lands. These microbes promote a healthier soil environment by enhancing nutrient cycling, soil water dynamics, organic matter decomposition, and nutrient solubilization and by suppressing soil-borne pathogens, etc. This chapter also provides a thorough understanding of the mechanisms employed by microbes, highlighting their role as a natural, economical, and eco-friendly solution to combat land degradation while supporting the global efforts to achieve food security and environmental sustainability.

13.2 Types of Land Degradation

Land degradation is a serious environmental threat that impairs the overall health and productivity of the land by natural or anthropogenic activities. Due to this alarming situation, land's ability to support life is reducing day by day, so the understanding of various types of land degradation could help to develop sustainable strategies to restore the degraded land. A structured classification of land degradation is yet to be developed; however, through various studies, it was suggested that land degradation is mainly done by these interconnected processes:

(a) Alteration in soil structure: involving soil erosion, drought, forest fires, soil compaction, and landslides.
(b) Decline in soil chemical properties: involving depletion of soil fertility and soil carbon through acidification, nutrient depletion, and salinization.
(c) Soil pollution mainly done by human activities such as dumping industrial waste, rapid urbanization, and agricultural-related activities (Fig. 13.1) (Upadhyay et al. 2023).

13.3 Alteration in Soil Structure

Transformation or alteration in soil structure is a global threat, mainly driven by forest fires, soil erosion, drought, etc. Forest fires possess detrimental effects on soil by degrading its physical, biological, and chemical attributes, such as soil water retention capacity, pore size, biodiversity loss, and sublimation of soil nutrients, such as nitrogen, phosphorus, etc. ((Certini 2005; Doerr and Cerdá 2005). Major repercussions of forest fire make the soil bare and highly susceptible toward water and wind erosion for many years (DeBano 2000; Ravi et al. 2009). Due to this soil erosion (loss of top soil) significantly restricts reforestation efforts for years. In addition to this, the burning of plant biomass causes the formation of harmful organic compounds known as polycyclic aromatic hydrocarbons (PAHs) (Kim et al. 2003).

Drought, another major type of degraded land caused by increasing human activities and global climate change, already covers approximately one-third of the Earth's terrestrial area and is increasing (Bridges 2019). In these areas, soil is found to be nutrient-poor with less organic matter, water deficit, and more saline due to the UV radiation and gusty winds (Ayangbenro and Babalola 2021). Under these extreme environmental stresses, plants suffer a range of biochemical, morphological, and physiological changes, which in turn impair their growth and productivity (Lugtenberg 2015).

13.4 Decline in Soil Chemical Properties

In order to decrease hunger and poverty, the Green Revolution (GR) was initiated in the 1960s by introducing the high-yielding varieties (HYVs) of rice and wheat. Initiatives taken by the government resulted in increased production of rice and wheat, but it caused a decrease in the production of other indigenous crops. Thus, GR and new farming techniques have significantly increased crop yield but gradually reduce the soil flora and fauna, soil fertility, and health due to the over usage of chemicals (fertilizers and pesticides) (Sadaf et al. 2024). Numerous environmental factors were also compromised, as seen by higher emissions of greenhouse gases (GHGs) during and after the GR, as well as land degradation and water pollution. According to Goyal et al. (2024), the exponential increment of human population is

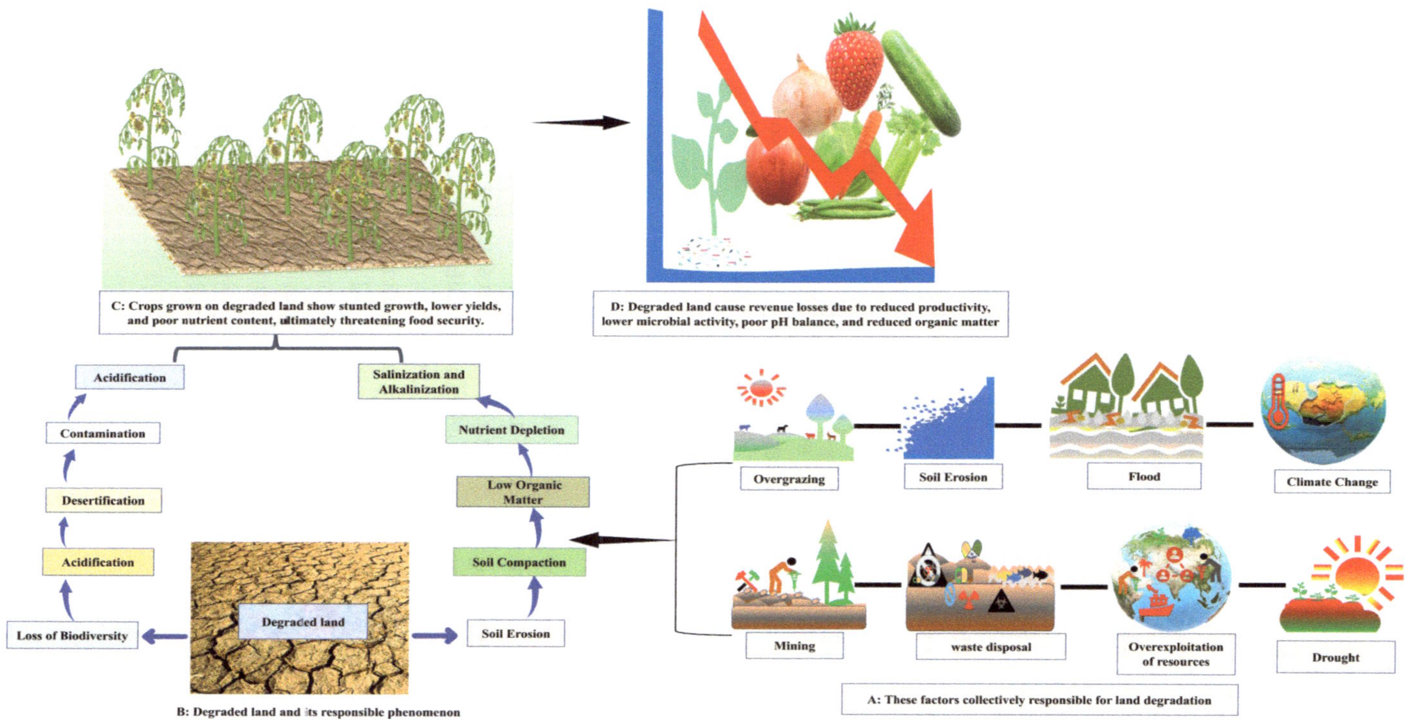

Fig. 13.1 Schematic representation of different factors involves un land degradation and its impact on crop productivity and soil fertility

responsible for driving a 70% enhancement in global food demand. Consequently, the vigorous cultivation of agricultural crops leads to depletion of soil nutrients and soil organic matter, which is ultimately responsible for land degradation (Kraaijvanger and Veldkamp 2015).

Soil salinity is another major cause of land degradation by degrading fertile soil of approximately 1–2% every year worldwide, with about 20% of agricultural land is already affected by salinization (Etesami and Beattie 2018; Munns and Tester 2008). The issue is predominantly appearing in arid and semiarid regions; however, it is also observed in some coastal areas like Netherlands because of the increase in sea levels. This issue arises from the increased concentration of sodium (Na^+) and chloride (Cl^-) ions while lowering the concentration of potassium (K^+) ions in the soil (Bothe 2012). Increased concentration of Na^+ damages plant cells and soil structure as well. Prolonged salinity stress for plants diminishes plants' ability to absorb moisture from soil, thereby creating the drought like situation for plants (Bothe 2012). Increased salinity of soil is linked with reduced soil organic carbon and microbial biodiversity. Soil salinization can significantly affect microbial respiration, mineralization, and enzymatic activities, which in turn are responsible for a decline in soil organic matter by stimulating the activity of carbon-degrading enzymes (AbuQamar et al. 2024; Morrissey et al. 2014; Setia et al. 2012). Consequently, soil fertility in those areas depleted dramatically. Therefore, there is an urgent need to develop bioremediation approaches for saline soil and cultivate salinity stress-tolerating plants to combat the challenges posed by saline soil effectively.

13.5 Soil Pollution

Over the last two to three decades, soil pollution has increased rapidly by various natural and anthropogenic activities and raising various environmental concerns about the Earth's terrestrial life. The natural factors responsible for soil pollution are volcanic eruptions, tsunamis, earthquakes, acid rainfall, etc., and there are lots of anthropogenic activities involving releasing tons of toxic chemicals used in agriculture and other industries, rapid urbanization, and dumping waste, such as various forms of plastic, medical, and radioactive waste (Al-Ansari and Al-Humaid 2025; Fang et al. 2025). These toxic compounds are not only harmful for short duration, but their longer shelf life made these compound riskier for the whole ecosystem of the planet. These vehement compounds are carcinogenic, teratogenic and mutagenic in nature (Koul and Taak 2018). In order to protect the Earth's ecosystem from the severity of these compounds, some strategies need to be developed to assess the threshold concentrations of these pollutants in the soil ecosystem.

13.6 Transformative Microbial Approaches for Restoring Degraded Lands

PBRMs can promote soil restoration in many of the cases mentioned in the above section. So, this section mainly focuses on the transformative strategies used by rhizospheric microorganisms for degraded land restoration.

13.7 PBRM Role in Transforming and Improving Soil Structure

Microorganisms have unique capabilities to adapt a wide range of extreme environmental conditions, such as high and low temperatures, salinity, drought, etc. Therefore, they can be the key player for providing essential nutrients to plants from the soil in different direct and indirect mechanisms while helping plants to enhance their tolerance to various stresses (biotic and abiotic) (Fig. 13.2) (Alsharif et al. 2020). Their role in dry areas is more commendable, as they have some unique abilities to upregulate genes related to osmoregulation and dormancy while downregulating the genes related to catabolism and nutrient cycling (Alsharif et al. 2020). PBRMs help the plants to combat drought stress by elevating sugar levels and chlorophyll content in the plant, like *Actinobacteria* produces trehalose sugar and stimulates plants' stress-related genes to develop drought tolerance while living in close association with plants (C. Chen et al. 2017; Vílchez et al. 2016). Consequently, these microorganisms can contribute to the recovery of drought-prone areas by improving soil stability and nutrient cycling. So, utilizing these organisms is a more sustainable way to restore the degraded land than replacing topsoil or adding organic matter (Fig. 13.2) (Chiquoine et al. 2016).

PBRMs have an essential role in restoring soil functions restoration wildfires. For many instances, these microorganisms have shown their potential in promoting plant growth by improving the nutrient content in fire-impacted soil (Aguirre-Monroy et al. 2019; Radhapriya et al. 2018). Even though forest fires reduce the overall biodiversity of the particular area, they somehow stimulate population of some bacterial genera from the *Bacillota, Actinomycetota, Bacteroidota,* and *Pseudomonadota phyla* having complex and unique interaction networks (García-Carmona et al. 2022; Lu et al. 2021). This survival instinct of microbes in this extreme condition reveals their high genetic plasticity with diverse metabolic profile that might provide a competitive edge over other genera (Baldrian 2016; Morán-López et al. 2015). Some of these genera, which belong to the order *Bacillales*, are potential indicators of wildfire severity, have the ability to degrade a large number of hydrocarbons generated by forest fires, and are termed as drought-tolerant copiotrophs (Ferrenberg et al. 2013). Some of the strains of the *Bacillales* order, such as *Bacillus, Paenibacillus, Lysinibacillus,* and *Psychrobacillus*, are studied extensively for their various plant growth-promoting properties by enhancing soil health and

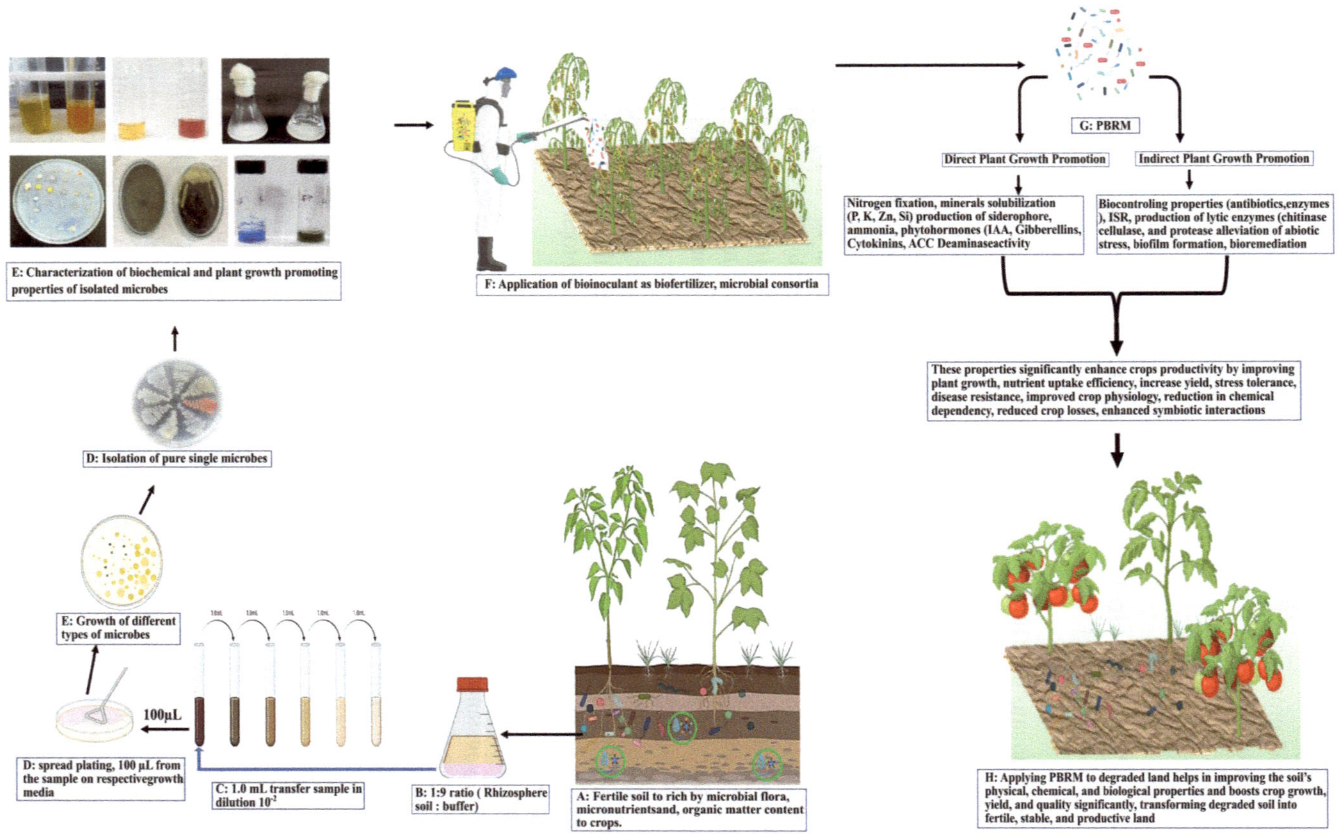

Fig. 13.2 Transformative approach of degraded land restoration via PBRMs

suppressing the activity of various plant pathogens (Harirchi et al. 2022; Pishchik et al. 2021; Xing et al. 2022). Among them, *Bacillus* and *Paenibacillus* strains have been reported for the role of mitigating salinity and drought stress on rice and wheat crops, respectively (Kumar et al. 2021; Samain et al. 2022).

Alternatively, in fire-affected lands, native soil microbial colonies can be enhanced via a biostimulation process with the addition of organic matter, oxygen, or specific plant species. This biostimulation increases the biodegradation ability of these microbes, which in turn removes PAH from burned soils, and this could be an effective, sustainable strategy for degraded land restoration by forest fires (Andreolli et al. 2015; Sales da Silva et al. 2020). The soil hydrophobicity generated after forest fires can be resolved by inoculating wax-degrading *Actinobacteria* strains in soil, making this approach more sustainable and economical (Roper 2006).

13.8 Impact of PBRMs on Soil Fertility

Soil fertility is adversely affected by intensive cultivation of crops, leading toward the depletion of soil organic matter and soil nutrients (e.g., soil NPK levels) at exponential rates. The increasing knowledge of PBRMs as an essential part of improving soil health and crop productivity has accelerated the adoption of many agronomic practices, such as the addition of organic matter extracted from plant and animals, which efficiently increase biological N fixation, and the recycling of nutrients within the soil ecosystem. These agricultural practices are designed to restore soil degradation and promote prolonged sustainability. In addition to this, utilization of biofertilizers and biopesticides supports beneficial microbial activity, while enriching soil with organic fertilizers such as manure, compost, and biochar contributes to its improved structure, fertility, and carbon sequestration (Fig. 13.2) (Singh and Seneviratne 2017). Various studies have also reported that soil nutrient use efficiency can be increase via inoculation of plant growth-promoting bacteria and fungi in combination with other organic fertilizers (Chouhan et al. 2023; Rashid et al. 2016). Therefore, biofertilizers and biopesticides in combination with organic fertilizers can provide a base to meet food demand for the increasing population while restoring soil fertility.

PBRMs are reported to improve soil fertility and plant growth by (i) increasing nutrient solubilization, such as phosphorous (P), potassium (K), zinc (Zn), and silicate (Si); (ii) biological nitrogen fixation; (iii) siderophore production; (iv) plant growth hormone production involving indole-3 acetic acid, gibberellins, cytokinin, ethylene, etc.; and (v) biocontrolling of various pathogens and antibiotic production (Fig. 13.2) (Hasan et al. 2024; Muhammad et al. 2025; Sun and Shahrajabian 2025). These rhizospheric microorganisms are attracted via plant root exudates, which act as a messenger and connect plant roots to rhizosphere microbes (Kawasaki et al. 2021). These root exudates belong to various classes such as sugars, amino acids, organic acids (malic acid, citric acid), phenolics (flavonoids, tannins, and coumarins), terpenoids, and secondary metabolites (strigolactones, phytoalexins, etc.). Among them, flavonoids, citric acid, and malic acid are reported to recruit

Rhizobium, Bacillus, and *Pseudomonas* (Rudrappa et al. 2008; Watson et al. 2015). Hence, plant root exudates have a role in promoting plant growth and development, elevating plant defense responses to abiotic stresses and diseases, and improving soil quality (He et al. 2024; Preece and Peñuelas 2020).

13.9 Bioremediation of Saline Soil

Bioremediation of saline soil can be treated by inoculating halophiles (salt-tolerant) microbes and arbuscular mycorrhizal fungi (AMF) (Chandrasekaran et al. 2014; Kumar Arora et al. 2020). The accumulation of salt-tolerant species in saline soil significantly contributes toward improving soil structure by upgrading soil nutrients, organic matter, pH, electrical conductivity, and ionic salts deposition (Kumar Arora et al. 2020). Additionally, halophiles indirectly improve soil salinity and plant growth by secreting secondary metabolites (volatiles and nonvolatiles), enzymes, and plant growth hormones (Kang et al. 2019; Khan et al. 2020). Plenty of studies have found that halophiles are able to produce osmoprotectants, such as proline (Pro) and glycine betaine (GB), and have the potential to alleviate salinity stress in plants (Tang et al. 2015; Zhang and Dai 2019). The Pro compound improves water absorption and boosts the antioxidant system, whereas GB plays a crucial role in osmotic regulation by managing the Na^+/K^+ ratio. This process effectively minimizes the buildup of harmful ions and enhances tolerance to salinity. Additional osmoprotectants, including sugars and sugar alcohols, contribute to ion homeostasis, enhance antioxidant mechanisms, and maintain cellular osmotic balance, thereby promoting tolerance to salinity (Zhang and Dai 2019). In addition to this, *Bacillus* sp. and *Arthrobacter* sp. were observed to enhance Pro accumulation in pepper plants (Sziderics et al. 2007). Conversely, *Pseudomonas pseudoalcaligenes* was noted for its significant production of GB in rice seedlings (Jha et al. 2011). In contrast to the individual inoculation with *Bacillus pumilus* or *P. pseudoalcaligenes*, the consortium of these offers enhanced protection by generating increased levels of antioxidants and osmoprotectants in plants subjected to high salt stress conditions. Canola (*B. napus*) plants subjected to salt stress and pre-inoculated with *P. putida* UW4 exhibited elevated levels of proteins associated with antioxidant activities, photosynthesis, and membrane transport (Cheng et al. 2012). Currently, many researchers are exploring the utilization of halophiles isolated from halophilic plant rhizosphere to promote plant development and salt tolerance in non-halophytic plants (Etesami and Beattie 2018). According to recent studies, these microbes alter plant genes that mitigate salinity stress (Pinedo et al. 2015; Yasmin et al. 2020). The interaction between plant stress responses, signaling compounds, and microbiome assembly allows phytomicrobiome manipulation to improve plant resilience in salt affected area, and therefore, this sector required more research to promote sustainable stress-resilient agriculture and soil improvement.

13.10 Degradation and Detoxification of Soil Pollutants by PBRMs

The process of bioremediation in polluted soils entails the disintegration of pollutants driven by the metabolic activities of indigenous microorganisms (Sales da Silva et al. 2020). These microbes also exert beneficial effects on the removal of pollutants indirectly via enhancing plant growth and augmenting contaminant bioavailability, known a phytoremediation. This function can be utilized to improve the effectiveness of phytoremediation (Wood et al. 2016). Microbe-based phytoremediation demonstrates its significance in the restoration of degraded soil by heavy metals, pesticides, and hydrocarbon contamination (Juwarkar 2012). This approach may facilitate improved recovery of degraded ecosystems, leading to a significant restoration of biodiversity (Juwarkar 2012). For instance, inoculating polluted soil with fungi at the time of vegetation serves two primary functions: it helps in the isolation of heavy metals from contaminated soils and promotes plant establishment in these polluted areas, which in turn enhances soil quality and health (Akhtar et al. 2020).

Efficient microbial degradation can be determined by the physicochemical properties and bioavailability of pollutants (Liang et al. 2022). The term bioavailability refers to the degree to which substances are physically and chemically accessible to microorganisms (Chaudhry et al. 2005; Mumtaz et al. 2024). Extreme (high and low) levels of pollutants can influence the results of remediation efforts, such as the optimal concentration for the removal of chromium (Cr) from soil by *Streptomyces* or chromium-specific soil microbes is approximately 1800 mg/L, at which these microbes remove the point where the removal efficiency reaches its peak (Huang et al. 2021; Wang et al. 2020). In context to this, complex organic pollutants are hydrophobic and less bioavailable, which makes them more difficult to degrade than simple organic pollutants (for instance, long-chain alkanes as opposed to short-chain alkanes). Although the simpler structure of pollutants demonstrates their higher bioavailability and degradation, they frequently show increased biological toxicity (Zhao et al. 2024). Cerqueira et al. reported that the characteristics and bioavailability of soil pollutants have a direct impact on the degradation efficiency of petroleum hydrocarbons, with simpler structures resulting in higher microbial degradation efficiency (Cerqueira et al. 2011). Additionally, fluctuations in soil environmental conditions can impact the effectiveness of microbial-based remediation of soil (Singh and Choudhary 2021). Temperature fluctuations can negatively impact microbial growth, adsorption, and degradation, which impairs the exchange of substances and energy both within and outside the cells, consequently affecting soil bioremediation efforts (Song et al. 2023). Moreover, suitable salt concentrations can aid in preserving osmotic pressure both within and outside cells during restoration of heavily polluted soil (Popescu and Dumitru 2009). For example, the removal rate of cadmium (Cd) was significantly higher by *Serratia marcescens* in the absence of sodium chloride, and there was a noted decline in the efficiency of heavy metal removal in increased concentration of sodium chloride (Li et al. 2013).

The initial findings on the community composition and structural characteristics of soil microbes have been succeeded by focusing on the diversity, structure, function, and regulatory mechanisms of microbial remediation across environmental conditions (Raiesi and Beheshti 2015; Yannikos et al. 2014). The results obtained from the prominent studies suggested that the understanding of bioremediation of contaminated soil has evolved drastically. This emphasizes the improved understanding of microbial remediation processes and helps establish a strong foundation for this research field (Chen et al. 2015). The primary research areas are mainly based on the bioremediation of inorganic pollutants (e.g., heavy metals) and organic pollutants, including PAHs. Factors include the introduction of heavy metals and PAHs into the soil due to mineral extraction, pesticide and fertilizer application, dumping, and accumulation of industrial waste (Rusănescu and Rusănescu 2023). Thus, to combat this issue, individual technologies may face limitations concerning their effectiveness, cost, and efficiency, explaining the urgent solutions that incorporate interdisciplinary approaches to explore the effectiveness of both microorganisms and other materials together for treating soil pollution.

13.11 Limitations of PBRMs in Restoration of Degraded Land

In spite of proving their potential candidature in degraded land restoration, PBRMs may face various challenges that need to be addressed to fully realize their benefits. The inoculated microbes are stressed by various challenges, such as:

(i) Environmental challenges, which involve their survival and sustainability in extreme weather conditions, temperature, pH, salinity, competition with native microbes or rhizocompetence, and soil variability.
(ii) Technical challenges involving effective strain selection, delivery mechanisms, shelf-life, and evaluation of the long-term impact of PBRMs on soil quality and plant growth.
(iii) Economic challenges such as high production cost, less incentives, and a lack of infrastructure for marketing these microbial products, restricting their availability and use.
(iv) Socioeconomic and policy barriers such as low awareness in farmers and other stakeholders, absence of regulatory laws and policies to promote the establishment of microbial technologies, and a wide gap between research organizations to local consumers, which is responsible for the non-adoption of technology by end users.

The understanding of these limitations aids in designing such strategies through which PBRMs can be utilized to their full potential in the restoration of degraded land. So, to harness the abilities of PBRMs, the following strategies may be employed:

(i) Focusing more on research and development of such specific microbial strains intended for a particular degraded soil type and implementation of cutting-edge omics technologies to get insights into microbial interactions and further developing consortia for field applications.
(ii) Enhancing microbial survival and proliferation, efficient delivery systems must be developed, for example, encapsulation or biochar-based carriers.
(iii) Formulating policies and subsidies for promoting the adoption of these microbial-based products with assurance of their safety and efficacy.
(iv) Running various awareness and training campaigns to educate more people about the advantages and proper usage of PBRMs and also promote community-driven initiatives to increase the utilization of these bio-products at the local level.
(v) Encouraging public-private partnerships by fostering collaborations among research institutions, government, and industries to propel innovations and funding opportunities to enhance the production and distribution of microbial technologies.
(vi) Establishing the standard protocols to track the long-term effect of PBRMs on soil health, biodiversity, and productivity, such as employing remote sensing and digital tools to assess the land restoration progress.

13.12 Conclusion and Future Perspectives

Land degradation is an alarming situation worldwide, and its restoration requires sustainable and eco-friendly approach that will combat the severe effects adjourned of deforestation, industrialization, rapid urbanization, and climate change. In this context, plant-beneficial rhizosphere microorganisms (PBRMs) have been considered as a key player in restoring degraded land by enhancing soil nutrient profiles and organic matter, improving plant growth, and fostering ecological equilibrium. These microorganisms are classified based on their function, such as plant growth promoters, mycorrhizal fungi, and nitrogen fixers. PBRMs hold significant potential for sustainable land management and ecological restoration processes. PBRMs significantly contribute toward improved soil fertility via enhancing nutrient availability through key processes, such as mineral solubilization, nitrogen fixation, and organic matter addition, which is critical for degraded lands. The bioavailability of these essential nutrients aids in restoring the nutrient balance in soil, which in turn creates a conducive environment for plant growth. Additionally, many bacterial strains are able to produce extracellular polysaccharides and biofilms, which aid in stabilizing soil aggregates, enhancing water retention capacity, reducing erosion, and ensuring better water management in drought-prone areas.

PBRM has the ability to promote overall plant growth and development by synthesizing several plant growth hormones, such as IAA, gibberellins, and cytokinins, which help stimulate root elongation and branching. This increment in roots helps plants take up water and nutrients from degraded soils. Furthermore, their biocontrolling properties, such as producing several antibiotics, hydrolytic enzymes, and secondary metabolites, are not only limited to enhancing crop productivity but also contribute toward the reintroduction of native vegetation, which is often susceptible to pathogenic attacks in degraded soils. Some microorganisms help restore degraded lands by chelating heavy metals that helps immobilize toxic metals, reducing their bioavailability, which in turn mitigates their harmful effects on plants. Similarly, halotolerant species of microbes can increase plant tolerance to salinity stress by releasing various osmoprotectants. Therefore, the increased vegetation facilitated by PBRMs promotes carbon sequestration and soil biodiversity proliferation, thereby enhancing overall ecosystem health.

In conclusion, adaptation of plant-beneficial rhizosphere microorganisms unveils a significant sustainable approach for the reclamation of degraded land impacted by both natural and other anthropogenic activities. Utilization of microbial consortia with complementary traits, such as biochar, organic manure, etc., synergistically improves nutrient cycling and stress tolerance, fosters plant growth, and provides biocontrol benefits. The amalgamation of these microbes into degraded land restoration practices aligns with sustainable development goals, particularly in land restoration, food security, and climate change mitigation. The extensive application of PBRMs across different agroecological zones unfolds their potential and makes way for future research work into more sustainable practices.

Acknowledgments The authors are highly thankful to SERB (EEQ/2021/001083), National Mission of Himalayan studies (NMHS), G.B. Pant National Institute of Himalayan Environment (NIHE), MOEF&CC (NIHE/NMHS/24-25/SC-XIII/MG/SL-14), Council of Science and Technology, Uttar Pradesh (CST-UP) (CST/D/Biotech-1280) and Banaras Hindu University, IoE (6031) incentives grant for providing support for applied research in plant-microbe interactions. BNS is thankful to RJP-PDF and BHU for providing support. AM is thankful to PMRF.

References

AbuQamar SF, El-Saadony MT, Saad AM, Desoky E-SM, Elrys AS, El-Mageed TAA, Semida WM, Abdelkhalik A, Mosa WFA, Al Kafaas SS, Naser S, Ibrahim EH, Alshamsi FMK, Mathew BT, El-Tarabily KA (2024) Halotolerant plant growth-promoting rhizobacteria improve soil fertility and plant salinity tolerance for sustainable agriculture—a review. Plant Stress 12:100482. https://doi.org/10.1016/j.stress.2024.100482

Aguirre-Monroy AM, Santana-Martínez JC, Dussán J (2019) *Lysinibacillus sphaericus* as a nutrient enhancer during fire-impacted soil replantation. Appl Environ Soil Sci 2019:1–8. https://doi.org/10.1155/2019/3075153

Akhtar O, Kehri HK, Zoomi I (2020) Arbuscular mycorrhiza and *Aspergillus terreus* inoculation along with compost amendment enhance the phytoremediation of Cr-rich technosol by *Solanum lycopersicum* under field conditions. Ecotoxicol Environ Saf 201:110869. https://doi.org/10.1016/j.ecoenv.2020.110869

Al-Ansari MM, Al-Humaid LA (2025) Sustainable approach to treat the hazardous medical waste polluted soil under greenhouse condition with *Jatropha integerrima*. J Taiwan Inst Chem Eng 166:105526. https://doi.org/10.1016/j.jtice.2024.105526

Alsharif W, Saad MM, Hirt H (2020) Desert microbes for boosting sustainable agriculture in extreme environments. Front Microbiol 11. https://doi.org/10.3389/fmicb.2020.01666

Andreolli M, Lampis S, Brignoli P, Vallini G (2015) Bioaugmentation and biostimulation as strategies for the bioremediation of a burned woodland soil contaminated by toxic hydrocarbons: A comparative study. J Environ Manag 153:121–131. https://doi.org/10.1016/j.jenvman.2015.02.007

Ayangbenro AS, Babalola OO (2021) Reclamation of arid and semi-arid soils: the role of plant growth-promoting archaea and bacteria. Curr Plant Biol 25:100173. https://doi.org/10.1016/j.cpb.2020.100173

Baldrian P (2016) Forest microbiome: diversity, complexity and dynamics. FEMS Microbiol Rev fuw040. https://doi.org/10.1093/femsre/fuw040

Bothe H (2012) Arbuscular mycorrhiza and salt tolerance of plants. Symbiosis 58(1–3):7–16. https://doi.org/10.1007/s13199-012-0196-9

Bridges EM (2019) In: Bridges EM, Hannam ID, Oldeman LR, de Vries FWTP, Scherr SJ, Sombatpanit S (eds) *Response to land degradation*. CRC Press. https://doi.org/10.1201/9780429187957

Cerqueira VS, Hollenbach EB, Maboni F, Vainstein MH, Camargo FAO, Peralba M, do C. R, Bento FM (2011) Biodegradation potential of oily sludge by pure and mixed bacterial cultures. Bioresour Technol 102(23):11003–11010. https://doi.org/10.1016/j.biortech.2011.09.074

Certini G (2005) Effects of fire on properties of forest soils: a review. Oecologia 143(1):1–10. https://doi.org/10.1007/s00442-004-1788-8

Chandrasekaran M, Boughattas S, Hu S, Oh S-H, Sa T (2014) A meta-analysis of arbuscular mycorrhizal effects on plants grown under salt stress. Mycorrhiza 24(8):611–625. https://doi.org/10.1007/s00572-014-0582-7

Chaudhry Q, Blom-Zandstra M, Gupta SK, Joner E (2005) Utilising the synergy between plants and rhizosphere microorganisms to enhance breakdown of organic pollutants in the environment (15 pp). Environ Sci Pollut Res—Int 12(1):34–48. https://doi.org/10.1065/espr2004.08.213

Chen M, Xu P, Zeng G, Yang C, Huang D, Zhang J (2015) Bioremediation of soils contaminated with polycyclic aromatic hydrocarbons, petroleum, pesticides, chlorophenols and heavy metals by composting: applications, microbes and future research needs. Biotechnol Adv 33(6):745–755. https://doi.org/10.1016/j.biotechadv.2015.05.003

Chen C, Xin K, Liu H, Cheng J, Shen X, Wang Y, Zhang L (2017) *Pantoea alhagi*, a novel endophytic bacterium with ability to improve growth and drought tolerance in wheat. Sci Rep 7(1):41564. https://doi.org/10.1038/srep41564

Cheng Z, Woody OZ, McConkey BJ, Glick BR (2012) Combined effects of the plant growth-promoting bacterium pseudomonas putida UW4 and salinity stress on the *Brassica napus* proteome. Appl Soil Ecol 61:255–263. https://doi.org/10.1016/j.apsoil.2011.10.006

Chiquoine LP, Abella SR, Bowker MA (2016) Rapidly restoring biological soil crusts and ecosystem functions in a severely disturbed desert ecosystem. Ecol Appl 26(4):1260–1272. https://doi.org/10.1002/15-0973

Chouhan GK, Mukherjee A, Singh S, Gaurav AK, Kumari B, Verma JP (2023) Developing eco-friendly endophytic bioinoculants for enhancing productivity and soil fertility in wheat. Agron J 115(4):1640–1652. https://doi.org/10.1002/agj2.21349

DeBano LF (2000) The role of fire and soil heating on water repellency in wildland environments: a review. J Hydrol 231–232:195–206. https://doi.org/10.1016/S0022-1694(00)00194-3

Doerr SH, Cerdá A (2005) Fire effects on soil system functioning: new insights and future challenges. Int J Wildland Fire 14(4):339. https://doi.org/10.1071/WF05094

Etesami H, Beattie GA (2018) Mining halophytes for plant growth-promoting halotolerant bacteria to enhance the salinity tolerance of non-halophytic crops. Front Microbiol 9. https://doi.org/10.3389/fmicb.2018.00148

Fang S, Hua C, Yang J, Liu F, Wang L, Wu D, Ren L (2025) Combined pollution of soil by heavy metals, microplastics, and pesticides: mechanisms and anthropogenic drivers. J Hazard Mater 485:136812. https://doi.org/10.1016/j.jhazmat.2024.136812

Ferrenberg S, O'Neill SP, Knelman JE, Todd B, Duggan S, Bradley D, Robinson T, Schmidt SK, Townsend AR, Williams MW, Cleveland CC, Melbourne BA, Jiang L, Nemergut DR (2013) Changes in assembly processes in soil bacterial communities following a wildfire disturbance. ISME J 7(6):1102–1111. https://doi.org/10.1038/ismej.2013.11

García-Carmona M, Lepinay C, García-Orenes F, Baldrian P, Arcenegui V, Cajthaml T, Mataix-Solera J (2022) Moss biocrust accelerates the recovery and resilience of soil microbial communities in fire-affected semi-arid Mediterranean soils. Sci Total Environ 846:157467. https://doi.org/10.1016/j.scitotenv.2022.157467

Goyal T, Mukherjee A, Chouhan GK, Gaurav AK, Kumar D, Abeysinghe S, Verma JP (2024) Impact of bacterial volatiles on the plant growth attributes and defense mechanism of rice seedling. Heliyon 10(8):e29692. https://doi.org/10.1016/j.heliyon.2024.e29692

Green JK, Seneviratne SI, Berg AM, Findell KL, Hagemann S, Lawrence DM, Gentine P (2019) Large influence of soil moisture on long-term terrestrial carbon uptake. Nature 565(7740):476–479. https://doi.org/10.1038/s41586-018-0848-x

Harirchi S, Sar T, Ramezani M, Aliyu H, Etemadifar Z, Nojoumi SA, Yazdian F, Awasthi MK, Taherzadeh MJ (2022) Bacillales: from taxonomy to biotechnological and industrial perspectives. Microorganisms 10(12):2355. https://doi.org/10.3390/microorganisms10122355

Hasan A, Tabassum B, Hashim M, Khan N (2024) Role of Plant Growth Promoting Rhizobacteria (PGPR) as a plant growth enhancer for sustainable agriculture: A review. Bacteria 3(2):59–75. https://doi.org/10.3390/bacteria3020005

He S, Li L, Lv M, Wang R, Wang L, Yu S, Gao Z, Li X (2024) PGPR: key to enhancing crop productivity and achieving sustainable agriculture. Curr Microbiol 81(11):377. https://doi.org/10.1007/s00284-024-03893-5

Huang C-C, Liang C-M, Yang T-I, Chen J-L, Wang W-K (2021) Shift of bacterial communities in heavy metal-contaminated agricultural land during a remediation process. PLoS One 16(7):e0255137. https://doi.org/10.1371/journal.pone.0255137

Hurni H, Giger M, Liniger H, Mekdaschi Studer R, Messerli P, Portner B, Schwilch G, Wolfgramm B, Breu T (2015) Soils, agriculture and food security: the interplay between ecosystem functioning and human well-being. Curr Opin Environ Sustain 15:25–34. https://doi.org/10.1016/j.cosust.2015.07.009

Jha Y, Subramanian RB, Patel S (2011) Combination of endophytic and rhizospheric plant growth promoting rhizobacteria in *Oryza sativa* shows higher accumulation of osmoprotectant against saline stress. Acta Physiol Plant 33(3):797–802. https://doi.org/10.1007/s11738-010-0604-9

Juwarkar AA (2012) Microbe-assisted phytoremediation for restoration of biodiversity of degraded lands: A sustainable solution. Proceedings of the National Academy of Sciences, India Section B: Biological Sciences. https://doi.org/10.1007/s40011-012-0098-x

Kang S-M, Shahzad R, Bilal S, Khan AL, Park Y-G, Lee K-E, Asaf S, Khan MA, Lee I-J (2019) Indole-3-acetic-acid and ACC deaminase producing Leclercia adecarboxylata MO1 improves *Solanum lycopersicum* L. growth and salinity stress tolerance by endogenous secondary metabolites regulation. BMC Microbiol 19(1):80. https://doi.org/10.1186/s12866-019-1450-6

Kawasaki A, Dennis PG, Forstner C, Raghavendra AKH, Mathesius U, Richardson AE, Delhaize E, Gilliham M, Watt M, Ryan PR (2021) Manipulating exudate composition from root apices shapes the microbiome throughout the root system. Plant Physiol 187(4):2279–2295. https://doi.org/10.1093/plphys/kiab337

Khademian R, Asghari B, Sedaghati B, Yaghoubian Y (2019) Plant beneficial rhizospheric microorganisms (PBRMs) mitigate deleterious effects of salinity in sesame (Sesamum indicum L.): physio-biochemical properties, fatty acids composition and secondary metabolites content. Ind Crop Prod 136:129–139. https://doi.org/10.1016/j.indcrop.2019.05.002

Khan MA, Asaf S, Khan AL, Adhikari A, Jan R, Ali S, Imran M, Kim K-M, Lee I-J (2020) Plant growth-promoting endophytic bacteria augment growth and salinity tolerance in rice plants. Plant Biol 22(5):850–862. https://doi.org/10.1111/plb.13124

Kim E-J, Oh J-E, Chang Y-S (2003) Effects of forest fire on the level and distribution of PCDD/Fs and PAHs in soil. Sci Total Environ 311(1–3):177–189. https://doi.org/10.1016/S0048-9697(03)00095-0

Koul B, Taak P (2018) Biotechnological strategies for effective remediation of polluted soils. Springer Singapore. https://doi.org/10.1007/978-981-13-2420-8

Kraaijvanger R, Veldkamp T (2015) Grain productivity, fertilizer response and nutrient balance of farming systems in Tigray, Ethiopia: A multi-perspective view in relation to soil fertility degradation. Land Degrad Dev 26(7):701–710. https://doi.org/10.1002/ldr.2330

Kumar Arora N, Fatima T, Mishra J, Mishra I, Verma S, Verma R, Verma M, Bhattacharya A, Verma P, Mishra P, Bharti C (2020) Halo-tolerant plant growth promoting rhizobacteria for improving productivity and remediation of saline soils. J Adv Res 26:69–82. https://doi.org/10.1016/j.jare.2020.07.003

Kumar A, Singh S, Mukherjee A, Rastogi RP, Verma JP (2021) Salt-tolerant plant growth-promoting *Bacillus pumilus* strain JPVS11 to enhance plant growth attributes of rice and improve soil health under salinity stress. Microbiol Res 242:126616. https://doi.org/10.1016/j.micres.2020.126616

Li C, Xu Y, Jiang W, Dong X, Wang D, Liu B (2013) Effect of NaCl on the heavy metal tolerance and bioaccumulation of *Zygosaccharomyces rouxii* and *Saccharomyces cerevisiae*. Bioresour Technol 143:46–52. https://doi.org/10.1016/j.biortech.2013.05.114

Liang G, Li S, Yu X, Bu Q, Qu H, Zhu H, Yao X, Lu A, Gong W (2022) Black carbon-mediated degradation of organic pollutants: a critical review. Process Saf Environ Prot 160:610–619. https://doi.org/10.1016/j.psep.2022.02.049

Lu M, Cheng S, Fang H, Xu M, Yang Y, Li Y, Zhang J, Müller C (2021) Organic nitrogen addition causes decoupling of microbial nitrogen cycles by stimulating gross nitrogen transformation in a temperate forest soil. Geoderma 385:114886. https://doi.org/10.1016/j.geoderma.2020.114886

Lugtenberg B (ed) (2015) Principles of plant-microbe interactions. Springer International Publishing. https://doi.org/10.1007/978-3-319-08575-3

Mahata S, Sharma VN (2021) The global problem of land degradation: a review. Nat Geographic J India 67(2):216–231. https://doi.org/10.48008/ngji.1773

Morán-López T, Fernández M, Alonso CL, Flores-Rentería D, Valladares F, Díaz M (2015) Effects of forest fragmentation on the oak–rodent mutualism. Oikos 124(11):1482–1491. https://doi.org/10.1111/oik.02061

Morrissey EM, Gillespie JL, Morina JC, Franklin RB (2014) Salinity affects microbial activity and soil organic matter content in tidal wetlands. Glob Chang Biol 20(4):1351–1362. https://doi.org/10.1111/gcb.12431

Muhammad H, Ijaz M, Sattar A, Ul-Allah S, Sher A, Asif M, Dilshad M, Mahmood K, Riaz MW, Zaheer MS, Rizwan M, Manoharadas S (2025) Synergistic effects of PGPRs and fertilizer amendments on improving the yield and productivity of Canola (*Brassica napus* L.). BMC Plant Biol 25(1):50. https://doi.org/10.1186/s12870-025-06062-3

Mumtaz MZ, Ahmad M, Etesami H, Mustafa A (2024) Editorial: mineral solubilizing microorganisms (MSM) and their applications in nutrient bioavailability, bioweathering and bioremediation, volume II. Front Microbiol 14. https://doi.org/10.3389/fmicb.2023.1345161

Munns R, Tester M (2008) Mechanisms of salinity tolerance. Annu Rev Plant Biol 59(1):651–681. https://doi.org/10.1146/annurev.arplant.59.032607.092911

Pinedo I, Ledger T, Greve M, Poupin MJ (2015) *Burkholderia phytofirmans* PsJN induces long-term metabolic and transcriptional changes involved in *Arabidopsis thaliana* salt tolerance. Front Plant Sci 6. https://doi.org/10.3389/fpls.2015.00466

Pishchik VN, Filippova PS, Mirskaya GV, Khomyakov YV, Vertebny VE, Dubovitskaya VI, Ostankova YV, Semenov AV, Chakrabarty D, Zuev EV, Chebotar VK (2021) Epiphytic PGPB *Bacillus megaterium* AFI1 and *Paenibacillus nicotianae* AFI2 improve wheat growth and antioxidant status under Ni stress. Plants 10(11):2334. https://doi.org/10.3390/plants10112334

Popescu G, Dumitru L (2009) Biosorption of some heavy metals from media with high salt concentrations by halophilic *Archaea*. Biotechnology & Biotechnological Equipment 23(sup1):791–795. https://doi.org/10.1080/13102818.2009.10818542

Preece C, Peñuelas J (2020) A return to the wild: root exudates and food security. Trends Plant Sci 25(1):14–21. https://doi.org/10.1016/j.tplants.2019.09.010

Quinton JN, Fiener P (2024) Soil erosion on arable land: an unresolved global environmental threat. Prog Phys Geog Earth Environ 48(1):136–161. https://doi.org/10.1177/03091333231216595

Radhapriya P, Ramachandran A, Palani P (2018) Indigenous plant growth-promoting bacteria enhance plant growth, biomass, and nutrient uptake in degraded forest plants. 3. Biotech 8(3):154. https://doi.org/10.1007/s13205-018-1179-1

Rahina K, Muhammad IC, Muhammad IA, Muhammad R, Obaid UR, Ayesha M, Muhammad AURK, Muhammad Z, Sarosh A (2018) Response of groundnut (*Arachis hypogaea* L.) to plant growth promoting Rhizobacteria in degraded soils. Afr J Agric Res 13(17):904–910. https://doi.org/10.5897/AJAR2018.13081

Raiesi F, Beheshti A (2015) Microbiological indicators of soil quality and degradation following conversion of native forests to continuous croplands. Ecol Indic 50:173–185. https://doi.org/10.1016/j.ecolind.2014.11.008

Rashid MI, Mujawar LH, Shahzad T, Almeelbi T, Ismail IMI, Oves M (2016) Bacteria and fungi can contribute to nutrients bioavailability and aggregate formation in degraded soils. Microbiol Res 183:26–41. https://doi.org/10.1016/j.micres.2015.11.007

Ravi S, D'Odorico P, Zobeck TM, Over TM (2009) The effect of fire-induced soil hydrophobicity on wind erosion in a semiarid grassland: experimental observations and theoretical framework. Geomorphology 105(1–2):80–86. https://doi.org/10.1016/j.geomorph.2007.12.010

Robinson DA, Hopmans JW, Filipovic V, van der Ploeg M, Lebron I, Jones SB, Reinsch S, Jarvis N, Tuller M (2019) Global environmental changes impact soil hydraulic functions through biophysical feedbacks. Glob Chang Biol 25(6):1895–1904. https://doi.org/10.1111/gcb.14626

Roper MM (2006) Potential for remediation of water repellent soils by inoculation with wax-degrading bacteria in South-Western Australia. Biologia 61(S19):S358–S362. https://doi.org/10.2478/s11756-006-0189-3

Rudrappa T, Czymmek KJ, Paré PW, Bais HP (2008) Root-secreted malic acid recruits beneficial soil bacteria. Plant Physiol 148(3):1547–1556. https://doi.org/10.1104/pp.108.127613

Rusănescu CO, Rusănescu M (2023) Application of fly ash obtained from the incineration of municipal solid waste in agriculture. Appl Sci 13(5):3246. https://doi.org/10.3390/app13053246

Sadaf T, Sehar S, Hassan S (2024) Agricultural revolutions and food security. In: Food security in the developing world. Wiley, pp 95–122. https://doi.org/10.1002/9781119265214.ch5

Sales da Silva IG, Gomes de Almeida FC, Padilha da Rocha e Silva NM, Casazza AA, Converti A, Asfora Sarubbo L (2020) Soil bioremediation: overview of technologies and trends. Energies 13(18):4664. https://doi.org/10.3390/en13184664

Samain E, Ernenwein C, Aussenac T, Selim S (2022) Effective and durable systemic wheat-induced resistance by a plant-growth-promoting rhizobacteria consortium of *Paenibacillus* sp. strain B2 and *Arthrobacter* spp. strain AA against *Zymoseptoria tritici* and drought stress. Physiol Mol Plant Pathol 119:101830. https://doi.org/10.1016/j.pmpp.2022.101830

Seneviratne SI, Corti T, Davin EL, Hirschi M, Jaeger EB, Lehner I, Orlowsky B, Teuling AJ (2010) Investigating soil moisture–climate interactions in a changing climate: A review. Earth Sci Rev 99(3–4):125–161. https://doi.org/10.1016/j.earscirev.2010.02.004

Setia R, Smith P, Marschner P, Gottschalk P, Baldock J, Verma V, Setia D, Smith J (2012) Simulation of salinity effects on past, present, and future soil organic carbon stocks. Environ Sci Technol 46(3):1624–1631. https://doi.org/10.1021/es2027345

Singh NK, Choudhary S (2021) Bacterial and archaeal diversity in oil fields and reservoirs and their potential role in hydrocarbon recovery and bioprospecting. Environ Sci Pollut Res 28(42):58819–58836. https://doi.org/10.1007/s11356-020-11705-z

Singh JS, Seneviratne G (eds) (2017) Agro-environmental sustainability. Springer International Publishing. https://doi.org/10.1007/978-3-319-49724-2

Song Q, Li X, Zhao Z, Li D, Li Y, Hou N (2023) Insights into ecotoxicity effects of PAHs and ecosystem responses of remediation strategies under cold stress: from PAHs degradation to ecological restoration regulated by signal molecular. Chem Eng J 475:146042. https://doi.org/10.1016/j.cej.2023.146042

Sun W, Shahrajabian MH (2025) Biostimulant and beyond: bacillus spp., the important Plant Growth-Promoting Rhizobacteria (PGPR)-based biostimulant for sustainable agriculture. Earth Syst Environ. https://doi.org/10.1007/s41748-024-00552-4

Sziderics AH, Rasche F, Trognitz F, Sessitsch A, Wilhelm E (2007) Bacterial endophytes contribute to abiotic stress adaptation in pepper plants (*Capsicum annuum* L.). Can J Microbiol 53(11):1195–1202. https://doi.org/10.1139/W07-082

Tang X, Mu X, Shao H, Wang H, Brestic M (2015) Global plant-responding mechanisms to salt stress: physiological and molecular levels and implications in biotechnology. Crit Rev Biotechnol 35(4):425–437. https://doi.org/10.3109/07388551.2014.889080

Ullah A, Bano A (2019) Role of PGPR in the reclamation and revegetation of saline land. Pak J Bot 51(1). https://doi.org/10.30848/PJB2019-1(43)

Upadhyay SK, Rajput VD, Kumari A, Espinosa-Saiz D, Menendez E, Minkina T, Dwivedi P, Mandzhieva S (2023) Plant growth-promoting rhizobacteria: a potential bio-asset for restoration of degraded soil and crop productivity with sustainable emerging techniques. Environ Geochem Health 45(12):9321–9344. https://doi.org/10.1007/s10653-022-01433-3

Vílchez JI, García-Fontana C, Román-Naranjo D, González-López J, Manzanera M (2016) Plant drought tolerance enhancement by Trehalose production of desiccation-tolerant microorganisms. Front Microbiol 7. https://doi.org/10.3389/fmicb.2016.01577

Wang H, Zhang H, Zhang X, Li Q, Cheng C, Shen H, Zhang Z (2020) Bioelectrochemical remediation of Cr (VI)/Cd (II)-contaminated soil in bipolar membrane microbial fuel cells. Environ Res 186:109582. https://doi.org/10.1016/j.envres.2020.109582

Watson BS, Bedair MF, Urbanczyk-Wochniak E, Huhman DV, Yang DS, Allen SN, Li W, Tang Y, Sumner LW (2015) Integrated metabolomics and transcriptomics reveal enhanced specialized metabolism in *Medicago truncatula* root border cells. Plant Physiol 167(4):1699–1716. https://doi.org/10.1104/pp.114.253054

Wood JL, Liu W, Tang C, E Franks A (2016) Microorganisms in heavy metal bioremediation: strategies for applying microbial-community engineering to remediate soils. AIMS Bioengin 3(2):211–229. https://doi.org/10.3934/bioeng.2016.2.211

Xing P, Zhao Y, Guan D, Li L, Zhao B, Ma M, Jiang X, Tian C, Cao F, Li J (2022) Effects of *Bradyrhizobium* co-inoculated with *Bacillus* and *Paenibacillus* on the structure and functional genes of soybean rhizobacteria community. Genes 13(11):1922. https://doi.org/10.3390/genes13111922

Yannikos N, Leinweber P, Helgason BL, Baum C, Walley FL, Van Rees KCJ (2014) Impact of Populus trees on the composition of organic matter and the soil microbial community in Orthic Gray Luvisols in Saskatchewan (Canada). Soil Biol Biochem 70:5–11. https://doi.org/10.1016/j.soilbio.2013.11.025

Yasmin H, Naeem S, Bakhtawar M, Jabeen Z, Nosheen A, Naz R, Keyani R, Mumtaz S, Hassan MN (2020) Halotolerant rhizobacteria *Pseudomonas pseudoalcaligenes* and Bacillus subtilis mediate systemic tolerance in hydroponically grown soybean (*Glycine max* L.) against salinity stress. PLoS One 15(4):e0231348. https://doi.org/10.1371/journal.pone.0231348

Zhang Q, Dai W (2019) Plant response to salinity stress. In: Stress physiology of Woody plants. CRC Press, pp 155–173. https://doi.org/10.1201/9780429190476-7

Zhao S, Yuan X-T, Wang X-H, Ai Y-J, Li F-P (2024) Research progress and hotspots in microbial remediation for polluted soils. Sustainability 16(17):7458. https://doi.org/10.3390/su16177458

Decision-Making in Forest Landscape Restoration

14

Bhimappa Honnappa Kittur, Manoj Kumar Jhariya,
Chandra Prakash Kala, Arnab Banerjee, and Abhishek Raj

Abstract

Despite recognizing the significance of forests, they are continuously disrupted by anthropogenic activities, though policies globally exist to control deforestation and land degradation stress more on prevention. However, framing of policy for halting deforestation and land degradation is insufficient; mandatory restoration efforts need to be undertaken to expedite the recovery of degraded forests. Forest landscape restoration stands out as the ultimate solution for revitalizing degraded forests. Selecting the right restoration approach is crucial, as missteps in restoration and management, such as incorrect site and species selection, using unhealthy seedlings, planting offseason, inadequate prevention of grazing and forest fires, and unresolved stakeholder conflicts, can hinder restoration success. Engaging all stakeholders in decision-making should be well-planned following preliminary studies. The chosen restoration approach must suit the site conditions, be cost-effective, and be feasibly completed within the allotted timeframe. In severely degraded or destroyed forests, where no hope in natural regeneration, active restoration would be a preferred option. Conversely, natural or passive restoration becomes a viable choice in areas where adequate mother trees for

B. H. Kittur · C. P. Kala
Faculty of Technical Forestry, Indian Institute of Forest Management, Bhopal, MP, India

M. K. Jhariya (✉)
Department of Farm Forestry, Sant Gahira Guru Vishwavidyalaya, Ambikapur, Chhattisgarh, India

A. Banerjee
Department of Environmental Science, Sant Gahira Guru Vishwavidyalaya, Ambikapur, Chhattisgarh, India

A. Raj
Pt. Deendayal Upadhyay College of Horticulture and Forestry, Dr. Rajendra Prasad Central Agriculture University, Samastipur, Bihar, India

V. C. Pandey (ed.), *Plantation-Based Land Restoration*,
https://doi.org/10.1007/978-981-95-2860-8_14

seed production and dispersal mechanisms exist. In areas where financial constraints and time limitations and less canopy gaps exist, partial interference through assisted means would be the ultimate choice and expedite the restoration process.

Keywords

Afforestation · Deforestation · Decision-making · Forest restoration · Active or passive restoration · Regeneration

14.1 Introduction

Forest cover decreased significantly in the last decade (2010–2020) by 7.84 million ha (FAO 2020) due to forest disturbances (forest fire, grazing, encroachment, landslides, forest produce collection, etc.), either natural or anthropogenic, poverty, population pressure, infrastructural development and intensive agriculture, and species invasion (Awasthi et al. 2022a), which causes an increase in CO_2 in the atmosphere (Bisht et al. 2023). The natural disturbances are vital (Lado et al. 2019; Kuuluvainen et al. 2021), and these disturbances catalyze for spontaneous change in the ecosystem (Renato et al. 2017; Thom et al. 2017). Conversely, anthropogenic disturbances cause adverse impact (Timothy and Ayten 2013), and these disturbances are changing in response to the advancement of human culture (Seidl et al. 2017). Among the various human-caused disturbances, forest fires have received much concern (Sarah 2020) and are increasingly reported in recent years (Martin et al. 2016; Rhett 2019; Rebecca et al. 2021). These disturbances effectively influence the fertility and stability of an ecosystem (Manral et al. 2022), initially causing the ecosystem to lose the microbial interactions, affecting the nutrient cycling and soil health and reducing the ecosystem productivity (Manral et al. 2022). With the development of time these changes and modifications in the ecosystems affect the ecosystem structure and functioning (Awasthi et al. 2022a, b). Hansen et al. (2013) reported global forest loss of 2.3 million hectares due to forest fire. A numerical estimate of global terrestrial ecosystems, the boreal forest, accounts for 29% (Lafleur et al. 2018) and is the best-studied ecosystem, which has undergone significant loss (65.2%) due to forest fire alone (Lafleur et al. 2018). In India, approximately 14,000 km^2 of forested areas were cleared in the last three decades due to hydroelectricity production, defense, and mining. Notably, the mining is accountable for the deforestation of 5000 km^2 and about disturbance of 115.3 km^2 *Shorea robusta* forest area and a loss of ~2.5 Tg of biomass that led to loss of 1.08 Tg C (~3.98 Tg of CO_2) over the 20 years (Mishra et al. 2022). Deforestation-induced warming and drying in the Amazon contribute to an extra $5.1 \pm 3.7\%$ decline in aboveground biomass. Similarly, biophysical impacts exacerbate carbon losses in the Congo by $3.8 \pm 2.5\%$ (Li et al. 2022). Similar studies highlighted anthropogenic deforestation across the globe (Li et al. 2015; Tan et al. 2022; Mon et al. 2023). Anthropogenic CO_2 emission and its related negative consequences on the environment were observed due to industrial development, mining, deforestation, and intensive agricultural practices. This unstoppable rising CO_2 concentration impairs key environmental services and its sustainability. Recently, the NOAA-based Mauna Loa Atmospheric Baseline Observatory has reported CO_2 concentration of about 419 ppm in 2021 along with 40 billion MT of CO_2 pollution every year in the environment. Therefore, it is crucial to halt the deforestation and implement management interventions aimed at fostering of forest restoration practices incorporating optimal silvicultural principles using ecological interactions.

14.2 Forest Landscape Restoration (FLR)

FLR (forest landscape restoration) is a continuous process that aims to work toward the proper functioning of ecosystem as well as the improvement of the socioeconomic condition of local community stakeholders under degraded habitat conditions that will move toward sustainable development (Fig. 14.1, WWF and IUCN 2020; Banerjee et al. 2023). Further, restoration is an approach that aims towards restoring the ecological services of forest ecosystem within a certain geographical area and providing diverse ecological outputs for human wellbeing (Van Oosten et al. 2014). Florida has gained global recognition for its successful restoration efforts—such as the Kissimmee River project—which have made significant contributions toward achieving the SDGs (Whalen et al. 2002), followed by restoration initiative at prairie ecosystem of the USA and the sustainable water management of the Netherlands (Brye et al. 2002; Mostert 2006). Although country's specific restoration program has been initiated, a proper institutional approach yet to be formed properly.

The FLR technique involves silvicultural interventions aimed at recovery of forest landscape. This includes introducing new species through seeding or planting, enhancing soil and water holding capacity, stabilizing slopes, etc. Overall, restoration is a recovery approach tailored to the type of disturbance and can be implemented across various scales, such as mining sites, landslides, grazed areas, and fire-affected forests (Macdonald et al. 2015). On the basis of the level of degradation, forest ecosystem has been classified in various ways, such as damaged, destroyed, and degraded forests. In case of degraded forest areas, they are characterized by partial removal of forest cover along with the presence of an undisturbed seed bank and soil conditions. In such sites, a "natural restoration" approach for halting the drivers of degradation and protection would be suitable for the natural regeneration (Ghazoul and Chazdon 2017). Damaged ecosystems are conditions where the soil system remains untouched while wide sources of seed dispersal are scanty (Di Sacco et al. 2021). Management of these ecosystems would be done through "assisted restoration" of native species with an objective to enhance the vegetation biodiversity. On the other hand, in case of destroyed forest, there is complete destruction of both function and structure through various biotic and abiotic factors. Reclamation or "reconstructive restoration" implies the application of vegetal cover for restoration purpose, where normal seeding procedure for reestablishment of vegetation becomes difficult due to presence of toxic pollutant substances

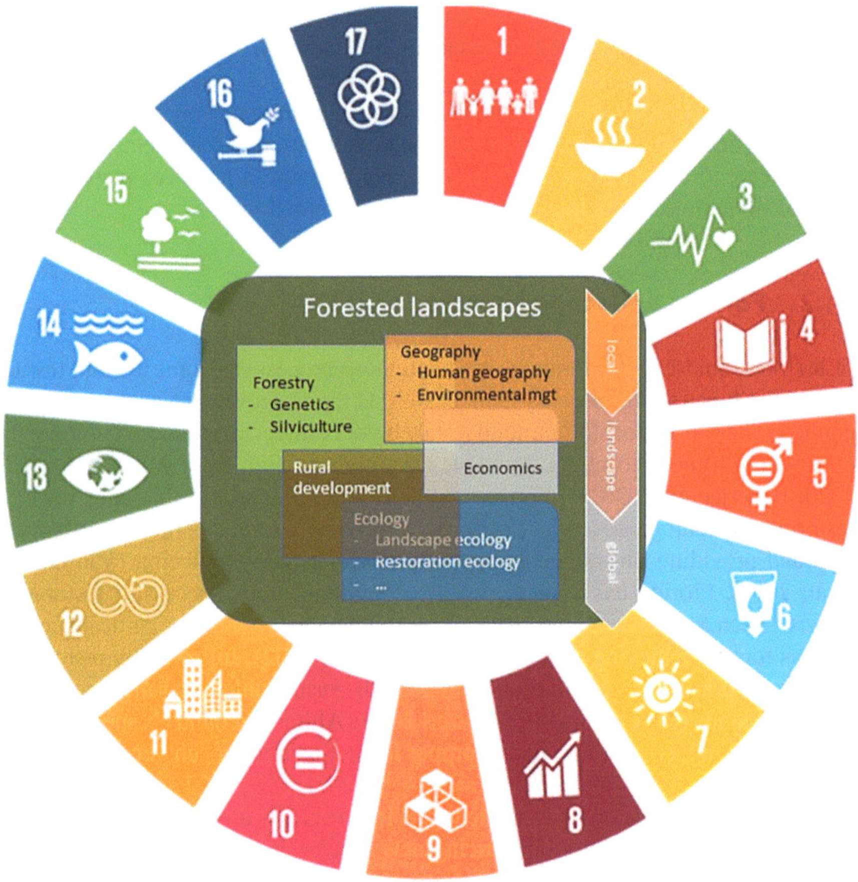

Fig. 14.1 A conceptual interconnection of forest landscapes to achieve Sustainable Development Goals (SDGs)

in the soil ecosystem (Wong 2003). Compare to natural restoration process, reconstructive approach through active restoration and assisted natural regeneration is mediated through planting of nursery grown seedlings followed by direct seed sowing methodology. However, the nature of restoration strategies varies according to the nature of the degraded forest, time frame, causal factors, as well as the existing conditions (Crouzeilles et al. 2016).

14.3 Forest Landscape Restoration to Meet Sustainable Development Goals (SDGs)

Forests play a pivotal role in attaining the SDGs. Forest vegetation provides a wide range of ecosystem goods and services to the inhabitants and stability to nature (Gosain et al. 2015). The UN recurrently stresses the significance of halting deforestation and forest degradation as crucial steps to achieve SDGs. For instance, the Billion Trees Afforestation Project restored the landscapes as well as created opportunities for employment and economic earning through various forest-related activities that have improved livelihoods in Pakistan (Goal 1; No Poverty; Ullah et al. 2022). It was also reported that restoring the potential area can store carbon up to 56–69% helping to mitigate climate change (Goal 13 Climate Action; Koch and Kaplan 2022). Furthermore, trees have a pivotal role in holding water and soil contaminants through rhizo-deposition (Goal 6; Clean Water and Sanitation; Bakshe and Jugade 2023). Forests provide fruits, roots, leaves, medicinal products, etc., serving as a source of food, especially during periods of famine (e.g., bamboo seeds) (Goal 2; Zero Hunger).

Although forests are vital for balancing the ecosystem and its functioning, the current situation is challenging because countries need a sustainable exploitation of resources, as suggested by the United Nations, in order to meet good environmental status and also to reduce or alleviate poverty. The sustainable forest utilization and management is a win-win situation that meets at least 8 of the 17 UN SDGs (Fig. 14.1) linking to reduce desertification, halt biodiversity loss, and reverse land degradation (Mahari et al. 2020). Hence, global target-bound restoration initiatives for emission reduction due to degradation and deforestation (Alexander et al. 2011), such as the Bonn Challenge (Lewis et al. 2019), the New York Forest Declaration (Suding et al. 2015), and regional initiatives, such as the Delhi declaration commitment on restoration of forest area up to 26 million hectares till 2030 (UNCCD 2016), and the 20 × 20 and AFR100 schemes aim toward restoration of 20 million and 100 million ha of forest areas, respectively, in Latin America and Africa (WRI 2023).

14.4 Landscape Restoration Through Sustainable Forest Management (SFM)

FLR-based approaches strengthen the goal for biodiversity conservation, food production, human well-being, livelihood promotion, and related social and cultural values. SFM including afforestation in the perspective of landscape management that promotes sustainability in global production and ecosystem health (Jhariya et al. 2019; Banerjee et al. 2020; Raj et al. 2023). Moreover, a landscape approach and its management through scientific forestry practices can restore land quality and its fertility, which are the basis of plant-soil health and its sustainable production (Stanturf 2021; Yusof et al. 2023; Machar 2020). A synergistic relation exists among various land use practices, which needs to be studied at ecological, economic, cultural, and social aspects under a viable landscape approach. Hence, SFM is a sustainable approach that works collectively for the betterment of the health of the ecosystem and environmental quality (Denier et al. 2015). Therefore, adopting SFM, including afforestation practices, can provide more ecosystem services with a greater level of sustainability (Jhariya et al. 2022, 2024). This is due to increasing biodiversity, enhancing land/soil fertility, greater C sequestration, and climate security that improve health, income, and sustainable food production.

14.4.1 Stakeholder Participation and Conflict Management in Restoration

Forest restoration is a multidimensional (political, bureaucratic, social, ecological, economic, etc.) approach and would receive equal importance. For instance, the political approach is most crucial during planning and execution (Baker et al. 2014). Politics brings negotiations over trade-offs to distribute scarce resources among different participating communities. On the other hand, the restoration approach is long term and time-consuming (Moreno-Mateos et al. 2020); decision-making in terms of site selection and economic and sociocultural aspects is obligatory, and prolonged discussion with stakeholders, comprising forest departments, village representatives, farmers, tribals, researchers, scientist, etc., plays prominent role (Fig. 14.2).

Generally, conflicts arise during land measurement and boundary marking, site selection and management, selection of plant species, control of grazing, etc. (Becerra et al. 2021). Therefore, the potential for solving conflicts for restoration must be emphasized, and resolving such conflicts requires many rounds of negotiations (Batie 2008). The conflicts also verse with the other factors, such as miscommunication, strained relationships, and fights for power (Folger et al. 2021). Sayer

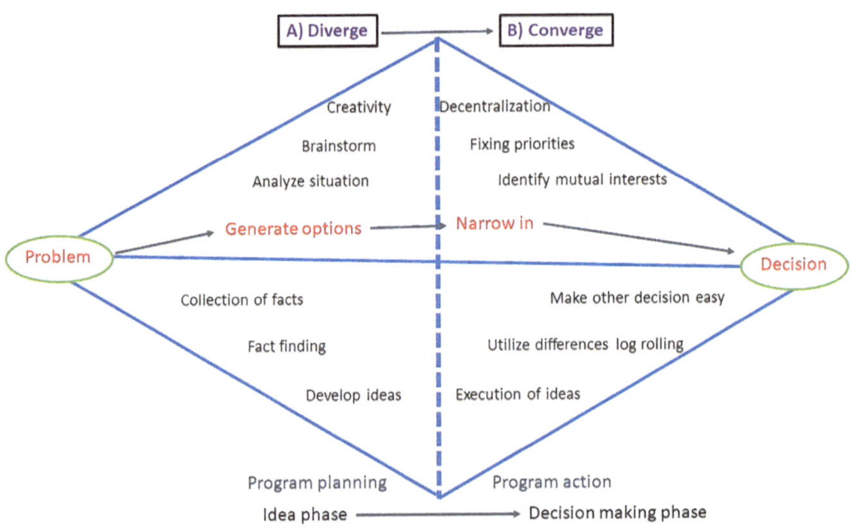

Fig. 14.2 Extension activity for conflict management in restoration approach (Modified from Jens 2007). The diagram depicts the transformation of forest extension from a divergent to a convergent by shifting the focus from scattered or disparate (problem) activities toward a unified and coordinated approach (decision). The conflict management aims to enhance integration, cooperation, and concerted efforts toward achieving shared objectives. Additionally, the diagram also illustrates a collective learning method for conflict resolution in restoration projects. Each situation must be analyzed, and actions are tailored for ideas of adaptive comanagement and knowledge negotiation for constructive problem-solving in restoration practices (Anthony 1978; Daniels and Walker 2001)

et al. (2019) reported similar instances where the restoration process fails because one person accepts work as "restoration"; another rejects the work as "degradation." Similarly, Brown (2005) also states that many individuals benefit while few face losses in terms of providing roads in personal land for reaching the restoration sites (Palmer et al. 2005) and for the prevention of the collection of fodder and firewood (Nepal and Weber 1995). Hence, conflict management plays an important role in successful restoration practice. Additionally, Jones and Dudley (2005) addressed the fundamental structure in the process of resolving the conflict raised in restoration projects, along with their negotiation principles, by using analytical tools. Further, they emphasized toward identification of proper indicators that would lead to capacity building and proper communication among the local stakeholders.

14.5 Decision-Making in Forest Restoration Project

The detailed information consisting of objectives of the project, implementation mode, and post restoration guidelines should be incorporated in the project management cycle and neatly presented to the stakeholders (Fig. 14.3). Decision on adaptation of a suitable approach is a challenging task, and models sometimes give ideal predictions in assessment of the sites and related activities. For instance, RESTS (Restoration Ecosystem Service Tool Selector) reflects the various criteria that become the key components of the assessment of ecosystem, scalability, time span, expenditure, and level of uncertainty, along with the benefit-cost ratio (Christin et al. 2016). SolVES (Social Values for Ecosystem Services) is another form that is based upon assessing the maps as well as providing an insight into the societal value of diverse ecosystem services (Sherrouse et al. 2014). ARIES (Artificial Intelligence

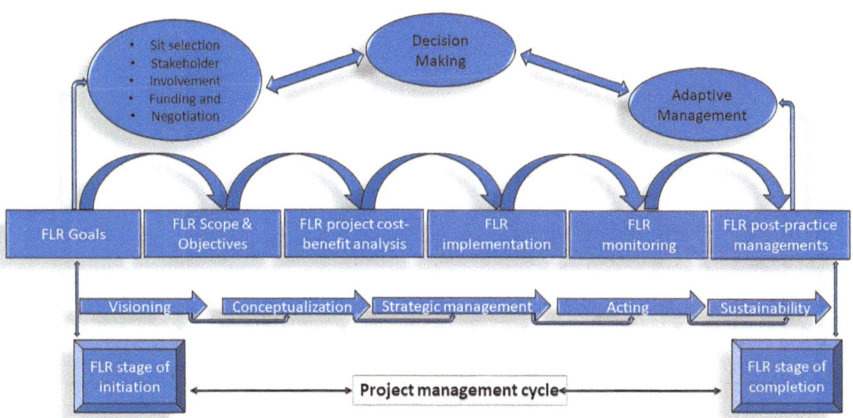

Fig. 14.3 Project Cycle Management (PCM) for FLR ration has four phases: visioning, conceptualization, strategic management, acting, and sustainability. PCM is an adaptive iterative process for moving from broad goals (a vision) and scope for improving the degraded landscapes through a conceptual stage. To ensure the sustainability of FLR, monitoring and long-term management must be integrated into the FLR project. (Modified from Stanturf et al. 2019)

for Ecosystem Services) is a model that provides a detailed analysis of ecosystem service production and their perceived use, along with associated complexity (Villa et al. 2014). However, the economic and ecological benefits anticipated to receive from restoration projects also determine the extent of the restoration project (Walters et al. 2021). Basically, the restoration economics is interdisciplinary in nature; undermining of economics in restoration projects would cause the greatest problem not only in monetary loss but also of time and work efficiency (Luz 2021). Economic aspects of uncertainty in restoration and the specificity of natural capital must be properly studied before undertaking the restoration project. Further, the ecological benefits arising from the restoration are also greatly important as these are imperative to society. Hence, while designing restoration projects, both the values (economic and ecological) must be equally considered.

14.5.1 Choice of Restoration Approach

The choice of forest restoration depends mainly on the locality factors and stakeholder requirements (Lazos-Chavero et al. 2016). Approaches to the restoration of degraded forests very much depend upon the remaining vegetation present in a habitat for natural regeneration purposes (Chazdon 2008). It was observed that under natural condition, restoration processes get increased under degraded conditions (Fig. 14.4). There is a stepwise approach regarding the degradation of ecosystem services associated with the natural, assisted natural, and active restoration approaches, linking with time, cost, and risk involved in the restoration approach

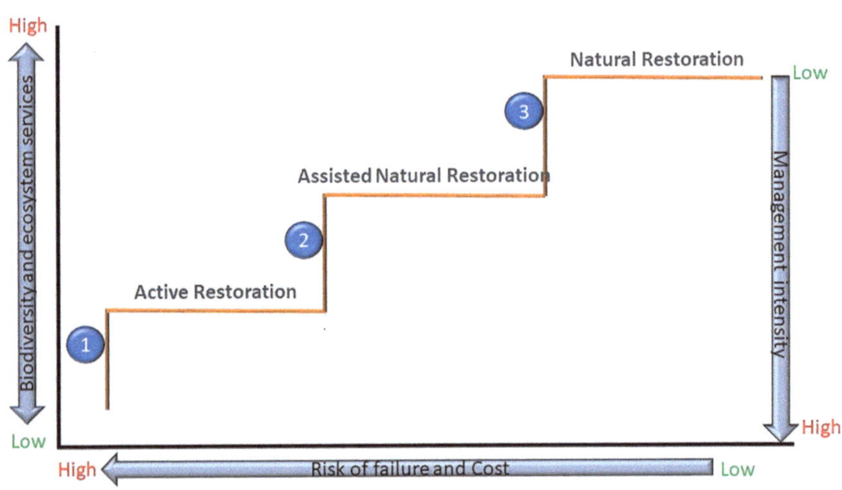

Fig. 14.4 The choice on restoration stairs for forest landscape restoration, depending on the state of degradation levels of biodiversity and ecosystem services, management intensity, and time (Revised from: Chazdon 2008; IIFM 2022). The ecosystem services increase as we move from active, assisted natural to natural, and conversely, the intensity of management is high in active and low in natural restoration approaches

(Fig. 14.4). In general, natural (passive) restoration is the best approach to achieve success of forest restoration compared to active and assisted natural restoration where the previous approach is not guaranteed (Ota et al. 2020). This would be exemplified through meta-analysis, which states that the natural restoration process is often successful in restoring tropical forest ecosystems for various bioresources as well as canopy structure and function (Crouzeilles et al. 2017). According to the research report, success in restoration in terms of vegetation structure and biodiversity appears to be 19–56% and 34–56%, respectively, higher under natural processes. They also reported that natural restoration of the forest holds minimal conservation value, revealing active restoration as the primary strategy.

Forest restoration attracts passive approach to strike a balance between the expenditure along with probable outcomes of ecosystem services. The anticipated results from passive restoration may often be perceived as uncertain. For instance, a survey was conducted in Brazil (Brancalion et al. 2016) to assess land allocation for active and passive restoration. The results revealed that active restoration has the highest level of land allocation (78.4%), followed by 7.4% land allocation in the assisted natural restoration process. Under passive restoration, areas reflect lesser percentage of canopy cover, species density, and lesser number of vegetal components. In general, restoration is an important aspect for bringing back the quality of degraded forest ecosystem. The aims of the restoration are to initiate and accelerate the recovery of land. Many factors are responsible for selecting best restoration approach suitable for a site.

14.6 Active Restoration

Active restoration assumes to be the most effective approach for target areas, achieving immediate results in terms of vegetation production. According to some research reports, active restoration was observed to be more stimulatory in comparison to passive one in terms of restoring forest structure (Trujillo-Miranda et al. 2018). In contrast, the ecological cost is more in active restoration in case of failure incidence. Such findings have been supported by Vyver et al. (2012) in their work in the Eastern Cape region of South Africa, where they underwent plantations of different-aged *Portulacaria afra, Pappea capensis,* and *Searsia longispina,* which were nursery propagated, followed by *Lycium ferocissimum* and *Rhigozum obovatum* as inter-canopy shrub in a degraded habitat. Comparative cost expenditure for plantation species was calculated between the normal procedure and the diverse species structure. Results revealed the unfeasibility of planting other canopy species, along with *P. afra,* in terms of net economic return.

It's worth highlighting that in areas where substantial degradation has occurred and there is no assurance of seed dispersal from the mother trees, in such areas, the only way for revegetation of the sites would be through an active approach (seeding or planting). Abiyu et al. (2016) conducted a field experiment on seed phenology at the northwest of Ethiopia. Results revealed better seed dispersal of *Olea europaea* in comparison to other species. Limitations in regeneration may be attributed toward a lack of suitable habitat as well as lesser dispersal ability. Considering these studies, it can be interpreted that natural regeneration is much more suitable to overcome the low dispersal ability of species and the reduction of negative effects of biotic and abiotic factors. However, in those areas, the ideal way would be the active restoration. Another advantage of the active restoration is the production of desired products by planting of fruit, leaves, timber, resin, and gum-yielding species (Program on Forests 2011).

In general, the choice of the restoration approach would be influenced by silvicultural techniques and local factors. For instance, the Indian Himalayan region is recurrently affected by natural and anthropogenic drivers and faces frequent cloud bursts, forest fires, landslides, and erosion. In addition, steep slopes, unconsolidated formations, eroded strata, a limited growing period, and shallow soils would cause degradation of forests. In order to rejuvenate such areas, a restoration approach of the degraded land of the village was done through public participation, as reported by Rawat et al. (2017). The common degraded land area was selected for the plantation of multiple species having multipurpose use through an active restoration process. Supported by appropriate bioengineering measures, proper capacity building of the local community stakeholders for effective management of the approach was taken care of. Various parameters, such as soil attributes, production of fodder, and growth attributes, followed by sequestration of carbon, were recorded in 4 years after restoration. A greater survival rate (96%), organic carbon (1.06%), total nitrogen (0.18%), total phosphorus (0.09%), and production of fodder biomass (9.568 tons/ha^{-1}/year^{-1}) from natural grass were reported. The success of a similar restoration approach, also quoted by Bindhy et al. (2018) in the same region, revealed that active restoration had a positive impact on ecology, economy, and livelihood improvement.

14.7 Passive Restoration

Passive restoration is best suited for areas with limited landscape degradation and the presence of sufficient mother trees proximal to the degraded sites (Table 14.1). For instance, the existence of a mother tree within the boundary of a 100 m area of restoration area promotes better regeneration as compared to other approaches of restoration. Mother tree presence is an important factor for recruitment density and the seed dispersal mechanism for zoochorous species. On the other hand, passive restoration was found to be fruitful in ecological invasion of woody plants under riparian ecosystem observed in Fynbos biomes of South Africa. A study was conducted beside the Berg River of the Western Cape involving fell and remove as well as fell and stack burn clearing treatments under passive and active restoration processes (Table 14.1). The same experiment was also repeated in greenhouse conditions for evaluation of germination and seed viability of native species. Results reveal zero germination of native species in area devoid of seeding application, which was attributed to the dominancy of herbs and graminoids, followed by the absence of propagules of indigenous species in the soil seed bank. Germination was

Table 14.1 Prominent cases of successful restoration approaches

Country/forest type	Landscape	Restoration type	Tree species	Reference
Southwest China	Agroforest and forest	Active restoration	*Alnus nepalensis, Betula alnoides, Choerospondias axillaris, Ligustrum lucidum, Lindera communis, Machilus yunnanensis,* and *Quercus acutissima*	Lu et al. (2017)
Yunnan, China	Forest	Assisted natural restoration	*Liatongus gagatinus, Onthophagus anguliceps, O. balthasari, O. diabolicus*	Sullivan et al. (2018)
Northwestern Ethiopia	Forests	Assisted natural regeneration	*Prunus Africana, Albizia gummifera, Hagenia abyssinica, Apodytes dimidata, Ekebrgia capensis*	Kassaye et al. (2023)
Ontario, Canada	Forests	Passive regeneration	*Hedera helix* and other herbs	Vasseur (2012)
Veracruz, Mexico	Forests	Active restoration	*Platanus Mexicana, Fraxinus uhdei, Liquidambar styraciflua, Juglans pyriformis, Quercus sapotifolia*	Toledo-Aceves et al. (2021)
Western cape, South Africa	Riparian ecosystems	Passive restoration	*Searsia angustifolia, Leonotis leonurus* and *Melianthus major*	Ruwanza et al. (2013)
Eastern cape, South Africa	Degraded thickets	Passive restoration	*(Portulacaria afra, Pappea capensis and nd Searsia longispina, (Lycium ferocissimum, and Rhigozum obovatum*	van der Vyver et al. (2012)
Northeastern Colombia	Degraded forest plantation	Active and passive restoration	*Azadirachta indica* (active restoration), *Croton leptostachyus* (passive restoration)	Restrepo et al. (2013)

low under both clearing treatments. However, some of the indigenous species reflected higher germination under fell and stack burn clearing treatment. Seasonal effects were also observed on seedling survival, as a lower number was recorded during the summer season due to soil moisture stress. On the other hand, the rate was higher under greenhouse conditions, and harvested seeds were found to be viable.

14.8 Assisted Natural Restoration

Assisted natural restoration is ideally suited for areas experiencing mild disturbance and aims to achieve successful restoration within a short period. This approach is best applied in areas with small canopy gaps and where seeding and planting are feasible, such as regions with gentle slopes, optimal soil depth, assured rainfall, and available labor resources. The ANR has a high scope due to its cost-effectiveness and a high success rate in restoration efforts. As a result, it has been widely adopted and applied in various restoration initiatives. This can be further substantiated in a case study reported from eastern Australia for a subtropical rainforest ecosystem (Uebel et al. 2017). The investigation revealed the positive outcome of the assisted regeneration process in three different land uses (grazing excluded plus assisted regeneration, grazing excluded, and grazed). The study involved assessment of the recovery of forest for a time span of 10 years after grazing, as well as 4–6 years of assisted regeneration. Assisted regeneration revealed betterment of canopy cover, an increase in indigenous species richness, and stem density as compared to non-assisted sites. This reveals that ANR not only has the capacity for upscale species composition and diversity with a little management intervention, such as filling gaps, weed removal, fire control, etc., but also a potential to fulfil ecosystem services (Kittur et al. 2014; Jhariya et al. 2014). For example, Yang et al. (2018) reported an evaluation of ANR potential. The study quantified important ecosystem services in ANR dominated by *Castanopsis carlesii* in comparison to other species. It also reduces surface runoff and sediment yield by more than half in comparison to other young plantation sites. A decline in dissolved organic carbon export was also reported by ANR. Aboveground biomass was found to be greater in young plantations as well as mature Chinese fir plantations. Hence, the potential of assisted natural restoration has gained broad acceptance as a standard practice for restoration. It is essential to conduct a preliminary evaluation and this assessment should consider budget constraints, project timelines, scope, stakeholder requirements, and other crucial factors before selecting the appropriate restoration approach.

14.9 Conclusion

Anthropogenic activities cause deforestation and degradation of forests not only diminishing the forest health and productivity but also disrupting the forest ecosystem functioning. Implementing prudent policy frameworks (e.g., Delhi Declaration, New York Forest Declaration) is crucial for halting deforestation and rejuvenating of forest ecosystem. However, the ultimate goal is to restore the landscape to the previous state in a short timeframe. FLR is an important strategy for improving the forest quality and maximizing of biodiversity along with delivery of precious ecosystem services. The choice of restoration approach must be evaluated on a pilot base considering site-specific social and ecological values. Active restoration techniques are most effective for degraded ecosystems, where the other approaches are seldom successful. Natural restoration is a simpler approach that follows natural processes but is time-consuming. Conversely, assisted natural restoration is often considered the best choice for partially disturbed landscapes. However, the decision-making process for landscape restoration involves consultation meetings with all stakeholders. Excluding villagers might achieve the objective of planting trees but may fail to address crucial needs of the villagers, such as improving groundwater tables, non-timber forest products (NTFPs), employment, etc. Conflicts often arise during the restoration process on boundary demarcation, resource sharing, firewood collection, grazing, etc. Managing such conflicts requires special attention and a conflict resolution mindset. The primary aim of decision-making in landscape restoration should be to improve the forest ecosystem health and augment resources for the local needs.

References

Abiyu A, Teketay D, Glatzel G, Gratzer G (2016) Seed production, seed dispersal and seedling establishment of two afromontane tree species in and around a church forest: implications for forest restoration. For Ecosyst 3:16. https://doi.org/10.1186/s40663-016-0076-5

Alexander S, Nelson CR, Aronson J, Lamb D, Cliquet A, Erwin KL, Finlayson CM, de Groot RS, Harris JA, Higgs ES, Hobbs RJ, Robin Lewis IIIRR, Martinez D, Murcia C (2011) Opportunities and challenges for ecological restoration within REDD+. Restor Ecol 19:683–689

Anthony O (1978) Theories of social conflict. Annu Rev Sociol 4:291–315

Awasthi P, Bargali K, Bargali SS, Jhariya MK (2022a) Structure and functioning of *Coriaria nepalensis* wall dominated Shrublands in degraded hills of Kumaun Himalaya. I. Dry matter dynamics. Land Degrad Dev 33(9):1474–1494. https://doi.org/10.1002/ldr.4235

Awasthi P, Bargali K, Bargali SS, Khatri K, Jhariya MK (2022b) Nutrient partitioning and dynamics in *Coriaria nepalensis* wall dominated Shrublands of Degraded Hills of Kumaun Himalaya. Front For Glob Change 5:913127. https://doi.org/10.3389/ffgc.2022.913127

Baker S, Eckerberg K, Zachrisson A (2014) Political science and ecological restoration. Environ Politics 23(3):509–524. https://doi.org/10.1080/09644016.2013.835201

Bakshe P, Jugade R (2023) Phytostabilization and rhizofiltration of toxic heavy metals by heavy metal accumulator plants for sustainable management of contaminated industrial sites: a comprehensive review. J Hazard Mater Adv 10:100293

Banerjee A, Jhariya MK, Yadav DK, Raj A (2020) Environmental and sustainable development through forestry and other resources. Apple Academic Press Inc., CRC Press—a Tayler and Francis Group, US & Canada. ISBN: 9781771888110. Pp. 400

Banerjee A, Jhariya MK, Bargali SS, Palit D (2023) Ecorestoration for sustainability, Wiley-Scrivener. ISBN: 9781119879718. pp. 1–441. https://onlinelibrary.wiley.com/doi/book/10.1002/9781119879954

Batie SS (2008) Fellows address: wicked problems and applied economics. Am J Agric Economic 90(5):1176–1191

Becerra L, Molendijk M, Porras N, Spijkers P, Reydon B, Morales J (2021) Fit-for-purpose applications in Colombia: defining land boundary conflicts between indigenous Sikuani and neighbouring settler farmers. Land 10(4):382

Bindhy P, Suman S, Om JR (2018) Ecological restoration and livelihood sustainability through mixed forestry in Himalaya: a case study of Kot Malla Village, Rudraprayag, Uttarakhand. In: Environmental concerns and sustainable development in Himalaya Publisher. Research India Press

Bisht S, Bargali SS, Bargali K, Rawat YS, Rawat GS (2023) Dry matter dynamics and carbon flux along riverine forests of Gori valley, Western Himalaya. Front Forest Glob Change 6. https://doi.org/10.3389/ffgc.2023.1206677

Brancalion PH, Schweizer D, Gaudare U, Mangueira JR, Lamonato F, Farah FT, Nave AG, Rodrigues RR (2016) Balancing economic costs and ecological outcomes of passive and active restoration in agricultural landscapes: the case of Brazil. Biotropica 48(6):856–867

Brown K (2005) Addressing trade-offs in forest landscape restoration. In: Mansourian S, Vallauri D, Dudley N (eds) Forest restoration in landscapes: beyond planting trees. Springer, New York, pp 59–62

Brye KR, Norman JM, Gower ST (2002) Assessing the progress of a tallgrass prairie restoration in southern Wisconsin. Am Midl Nat 148(2):218–235
Chazdon RL (2008) Beyond deforestation: restoring forests and ecosystem services on degraded lands. Science 320(5882):1458–1460
Christin ZL, Bagstad KJ, Verdone MA (2016) A decision framework for identifying models to estimate forest ecosystem services gains from restoration. For Ecosyst 3:3. https://doi.org/10.1186/s40663-016-0062-y
Crouzeilles R, Curran M, Ferreira MS, Lindenmayer DB, Grelle CEV, Benayas JMR (2016) A global meta-analysis on the ecological drivers of forest restoration success. Nat Commun 7:11666. https://doi.org/10.1038/ncomms11666
Crouzeilles R, Ferreira MS, Chazdon RL, Lindenmayer DB, Sansevero JBB, Monteiro L, Iribarrem A, Latawiec AE, Strassburg BBN (2017) Ecological restoration success is higher for natural regeneration than for active restoration in tropical forests. Sci Adv 3(11):e1701345
Daniels SE, Walker GB (2001) Working through environmental conflict. The Collaborative Learning Approach, Preager
Di Sacco A, Hardwick KA, Blakesley D, Brancalion PH, Breman E, Cecilio Rebola L, Chomba S, Dixon K, Elliott S, Ruyonga G, Shaw K (2021) Ten golden rules for reforestation to optimize carbon sequestration, biodiversity recovery and livelihood benefits. Glob Chang Biol 27(7):1328–1348
Denier L, Scherr S, Shames S, Chatterton P, Hovani L, Stam N (2015) The little sustainable landscapes book: achieving sustainable development through integrated landscape management. Global Canopy Programme, Oxford
FAO (2020) Global Forest Resources Assessment 2020: Main report. Rome. https://doi.org/10.4060/ca9825en
Folger JP, Poole MS, Stutman RK (2021) Working through conflict: strategies for relationships, groups, and organizations. Routledge
Ghazoul J, Chazdon R (2017) Degradation and recovery in changing forest landscapes: a multi-scale conceptual framework. Annu Rev Environ Resour 42:161–188
Gosain BG, Negi GCS, Dhyani PP, Bargali SS, Saxena R (2015) Ecosystem services of forests: carbon stock in vegetation and soil components in a watershed of Kumaun Himalaya, India. Int J Ecol Environ Sci 41(3–4):177–188
Hansen MC, Potapov PV, Moore R, Hancher M, Turubanova SA, Tyukavina A, Thau D, Stehman SV, Goetz SJ, Loveland TR, Kommareddy A, Egorov A, Chini L, Justice CO, Townshend JRG (2013) High-resolution global maps of 21st-century forest cover change. Science 342(6160):850–853. https://doi.org/10.1126/science.1244693
IIFM (2022) Mid-term evaluation of forestry plantations funded by the National Mission for clean Ganga. Report Submitted to NMCG. Indian Institute of Forest Management (IIFM), Bhopal
Jens E (2007) Conflict and forest landscape restoration—a collaborative learning approach. IUFRO Conference on Forest Landscape Restoration-Seoul, Korea May 14–19:2007
Jhariya MK, Bargali SS, Swamy SL, Kittur B, Bargali K, Pawar GV (2014) Impact of forest fire on biomass and carbon storage pattern of tropical deciduous forests in Bhoramdeo wildlife sanctuary, Chhattisgarh. Int J Ecol Environ Sci 40(1):57–74
Jhariya MK, Yadav DK, Banerjee A, Raj A, Meena RS (2019) Sustainable forestry under changing climate. In: Jhariya MK et al (eds) Sustainable Agriculture. Forest and Environmental Management. Springer, Singapore, ISBN 978-981-13-6829-5, pp 285–325
Jhariya MK, Meena RS, Banerjee A, Meena SN (2022) Natural resources conservation and advances for sustainability. Elsevier. ISBN: 9780128229767. https://doi.org/10.1016/C2019-0-03763-6
Jhariya MK, Meena RS, Banerjee A, Kumar S, Raj A (2024) Agroforestry for carbon and ecosystem management. Elsevier. ISBN: 9780323953931, eBook ISBN: 9780323958752. Pp. 434
Jones S, Dudley N (2005) Negotiations and conflict management. In: Mansourian S, Vallauri D, Dudley N (eds) Forest restoration in landscapes: beyond planting trees. Springer, New York
Kassaye M, Nigatu A, Gedamu B, Derebe Y, Tamer M, Ayne F, Emiru E, Belay B (2023) Assisted natural regeneration of selected tree species through silviculture interventions as a tool for restoring degraded natural forests in northwestern Ethiopia. For Sci Technol 19(3):232–240

Kittur B, Swamy SL, Bargali SS, Jhariya MK (2014) Wildland fires and moist deciduous forests of Chhattisgarh, India: divergent component assessment. J For Res 25(4):857–866
Koch A, Kaplan JO (2022) Tropical forest restoration under future climate change. Nat Clim Chang 12:279–283. https://doi.org/10.1038/s41558-022-01289-6
Kuuluvainen T, Angelstam P, Frelich L, Jõgiste K, Koivula M, Kubota Y, Lafleur B, Macdonald E (2021) Natural disturbance-based forest management: moving beyond retention and continuous-cover forestry. Front For Glob Change 4:629020. https://doi.org/10.3389/ffgc.2021.629020
Lado K, Thomas AN, Janez K (2019) Effects of disturbance on understory vegetation across Slovenian forest ecosystems. Forests 10:1048. https://doi.org/10.3390/f10111048
Lafleur B, Fenton NJ, Simard M, Leduc A, Pare D, Valeria O, Bergeron Y (2018) Ecosystem management in paludified boreal forests: enhancing wood production, biodiversity, and carbon sequestration at the landscape level. For Ecosyst 5:27. https://doi.org/10.1186/s40663-018-0145-z
Lazos-Chavero E, Zinda J, Bennett-Curry A, Balvanera P, Bloomfield G, Lindell C, Negra C (2016) Stakeholders and tropical reforestation: challenges, trade-offs, and strategies in dynamic environments. Biotropica 48(6):900–914. https://www.jstor.org/stable/48576586
Lewis SL, Wheeler CE, Mitchard ETA, Koch A (2019) Restoring natural forests is the best way to remove atmospheric carbon. Nature 568:25–28
Li M, De Pinto A, Ulimwengu JM, You L, Robertson RD (2015) Impacts of road expansion on deforestation and biological carbon loss in the Democratic Republic of Congo. Environ Resour Econ 60:433–469
Li Y, Brando PM, Morton DC, Lawrence DM, Yang H, Randerson JT (2022) Deforestation-induced climate change reduces carbon storage in remaining tropical forests. Nat Commun 13:1964. https://doi.org/10.1038/s41467-022-29601-0
Lu Y, Ranjitkar S, Harrison RD, Xu J, Ou X, Ma X, He J (2017) Selection of native tree species for subtropical forest restoration in Southwest China. PLoS One 12(1):e0170418. https://doi.org/10.1371/journal.pone.0170418
Luz MLA (2021) Is economics of restoration helping with decision-making challenges? Insights guided by bibliometrics. Environ Develop 40:100674
Macdonald SE, Landhäusser SM, Skousen J, Franklin J, Frouz J, Hall S, Jacobs DF, Quideau S (2015) Forest restoration following surface mining disturbance: challenges and solutions. New For 46:703–732
Machar I (2020) Sustainable landscape management and planning. Sustainability 12(6):2354
Mahari WAW, Azwar E, Li Y, Wang Y, Peng W, Ma NL, Yang H, Rinklebe J, Lam SS, Sonne C (2020) Deforestation of rainforests requires active use of UN'S sustainable development goals. Sci Total Environ 742:140681
Manral V, Bargali K, Bargali SS, Jhariya MK, Padalia K (2022) Relationships between soil and microbial biomass properties and annual flux of nutrients in central Himalayan forests, India. Land Degrad Dev 33(12):2014–2025. https://doi.org/10.1002/ldr.4283
Martin D, Tomida M, Meacham B (2016) Environmental impact of fire. Fire Sci Rev 5:5. https://doi.org/10.1186/s40038-016-0014-1
Mishra A, Swamy SL, Thakur TK, Kumar A, Pandey M (2022) Impact of coal mining on land use changes, deforestation, biomass, and C losses in Central India: implications for offsetting CO_2 emissions. Land Degrad Dev 33(18):3731–3747
Mon SM, Okuda T, Yamada T, Thant AM, Shin T, Chew WC, Mandal MSH, Shigematsu C (2023) Can commercialization of non-timber forest product (NTFP) reduce deforestation in Myanmar? Tropics 31(4):81–93
Moreno-Mateos D, Alberdi A, Morriën E, van der Putten WH, Rodríguez-Uña A, Montoya D (2020) The long-term restoration of ecosystem complexity. Nat Ecol Evol 4(5):676–685
Mostert E (2006) Integrated water resources management in The Netherlands: how concepts function. J Contemp Water Res Educ 135(1):19–27
Nepal SK, Weber KE (1995) Prospects for coexistence: wildlife and local people. Ambio 24(4):238–245

Ota L, Chazdon RL, Herbohn J, Gregorio N, Mukul SA, Wilson SJ (2020) Achieving quality forest and landscape restoration in the tropics. Forests 11(8):820

Palmer MA, Bernhardt ES, Allan JD, Lake PS, Alexander G, Brooks S, Carr J, Clayton S, Dahm CN, Follstad Shah J, Galat DL (2005) Standards for ecologically successful river restoration. J Appl Ecol 42(2):208–217

Program on Forests (2011) Investing in trees and landscape restoration in Africa: overview. Program for Forests (PROFOR). Assessed on 26 Nov 2023, Washington, DC

Raj A, Jhariya MK, Banerjee A, Nema S, Bargali K (2023) Land and environmental management through forestry. Wiley-Scrivener Publishing. ISBN: 978-1-119-91050-3. pp. 400

Rawat LS, Maikhuri RK, Dhyani D, Bahuguna YM, Pharswan DS (2017) Ecological restoration of village common degraded land through participatory approach for biodiversity conservation and socio-economic development in Indian Himalayan region. Acta Ecol Sin 37(4):240–252

Rebecca CS, Randi J, Sander V (2021) Overwintering fires in boreal forests. Nature 593:399–404

Renato C, Mariana SF, Robin LC, David BL, Jerônimo BB, Sansevero LM, Alvaro I, Agnieszka EL, Bernardo BNS (2017) Ecological restoration success is higher for natural regeneration than for active restoration in tropical forests. Sci Adv 3(11)

Restrepo MF, Florez CP, Osorio NW, León JD (2013) Passive and active restoration strategies to activate soil biogeochemical nutrient cycles in a degraded tropical dry land. International Scholarly Research Notices, 2013. Article ID 461984. https://doi.org/10.1155/2013/461984

Rhett AB (2019) The year rainforests burned 2019. Mongabay Series: Glob for. Accessed on 29th June 2021

Ruwanza S, Gaertner M, Esler KJ, Richardson DM (2013) The effectiveness of active and passive restoration on recovery of indigenous vegetation in riparian zones in the Western cape, South Africa: a preliminary assessment. S Afr J Bot 88:132–141

Sarah R (2020) Intense boreal Forest fires a climate concern. World Resources Institute. Accessed on 03 July 2021. https://www.wri.org/insights/intense-boreal-forest-fires-climate-concern

Sayer J, Sheil D, Galloway G, Riggs RA, Mewett G, MacDicken KG, Arts BJM, Boedhihartono AK, Langston J, Edwards DP (2019) SDG 15 life on land–the central role of forests in sustainable development. In: Sustainable development goals: their impacts on forest and people. Cambridge University Press, pp 482–509

Seidl R, Thom D, Kautz M, Martin-Benito D, Peltoniemi M, Vacchiano G, Wild J, Ascoli D, Petr M, Honkaniemi J, Lexer MJ, Trotsiuk V, Mairota P, Svoboda M, Fabrika M, Nagel TA, Reyer CPO (2017) Forest disturbances under climate change. Nat Clim Chang 7:395–402. https://doi.org/10.1038/nclimate3303

Sherrouse BC, Semmens DJ, Clement JM (2014) An application of Social Values for Ecosystem Services (SolVES) to three national forests in Colorado and Wyoming. Ecol Indic 36:68–79

Stanturf JA (2021) Forest landscape restoration: building on the past for future success. Restor Ecol 29(4):e13349

Stanturf JA, Kleine M, Mansourian S, Parrotta J, Madsen P, Kant P, Burns J, Bolte A (2019) Implementing forest landscape restoration under the Bonn challenge: a systematic approach. Ann For Sci 76:50. https://doi.org/10.1007/s13595-019-0833-z

Sullivan CD, Slade EM, Bai M, Shi K, Riordan P (2018) Evidence of forest restoration success and the conservation value of community-owned forests in Southwest China using dung beetles as indicators. PLoS One 13(11):e0204764. https://doi.org/10.1371/journal.pone.0204764

Suding K, Higgs E, Palmer M, Callicott JB, Anderson CB, Baker M, Gutrich JJ, Hondula KL, Lafevor MC, Larson BMH, Randall A, Ruhl JB, Schwartz KZS (2015) Committing to ecological restoration: efforts around the globe need legal and policy clarification. Science 348:638–640

Tan YL, Chen JE, Yiew TH, Habibullah MS (2022) Habitat change and biodiversity loss in south and southeast Asian countries. Environ Sci Pollut Res 29(42):63260–63276

Timothy OR, Ayten E (2013) Emerging threats to forests: resilience and strategies at system scale. Am J Plant Sci 4(3A):29506. https://doi.org/10.4236/ajps.2013.43a093

Thom D, Rammer W, Seidl R (2017) Disturbances catalyze the adaptation of forest ecosystems to changing climate conditions. Glob Chang Biol 23(1):269–282. https://doi.org/10.1111/gcb.13506

Toledo-Aceves T, Trujillo-Miranda AL, López-Barrera F (2021) Tree regeneration in active and passive cloud forest restoration: functional groups and timber species. For Ecol Manag 489:119050

Trujillo-Miranda AL, Toledo-Aceves T, López-Barrera F, Gerez-Fernández P (2018) Active versus passive restoration: recovery of cloud forest structure, diversity and soil condition in abandoned pastures. Ecol Eng 117:50–61

Uebel K, Wilson KA, Shoo LP (2017) Assisted natural regeneration accelerates recovery of highly disturbed rainforest. Ecol Manage Restor 18(3):231–238

Ullah A, Zeb A, Saqib SE, Kächele H (2022) Landscape co-management and livelihood sustainability: lessons learned from the billion trees afforestation project in Pakistan. Land Use Policy 115:106034

UNCCD (2016) Land degradation neutrality, target setting Programme, land degradation neutrality target setting. A technical guide. Draft for consultation during the Land Degradation Neutrality Target Setting Programme inception phase. Accessed on 13th Dec 2023

Van Oosten C, Gunarso P, Koesoetjahjo I, Wiersum F (2014) Governing forest landscape restoration: cases from Indonesia. Forests 5:1143

van der Vyver ML, Cowling RM, Campbell EE, Difford M (2012) Active restoration of woody canopy dominants in degraded south African semi-arid thicket is neither ecologically nor economically feasible. Appl Veg Sci 15(1):26–34

Vasseur L (2012) Restoration of deciduous forests. Nat Educ Knowl 3(12):1

Villa F, Bagstad KJ, Voight B, Johnson GW, Portela R, Honzák M, Batker D (2014) A methodology for adaptable and robust ecosystem services assessment. PLoS One 9:e91001

Walters G, Baruah M, Karambiri M, Adjei POW, Samb C, Barrow E (2021) The power of choice: how institutional selection influences restoration success in Africa. Land Use Policy 104:104090

Whalen PJ, Toth LA, Koebel JW, Strayer PK (2002) Kissimmee River restoration: a case study. Water Sci Technol 45(11):55–62

WRI (2023) Initiative 20x20. www.wri.org/our-work/project/initiative-20x20/about-initiative-20x20#project-tabs. Assessed on 26 Nov 2023

WWF and IUCN (2020) Forests Reborn, A workshop on forest restoration. In Proceedings of the WWF/IUCN International Workshop on Forest Restoration, Segovia, Spain, 3–5 July 2020

Wong MH (2003) Ecological restoration of mine degraded soils, with emphasis on metal contaminated soils. Chemosphere 50(6):775–780

Yang Y, Wang L, Yang Z, Xu C, Xie J, Chen G, Lin C, Guo J, Liu X, Xiong D, Lin W (2018) Large ecosystem service benefits of assisted natural regeneration. J Geophys Res Biogeo 123(2):676–687

Yusof NNM, Hatta SKM, Kamarudin SH, Jamil NM, Supardan SN, Suratman MN (2023) Forest landscape restoration for environmental management. Land and environmental management through forestry, pp 161–199

MIX
Papier aus verantwortungsvollen Quellen
Paper from responsible sources
FSC® C105338
www.fsc.org

If you have any concerns about our products,
you can contact us on
ProductSafety@springernature.com

In case Publisher is established outside the EU,
the EU authorized representative is:
Springer Nature Customer Service Center GmbH
Europaplatz 3, 69115 Heidelberg, Germany

Printed by Libri Plureos GmbH
in Hamburg, Germany